Chemical Water and
Wastewater Treatment VII

Chemical Water and Wastewater Treatment VII

Proceedings of the
10th Gothenburg Symposium 2002
June 17–19, 2002
Gothenburg, Sweden

Edited by
Hermann H. Hahn
Erhard Hoffmann
Hallvard Ødegaard

Publishing

Published by IWA Publishing, Alliance House, 12 Caxton Street, London SW1H 0QS, UK

Telephone: +44 (0) 20 7654 5500; Fax: +44 (0) 20 7654 5555
Email: publications@iwap.co.uk; Web: www.iwapublishing.com

First published 2002
© 2002 IWA Publishing

Printed by TJ International (Ltd), Padstow, Cornwall, UK

British Library Cataloguing in Publication Data
A CIP catalogue record for this book is available from the British Library

Library of Congress Cataloging- in-Publication Data
A catalog record for this book is available from the Library of Congress

ISBN 1 84339 009 4

Preface

Environment, which includes the repair and protection of air, water and soil, does not register anymore as the number one problem in many of our countries. There appear to be other problems that have caught the attention of the multitudes, such as the fight against poverty and problems that result from many more people demanding a higher standard of living. Does this mean that the topics addressed in our symposia have lost their relevance?

Treating potable and polluted waters to provide wholesome water to all people and to dispose of all spent waters safely, though intricately connected with environmental problems, was and still is one of the most important tasks of our day. Official documents of the United Nations state that more than 1.2 billion people suffer from insufficient water supply and an even larger number of people, up to 4 billion, lack hygienic disposal of waste and wastewater. And these numbers are likely to increase over the years to come.

Water technology has thus become an economic factor, an export item of the highest priority. And with it comes the necessary know-how transfer; an item that has been the key issue of the Gothenburg symposia from their very beginning. With the challenges the world faces today, these symposia are needed more than ever before, as illustrated by the number of papers offered for presentation and the ever growing audience at each conference event.

As to the content of this book, the symposium proceedings have traditionally been sufficiently focused to attract high interest from many specialists. As is the case with other international events, it seems that such topical programs draw proportionately larger audiences. On the other hand, each Gothenburg Symposium has also been open to new ideas, new concepts and new technologies. Most noticeable to us, the editors, is the increasing geographic spread of the contributing authors. The original Scandinavian and Western European joint venture has now become a truly international affair with nearly equal input from all continents, and with the number of reports from developing countries increasing most noticeably.

If one would have presented this year's program of the symposium to the initiators of the 1984 conference, they might have expressed surprise at some of the presentations in view of the aim of these symposia to provide solutions to real world problems. Who would have thought that composite inorganic and organic flocculants might become available for practical applications in such short time, or that granulated iron hydroxide would be used for copper control in roof run-off? The development of the field of chemical water and wastewater treatment is impressive and the Gothenburg symposia have played a significant role in this process.

For the editorial team, Hallvard Ødegaard set the stage of each event by structuring the organizing committee's program work, and Erhard Hoffmann cooperated successfully with Karin Knisely to provide an understandable, attractive and well-formulated manuscript to our new publisher, the IWA. All of us are, as always, indebted to the authors of outstanding papers, which are the basis of any successful conference.

H.H. Hahn
E. Hoffmann
H. Ødegaard

Karlsruhe and Trondheim
March 2002

Members of the Organising Committee

H.H. Hahn, University Fridericiana, Germany

T. Hedberg, Chalmers University of Technology, Sweden

L. Johansson, Kemira Kemwater, Sweden

H. Ødegaard, Norwegian University of Science and Technology, Norway

Members of the Scientific Committee

Prof. E. Arvin, Denmark

Prof. T. Asano, USA

Prof. B. Balmér, Sweden

Prof. M. Boller, Switzerland

Dr. N. Booker, Australia

Prof. J.C. van Dijk, The Netherlands

Dr. P. Dolejs, Czech Republic

Prof. S.S. Ferreira Filho, Brazil

Prof. J. Fettig, Germany

Prof. N. Graham, England/Hong Kong

Prof. P.M. Huck, Canada

Prof. B. Jimenez Cisneros, Mexico

Mr. I. Karlsson, Sweden

Prof. R. Mujeriego, Spain

Prof. H. Ødegaard, Norway

Prof. M. Ottaviani, Italy

Prof. R. Pujol, France

Mr. F. Rogalla, USA

Prof. H. X. Tang, China

Prof. T. Tuhkanen, Finnland

Prof. Y. Watanabe, Japan

Contents

Coagulation / Flocculation Mechanisms

Flocculation and Floc Separation

Chemical Dosing Control

Wastewater Treatment

Sludge Treatment

Enhanced Removal and Reuse

Coagulation / Flocculation Mechanisms

H.H. Hahn, E. Hoffmann, H. Ødegaard (Eds.)
Chemical Water and Wastewater Treatment VII, pp. 3-16
© IWA Publishing, London
ISBN: 1 84339 009 4

Production of the Coagulation Agent PAX-14. Contents of Polyaluminium Chloride Compounds

E. G. Søgaard

Aalborg University, Department of Chemistry and Applied Engineering Science, Esbjerg, Denmark.
email: egs@aue.auc.dk

Abstract

The nature of the polyaluminium chloride species in the synthesis of a coagulation agent, PAX-14, produced by Kemira Miljø A/S, Esbjerg was investigated.

Testing methods ranged from simple pH, viscosity, conductivity, refractory index, basicity and density tests to measurements of zetapotential, Ferron test, ESI-MS and Al-27 NMR.

The results indicate that shortly after having reached the final process temperature the batch content of PAX 14 does not change its contents during the next couple of hours at fixed conditions. After filtration the solution is stable for several years. No particles with a size over 50 nm could be detected. The precipitating agency towards phosphates was not improved during the period after the batch had reached its final temperature.

These results indicate that the production time may be reduced by a couple of hours.

With the help of Al-27 NMR we found no contents of known polyaluminium compounds larger than octahedral dimers, $[Al_2(OH)_2(H_2O)_8]^{4+}$ and possibly also octahedral trimers, $[Al_3(OH)_4(H_2O)_9]^{5+}$. After dilution a tetrahedral polyaluminum compound with thirteen nuclei, $[Al_{13}O_4(OH)_{24}(H_2O)_{12}]^{7+}$, known from literature was identified. A simultaneous decrease in dimer content indicated that dimers are precursors of this polyaluminium compound. Aging of diluted samples increases the content of Al_{13}.

[3]

ESI-MS indicated, that the production of polyaluminium compounds were produced together with dimers and trimers from solutions when they were diluted and subsequently exposed to a vacuum.

The conventional use of the Ferron test to determine polyaluminium species content may overestimate the percentage of Al_{13} compounds because it misidentifies dimers and trimers as Al_{13} in undiluted PAX 14.

Introduction

PAX 14 is one of a wide range of coagulation agents that is used mainly to precipitate phosphates and for flocculation purposes in wastewater treatment plants. In certain water treatment plants PAX 14 is used for the precipitation of humic substances in the production of drinkable water. It is also used as a coagulation agent for different substances in industrial waste waters or reuse waters.

PAX is an abbreviation for polyaluminium compounds that are defined as hydroxide or oxyhydroxide species with more than one aluminium nucleus. These polyaluminium compounds have been studied in laboratory for several years and many species have been proposed to be the principal components present in aqueous solutions at different pHs up to a pH just before the precipitation of aluminium hydroxide (Akitt et al., 1972; Akitt and Farthing, 1978, 1981; Akitt and Elders, 1988; Bertsch et al., 1986a, 1986b, 1987, 1989; Bottero et al., 1980, 1987; Furrer et al., 1992a, 1992b; Thompson et al., 1987). Most of the knowledge about polyaluminium compounds stems from X-ray diffraction spectroscopy investigations of powders crystallised from aqueous solution by help of addition of sodium or potassium sulphates, selenates or hydroxides (Karlsson, 1998).

One of the recent papers on aluminium compounds supports previous work on the existence of at least five different polyaluminium and polycationic compounds besides the dimer, $Al_2(OH)_2^{4+}$, and the trimer, $Al_3(OH)_4^{5+}$ (Allouche et al., 2001). Two of these five polyaluminium polycationic compounds have the ε-Keggin structure, $Al_{13}O_4(OH)_{24}(H_2O)_{12}^{7+}$ and [6+]. They are denoted as Al_{13} ε-Johansson[7+] and Al_{13} ε-Johansson[6+] respectively after their discoverer G. Johansson (Johansson, 1960). A third compound has the δ-Keggin $Al_{13}(O_4(OH)_{24}(H_2O)_{12}^{7+}$ structure and is referred as δ-Nazar[7+] (Allouche et al., 2001). The fourth has a polycationic structure with 15 positive charges, $Al_{13}(OH)_{24}(H_2O)_{24}^{15+}$, and is referred as Al_{13} Mögel[15+] (Seichter et al., 1998; Karlsson, 1998). These four compounds are also referred generally as Al_{13}. The final one, $Al_{30}O_8(OH)_{56}(H_2O)_{24}^{18+}$, the largest of all polycationic compounds discovered until now, is denoted as Al_{30} δ-Tauelle[18+] (Rowsell et al., 2000, Allouche et al., 2001).

The ratio of hydroxide to aluminium, $r = OH/Al$, is an important parameter which is closely connected with the creation of polyaluminium compounds in aqueous solution. The solutions are stable during precipitation up to $r = 2.46$, but also already at the level of r~1.0 corresponding to a pH~3.9. The most familiar polyaluminum compound, Al_{13} ε-Johansson^{7+}, is detected with the aid of Al-27 NMR on laboratory samples.

The synthesis of PAX 14, produced by Kemira Miljø A/S, Esbjerg, was investigated for primary products of polyaluminium chloride species. Even if the pH of the final solution of the production is about 1.2, the product belongs to a large family of so-called prepolymerized inorganic coagulants (Ødegaard *et al.*, 1990). A thermodynamic calculation, based on an equilibrium constant for the formation of Al_{13}, ε-Johannson^{7+}, shows that no detectable Al_{13} could be present at that very low pH (Furrer *et al.*, 1992a). The objective of the current investigation was to find out which aluminium compounds are present in PAX 14 during and after production.

Materials and sampling

Aluminium hydroxide, $Al(OH)_3$, synthesized from bauxite, hydrochloric acid, and water were used as bulk chemicals to produce the final product, PAX 14. The production was performed in a stirred batch of 9.7 tons with increasing temperature up to 160 °C and with the corresponding pressure reaching approximately 5 bar.

Samples of the batch were collected from start to end of the production period and they were tested in a laboratory on site immediately after sampling. After stabilisation of the product for several hours, samples were tested again. Other production run samples were tested after 2-3 years. Characteristics of the samples were compared with properties of samples prepared in laboratory. Laboratory samples with OH/Al ratios between 1.0 and 2.4 were produced as described by Furrer *et al.* (1992b).

Methodology

The testing methods included simple electrochemical pH, chloride and conductivity measurements. Other methods were also performed in a laboratory on site and included measurements of viscosity, refractory index, aluminium contents, basicity and density. Measurements of zeta potential, Ferron test, electrospray ionisation mass spectrometry (ESI-MS) and Al-27 nuclear magnetic resonance spectroscopy (Al-27 NMR) were carried out in university laboratories. An introductory description of ESI-MS is given Kebarle and Tang (1993).

Aluminium content was determined by back titration of a surplus of EDTA with a solution of $ZnSO_4$ and with dithizone as indicator. *Basicity* was measured by back titration of surplus of sulphuric acid in the presence of sodium fluoride. A Malvern ZetaMaster S (Version PCS: v 1.27) measured *zeta potential* a few seconds after 1:100 dilution of a sample with water. The *Ferron test* was performed according to the instructions provided by Kemira (Kujala and Mikkonen, 1997). *Al-27 NMR* was performed with a Varian Vinity INOVA 300. The apparatus for

ESI-MS was a Finnigan TSQ 700 triple quadropole mass spectrometer with a nanoelectrospray source. Samples were diluted 1:100 in either water or methanol. Some samples were tested without dilution.

Results and discussion

Figures 1a and 1b show the time dependency of temperature and pressure, respectively, for two batches. After 100 min the curves are almost identical. The dots correspond to sampling points. While the final temperature was reached at 150 min, the pressure of the batch continued to increase by about 10% probably due to the establishment of a final temperature equilibrium.

Figure 2a shows that pH increased as expected during the dissolving process and remained approximately constant after 100 min. Dissolved aluminium increased up to 7 % and reached its final value after 150 min (Figure 2b). The

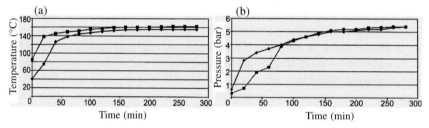

Figure 1. Temperature (a) and pressure (b) as a function of time shown for two batch productions of PAX14

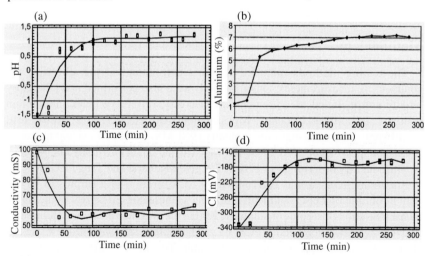

Figure 2. Various parameters in a batch production of PAX 14 as a function of time (a) pH, (b) % Aluminium, (c) Conductivity, and (d) Selective electrode measurements of chloride content

solutions are unstable without filtration and a precipitate shows up after about 12 hours. More than 80% of the total amount of aluminium is dissolved during one hour (Figure 2b).

Conductivity decreased during the dissolving process as the number of H^+ ions decreased (Figure 2c). An increasing inaccessibility of chloride ions measured by the chloride selective electrode will also decrease conductivity (Figure 2d). The precise reason for this decreasing amount of measurable Cl^- is unknown. Chloride ions are not a part of the polyaluminium compound systems. Their inaccessibility could be due to some interaction with not yet dissolved aluminium colloids in the batch solution where they are present in the diffuse double layers covering the colloids. Later they can be trapped in the filtration system together with the particles.

The ratio, OH/Al, calculated from basicity by help of the expression OH/Al = basicity •0.03, is considered to be a measure of the degree of polymerisation since OH^- is expected to be a part of the ligand system around the aluminium ions. The OH/Al ratio increases during the whole period of production, supporting the fact that some polymerisation reactions could take place (Figure 3). However, 85 % of the final basicity is produced before 150 min.

Figures 4a and 4b show the increasing refractory index and the increasing viscosity during the PAX 14 production process. After 100 min there were only small changes in refractory index and viscosity, indicating that the batch had reached

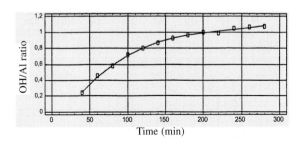

Figure 3. OH/Al ratio in a batch production of PAX 14 as a function of time

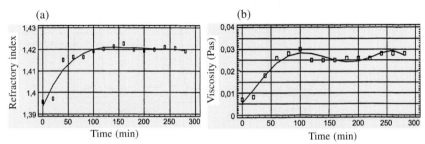

Figure 4. (a) Refractory index and (b) Viscosity as a function of time in a batch production of PAX 14

its final condition. These results indicate that shortly after the batch reaches its final process temperature, the batch content does not change significantly during the next couple of hours at fixed conditions.

Zeta potential measurements were performed in order to find out if the batch contained any colloidal charged particles, which could act as an indicator of the presence of polycationic polyaluminium compounds increasing during the process.

The zeta potential increased from values below 20 mV to relatively high values above 60 mV and stabilised a little above 60 mV after 180 min (Figure 5). Even if the samples were diluted before measurement we know from later results that a few seconds is not enough to create significant amounts of polyaluminium compounds. Thus we consider the zeta potentials to represent species already present

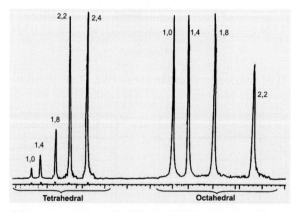

Figure 5. Zetapotential as a function of time in a batch production of PAX 14

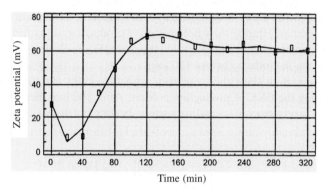

Figure 6. Five laboratory samples with different OH/Al ratios show increasing amounts of the tetrahedral part of Al_{13} and decreasing amounts of the octahedral $Al(H_2O)_6^{3+}$ as a function of increasing OH/Al ratio. For the sample with OH/Al = 2.4 the octahedral line had disappeared

in the samples. Therefore we expected to have highly positively charged species in solution. They could either be colloidal particles of not yet dissolved aluminium hydroxide crystallites or highly charged polyaluminium compounds produced during a reaction in the batch.

After production the batch is filtered by filter cake filtration for one hour. After filtration the solution is stable for several years. No particles with a size over 50 nm could be detected by Malvern Master Sizer.

Another series of experiments with unfiltered samples from the production showed that the ability to precipitate phosphates was not improved during the period after the batch had reached its final temperature. This agency of cake filtrated samples also remained on the same level. Therefore the production time in principal could be diminished to about 150 min.

Al-27 NMR was performed on 5 samples prepared in laboratory with OH/Al ratios varying from 1.0 to 2.4 and all with an aluminium concentration of 0.1 M. Two lines were observed on the spectra, one at about zero ppm and one at about 63 ppm, identified as the octahedral monomeric $Al(H_2O)_6^{+++}$ and the tetrahedral part of Al_{13} compound ε-Johannson^{7+}, respectively. The five spectra are combined in Figure 6 to show how the Al_{13} compound increases at the expense of the monomers as the ratio OH/Al increases.

This result was well known and expected. At OH/Al = 2.4 we observed that the line corresponding to the monomeric substance had disappeared. All aluminium in this sample was transformed into ε-Johannson^{7+}.

Special attention should be given to the "foot", that becomes broader in the last part of the integral curve in Figure 7b. This "tail" belongs to the twelve octahedral coordinated Al ions in Al_{13} ε-Johannson^{7+}. Only the tetrahedral coordinated centrally placed aluminium ion gives rise to the line at 63 ppm. For comparison the spectrum of the sample with a OH/Al ratio of 1.0 and 2.4 are given in Figures 7a and 7b, respectively.

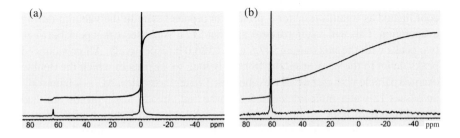

Figure 7. Comparison of two Al-27 NMR spectra with OH/Al ratio (a) 1.0 and (b) 2.4. A very broad foot in the spectrum of (b) indicates the existence of the 12 octahedral aluminium ions surrounding the tetrahedral part of Al_{13}. Only the central tetrahedral aluminium ion contributes to the peak at 63 ppm

With the aid of Al-27 NMR we found no presence of known polyaluminium compounds larger than octahedral dimers, $[Al_2(OH)_2(H_2O)_8]^{4+}$, and possibly also octahedral trimers, $[Al_3(OH)_4(H_2O)_9]^{5+}$ in the final and aged sample from the batch production of PAX 14.

These oligomeric compounds were identified as a "shoulder" left of the main peak corresponding to the large amount of the monomer $Al(H_2O)_6$ (see Fig. 8a). However, after dilution with water to 15 % and further down to 3 % the aforementioned tetrahedral polyaluminum compound with thirteen nuclei, $[Al_{13}O_4(OH)_{24}(H2O)_{12}]^{7+}$, known as ε-Johannson[7+] was identified at 63 ppm. A simultaneous decrease in dimers or trimers indicated that dimers are precursors of this polyaluminium compound. Aging of a diluted sample increased the content of Al_{13}. Figures 8a and 8b show the spectrum of undiluted PAX 14 with its oligomeric shoulder and a 4 % diluted sample of PAX 14 aged for nine days.

Figure 9 shows peak-areas of monomer, oligomer and polymer as a function of time indicating that the oligomers, which are dimers and trimers, could be precursors of Al_{13} when the sample is diluted. Compared to spectra of samples prepared in laboratory these spectra indicate that the PAX 14 prepolymerisation process of Kemira produces small amounts of the oligomeric compounds $[Al_2(OH)_2(H_2O)_8]^{4+}$ and $[Al_3(OH)_4(H_2O)]^{5+}$.

ESI-MS could also be used to detect the production of the polyaluminium compound, Al_{13} ε-Johannson[7+], together with dimers and trimers from the final PAX 14 solution when it was diluted and subsequently exposed to a vacuum. Figure 10a shows the ESI-MS spectrum of a 1:100 diluted sample of PAX 14 with m/z-values between 50 u and 2000 u. The area between 50 u and 670 u is shown in more detail in Figure 10b.

From this figure it is observed that the mutual distance between the most pronounced peaks is $\Delta m/z = 9$ u corresponding to water molecules escaping from a double positive charged particle. Other peaks with lower intensity show distances of 6 u corresponding to triple charged particles. The mass of a Al_{13} ε-Johannson[7+] particle is 1039.1 u. A double charged particle can arise if water molecules coordinated as ligands transfer 5 protons in reduced state to the vacuum during evaporation. This will result in a peak at about 517 u We observe that peak between two peaks with the m/z values 507.6 u and 526.0 u respectively. All pronounced peaks down to the peak at m/z = 489.8 u belong to a series in which the double charged particle with m/z = 517 u evaporates 3 water molecules. Al_{13} ε-Johannson[7+] possesses 12 water molecules and 5 of these had already lost a proton and had turned into hydroxide groups. So, at m/z = 489.8 u we observe a peak from a particle with only four intact water molecules left. However, we also find a peak with m/z = 462.8 u corresponding to the same substance with three more water molecules lost. The next pronounced peak is at m/z = 417.7 u and at m/z = 408.9 u. If they were parts of the same compound system four and five more water molecules should be lost. That would involve hydroxy groups in the bridge systems between aluminium ions. Another similar possibility is that these two peaks belong

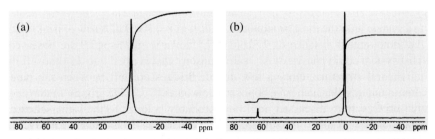

Figure 8. Comparison of two Al-27 NMR spectra of (a) undiluted PAX 14 and (b) a 4 % diluted sample after 9 days. Notice the "shoulder" left of the main peak at zero ppm. This "shoulder" belongs to oligomers like the dimers and the trimers that were produced in the process

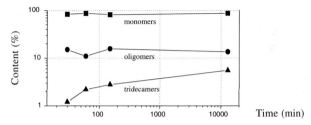

Figure 9. In double logarithmic scale the content of monomers, oligomers and tridecamers as a function of time in a 4 %-diluted PAX 14 sample. The amount of tridecamer grows at the expense of oligomers that act as precursors of Al_{13}

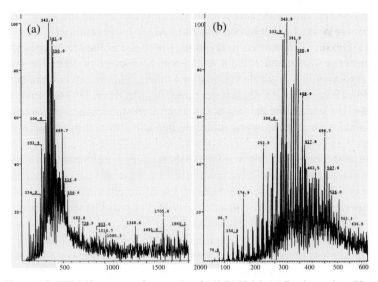

Figure 10. ESI-MS spectra of a sample of 1% PAX 14. (a) Region m/z = 50 u – 2000 u, (b) Region m/z = 50 u – 670 u

to a system with the most pronounced peak at m/z = 390.9 u. A row of peaks with the same mutual distance of 9 from 417.7 down to m/z = 363.9 are observed. That system could consist of Al_{13} ε-Johannson[7+] that is dried into an almost fully tetrahedral structure. Only a few double bridges are left between the outer aluminium ions that also have bonds to a few outer hydroxide groups. In that case the whole system discussed until now probably belongs to the same series of compounds with some "missing links" between m/z = 489.8 u and m/z = 417.7 u. This series consists of compounds ranging from $AlO_4(Al_{12}O_x)(OH)_y$ to Al_{13} ε-Johansson with some extra hydrogen bonded water molecules.

The spectrum of another system with only single positive charged compounds begins with the most pronounced peak at m/z = 342.9 u. Water molecules are evaporated in a unbroken series down to m/z = 234.9. This last peak could belong to a combination of a dimer and a trimer that are connected probably through an oxide or a hydroxide bridge, $Al_5O_2(OH)_4+$. This system must have been reduced through the pick-up of electrons at the cathode. Seven peaks belong to that series corresponding to the pick-up of six water molecules until m/z = 342.9. It is not known why this cluster with five aluminium atoms should be of special stable origin to be able to survive in vacuum. However, it indicates the pre-existence of dimers and trimers in PAX 14.

A system with a mutual spectral interpeak distance of 6 u and therefore with a triple positive charge consists of 11 peaks from m/z = 302.9 u to m/z = 356.8 u. It has its most pronounced peaks at 326.6 u and 332.9 u. The lowest observed molecular weight in this system is 908.7 u corresponding to Al_{13} ε-Johansson that has lost 7 water molecules and 4 hydrogen atoms to obtain the triple charge. Therefore the peak at 338.9 u corresponds to Al_{13} ε-Johansson itself.

A few not very pronounced peaks in the lower end of the spectrum with m/z values between 78.8 u and 150.8 u and a mutual distance of 18 u. The peak at 78.8 u corresponds to a single charged monomeric aluminium compound with two hydroxide groups and a single water molecule, $Al(OH)_2H_2O^+$. Four more water molecules are present in the heaviest species corresponding to an octahedral coordination with an extra water molecule bonded with hydrogen bonds to the complex.

In the total spectrum no important peaks seem to involve chloride ions. Chlorine has two isotopes that would have given rise to at least two peaks with a mutual distance of 2 u and with intensity ratio of 3:1. More chlorine ions present on a certain species should give rise to a dense system of peaks. If they are present in the spectrum they are not very pronounced.

The ferron test is a spectrophotometric method to distinguish among three species of aluminium contents, monomers, Al_{13} polymers and larger clusters of polymers present together in an aqueous solution. An equilibrium between Al^{3+} and 8-hydroxy-7-iodoquinoline-5-sulfonic acid (ferron) and their complexation

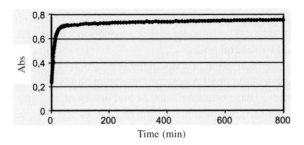

Figure 11. Diagram of ferron test of a sample with OH/Al = 1.8. Absorbance at $\lambda = 370$ nm as a function of time (min)

compound in a buffer solution is the basis for the method. The method uses the different chemical kinetic rates at which monomeric aluminium and polycationic aluminium clusters can undergo complexation towards equilibrium. The model is that the polycationic compounds are slowly disintegrated while they produce monomeric aluminium that goes into the equilibrium with ferron. The ferron test has been studied by several authors (Gessner and Winzer, 1979; Bertsch *et al.*, 1989; Parker and Bertsch, 1992a, 1992b; Shen and Dempsey, 1998)

Figure 11 shows the absorbance of ferron complexes as a function of time for a sample that is known to consist of both mononuclear aluminium and polynuclear Al_{13} from the Al-27 NMR investigations. In this figure the absorbance before $t = 2$ min stems from monomeric Al. The monomeric aluminium gives an absorbance of 0.3 and makes up about 40 % of the sample. A more detailed look at the curve, however, indicates that the absorbance is already increasing linearly with time after 0.5 min and an extrapolation to zero shows an initial absorbance of 0.21. According to the very detailed work of Gessner and Winzern the halftime of the first order reaction between monomeric aluminium and ferron is only 0.3 min and the reaction terminates after 2 min. However, for the reaction between oligomers (e.g. as dimers) and ferron, halftime is 3.0 min whereas it is 9.5 min between the Al_{13} polycationic aluminium and ferron. This means that with the prescribed concentrations for the test performed according to Kemira's standards, the method will overestimate the number of monomers when large numbers of polymers, especially in the form of oligomers, are present. With this method any oligomers that may be present will be determined as Al_{13} polymers. This is exactly what happened in our analysis of PAX 14 with the ferron test. The 5-10 % Al_{13} in PAX 14 that was calculated from the ferron test was in fact due to the existence of oligomeric substances in PAX 14 as shown in Al-27 NMR.

Conclusions

- The finding that there were only small changes in refractory index and viscosity after 100 min suggest that production time of PAX 14 can be decreased by several hours.
- The pre-polymerisation process in the production of PAX 14 is in reality a production of the small polyaluminium compounds in the form of dimers and trimers. These oligomers act as precursors of Al_{13} (polyaluminium chloride) diluted in aqueous solution of PAX 14 at a pH greater than 3.5.
- ESI-MS could help to reveal the presence of the most stable clusters in drying PAX 14 exposed to a vacuum.
- Results obtained using the ferron test on PAX 14 indicating the presence of the polyaluminium compound Al_{13} are in fact due to the presence of oligomers.

Acknowledgements

The author wants to express his thanks to several contributors to this work. Thanks to 16 students who participated in different areas of the work as a part of their education in colloid and interface science leading to a Diploma or Master's Degree in Chemical Engineering Science at Aalborg University Esbjerg. Thanks to Professor H. J. Jacobsen, Instrument Center for Solid State NMR Spectroscopy, University of Aarhus for his help with Al-27 NMR measurements. Thanks to post-doc Thomas J. D. Jørgensen, Dept. of Biochemistry and Molecular Biology, Southern Danish University, Odense, for his help with ESI-MS. Also thanks to Senior Researcher Armand Masion from J. Y. Bottero´s group at CEREGE, Aix-en-Provence Marseille for discussions about polyaluminium compounds. Thanks to assistant professor Alexei Soloviev, Dept. of Chemistry and Applied Engineering Science, Aalborg University Esbjerg and his wife Liudmila Solovieva for their help with the manuscript and finally thanks to Kemira Miljø, Esbjerg and Kemira Kemwater, Helsingborg for cooperation and support with sampling and testing.

References

Akitt, J.W., Greenwood, N.N., Khandelwal, B.L., and Lester, G.D. (1972) ^{27}Al Nuclear Magnetic Resonance Studies of the Hydrolysis and Polymerisation of the Hexa-aquo-aluminium(III) Cation, *J.C.S. Dalton Transactions*, 604-610

Akitt, J.W., and Farthing, A. (1978) New ^{27}Al NMR Studies of the Hydrolysis of the Aluminium(III) Cation, *J. Magnetic Resonance* **32**, 345-352

Akitt, J.W., and Farthing, A. (1981) Aluminium-27 Nuclear Magnetic Resonance Studies of the Hydrolysis of Aluminium(III). Part 5. Slow Hydrolysis using Aluminium Metal, *J.C.S. Dalton Transactions*, 1624-1628

Akitt, J.W., and Elders, J.M. (1988) Multinuclear Magnetic Resonance Studies of the Hydrolysis of Aluminium(III). Part 8, Base Hydrolysis monitored at Very High Magnetic Field, *J.C.S. Dalton Transactions*, 1347-1355

Allouche, L. Gérardin, C., Loiseau, T. Férey, G., and Taulelle, F. (2000), Al30: A Giant Aluminium Polycation, *Angew. Chem. Int. Ed.* **39**, No3, 511-514

Allouche, L., Huguenard, C., and Taulelle, F. (2001) 3QMAS of three aluminium polycations: space group consistency between NMR and XRD. *J.Phys. and Chem. of Solids* 62, 1525-1531

Bertsch, P.M., Thomas G.W., and Barnhisel, R.I. (1986) Characterization of Hydroxy-Aluminium Solutions by Aluminium-27 Nuclear Magnetic Resonance Spectroscopy, *Soil Sci. Soc. Am. J.* **50**, 825-830

Bertsch, P.M., Layton, J.W., and Barnhisel, R. I. (1986) Speciation of Hydroxy-Aluminium Solutions by Wet Chemical and Aluminium-27 NMR Methods, Soil Sci. Soc. Am. J. **50**, 1449-1454

Bertsch, P.M., (1987) Conditions for Al_{13} Polymer Formation in Partially Neutralized Aluminium Solutions, *Soil Sci. Soc. Am. J.* **51,** 826-828

Bertsch, P.M., Anderson, M.A., and Layton, J.W. (1989) Aluminium-27 Nuclear Magnetic Resonance Studies of Ferron-Hydroxy-Polynuclear Al Interactions, *Magnetic Resonance in Chemistry* **27**, 283-287

Bottero, J.Y. Cases, J.M. Fiessinger, F., and Poirier, J.E. (1980) Studies of Hydrolyzed Aluminium Chloride Solutions. 1. Nature of Aluminium Species and Composition of Aqueous Solutions, *J. Phys. Chem.* **84**, 2933-2939

Bottero, J.Y. Axelos, D., Tchoubar, J.M., Cases, M., Fripiat, J.J., and Fiessinger, F. (1987) Mechanism of Formation of Aluminium Trihydroxide from Keggin Al_{13} Polymers, *J. Colloid and Interface Science* **117**, 47-57

Furrer, G. Trusch, B., and Müller C. (1992) The formation of polynuclear Al13 under simulated natural conditions, *Geochimica et Cosmochimica Acta* **56**, 3831-3838.

Furrer, G. Ludwig, C. and Schindler, P.W. (1992) On the Chemistry of the Keggin Al_{13} Polymer, *J. Colloid and Interface Science* **149**, 56-67

Gessner, W., and Winzer, M. (1979) Über das Verhalten von Aluminiumsalzen mit unterschiedlich hoch kondensierten Al-oxo-Kationen bei der Reaktion mit Ferron [8-Hydroxy-(7)-iodchinolinsulfonsäure-(5)], *Z. anorg. Allg. Chem.* **452**, 151-156

Karlson, M. (1998), Structure Studies of Aluminium(III) Complexes in Solids, in Solutions at the solid/water Interface. Thesis, Umeå Universitet, Solfjädern Offset AB, Umeå 1998

Kebarle, P., and Tang, L. (1993) From Ions in Solution to Ions in the Gas Phase, The Mechanism of Electrospray Mass Spectrometry, *Analytical Chemistry* **65**, 972A- 986A

Ødegaard, H., Fettig, J., and Ratnaweera, C. (1990) Coagulation with Prepolymerized Metal Salts. In: Chemical Water and Wastewater Treatment IV, Hahn, H. H. and Klute, R. (Eds), Springer Verlag, New York, 189-220

Parker, D.R., and Bertsch, P.M. (1992) Identification and Quantification of the "Al_{13}" Tridecameric Polycation Using Ferron, *Environ. Sci. Technol.* 26, 908-914

Parker, D.R., and Bertsch, P.M. (1992) Formation of the Al_{13} Tridecameric Polycation under Diverse Synthesis Conditions, *Environ. Sci. Technol.* 26, 914-921

Rowsell, J., and Nazar, L. F. (2000) Speciation and Thermal Transformation in Alumina Sols: Structures of the Polyhydroxyoxoaluminium Cluster $[Al_{30}O_8(OH)_{56}(H_2O)_{26}]^{18+}$ and its δ-Keggin Moieté, *J. Am. Chem. Soc.* 122, 3777-3778

Seichter, W., Mögel, H.-J., Brand, P., and Saleh, D. (1998) Crystal Structure and Formation of the Aluminium Hydroxide Chloride $[Al_{13}(OH)_{24}(H_2O)_{24}]Cl_{15}\cdot 13H_2O$, *Eur. J. Inorg. Chem.* 795-797

Shen, Yun-Hwei, and Dempsey, B.A. (1998) Synthesis and speciation of polyaluminium chloride for water treatment, *Environmental International* **24**, 899-910

Thompson, A.R., Kunwar, A.C., Gutowsky, H.S., and Oldfield, E. (1987) Oxygen-17 and Aluminium-27 Nuclear Magnetic Resonance Spectroscopic Investigations of Aluminium(III) Hydrolysis Products, *J. Chem. Soc. Dalton Trans* 2317-2322

H.H. Hahn, E. Hoffmann, H. Ødegaard (Eds.)
Chemical Water and Wastewater Treatment VII, pp. 17-28
© IWA Publishing, London
ISBN: 1 84339 009 4

The Characteristics of Composite Flocculants Synthesized with Inorganic Polyaluminum and Organic Polymers

H. Tang and B. Shi*

*State Key Laboratory of Environmental Aquatic Chemistry (SKLEAC), Chinese Academy of Sciences, Beijing, China
email: hxtang@public.bta.net.cn

Abstract

In order to improve the aggregating property of flocculants, inorganic polyaluminum was synthesized with organic polymer additives. Cationic, anionic and nonionic polymers were used. Various characteristics such as chemical speciation, charge transformation, surface adsorption, microscopic configuration, and coagulation of composite flocculants were examined and compared among the composites and PAC and among the composites themselves.

The efficiency of composites was significantly higher than that of polyaluminum. The cationic polymers enhanced the charge neutralization power and the anionic polymer improved aggregation without noticeably decreasing the charge.

Introduction

Inorganic polymer flocculants such as polyaluminum (PAC) and polyiron (PFC) are now widely used in water and wastewater treatment facilities. Their efficiency is significantly higher than that of the traditional metal salt coagulants, but still much lower than that of organic polymer flocculants. The basic requisite characteristics of any coagulant or flocculant are the charge-neutralization and bridge-aggregation abilities for the removal of fine particles in raw water. While inorganic polymers are superior to traditional coagulants with regard to these characteristics, they are still inferior to the organic polymers. There are two ways to enhance the effectiveness of polyaluminum and polyiron: increase the proportion of the most efficient species in their original composition or add some other components to produce new composite flocculants.

[17]

Concerning the first method, some investigators have tried to increase the percentages of tridecameric polynuclear species (Al_{13}) in commercial polyaluminum products. It has been assumed that $AlO_4Al_{12}(OH)_{24}^{7+}$ in PAC may be the water soluble polymer with the most powerful coagulating and aggregating capabilities. This species is considered to be an artifact of special synthesis procedures (Bertsch and Parker, 1995) and in laboratory they can be prepared to above 80 % of the total Al species. However, only 40 to 50 % is possible when commercial PAC is produced by industrial manufacturing processes. One of the objectives of our current work was to develop the industrial technology to produce flocculants with higher Al_{13} content. Moreover, a new type of PAC consisting entirely of Al_{13} has been designed to be produced by nanometer technology.

The second method of improving polyaluminum and polyiron by adding other additive components to make more efficient inorganic polymer composites has been in practice for years. The main goals of this method are to increase the molecular size and enhance the aggregating power of the flocculants. The additive frequently used is silicate or polysilicate. The composites with anionic silicates may improve stability and aggregation, but their merits would be partly offset in the charge neutralizing ability of the original cationic inorganic polymer flocculants (Tang *et al.*, 1998). It is logical to combine inorganic polymer with some cationic organic polymer flocculant that should strengthen both the aggregating and charge-neutralizing capabilities. Alternatively, another strategy might be to produce an anionic polymer with strong aggregating power but weak charge. In fact, dosing the organic polymer flocculants in situ with inorganic coagulants in separate steps has been a frequently used procedure in our existing water plant. Combining these chemicals into one prescription would be convenient for the dosing facilities and should improve the efficiency of the technology for coagulation and flocculation.

This paper summarizes our laboratory study on the preparation, characteristics and effectiveness of composites made from inorganic and various organic polymer flocculants (Shi and Tang, 1999; Shi and Tang 2000).

Material and Procedures

Flocculants

The flocculants used are listed in Table 1. The inorganic polymer polyaluminum chloride (PAC) flocculants were prepared in our laboratory with slow micro-titration. A predetermined amount of 0.5 mol/L NaOH solution was injected into a 1.0 mol/L $AlCl_3$ solution with a 655 Dosmat micro-buret at a rate of 0.04 mL/min under magnetic stirring. The basicities i.e. the molar ratio of NaOH/Al for aliquots were 0.5, 1.0, 1.5, 2.0, respectively, and their ageing time was 24 hours. Another

Table 1. Flocculants for composing

Inorganic	Name	Manufacturer	Form	Al₂O₃%	Basicity
	PAC	Laboratory	liquor	10	0.5, 1.0, 1.5, 2.0
	CPAC	Tangshan, China	powder	32	1.4

Organic	Name	Manufacturer	Form	Viscosity (1% water Solution), cps	
Cationic	C109	Sanfloc, Japan	powder	2,000	
	Chitosan	Laboratory	particle	(MW 300,000)	
Anionic	AH200P	Sanfloc, Japan	powder	5,500	
	AN910SH	PAM, France	particle		
Nonionic	N505P	Sanfloc, Japan	powder	4,500	

inorganic polymer (CPAC) is a commercial product of spray-dried powder made in China. Chitosan is a weak cationic organic polymer flocculant produced in laboratory by extraction from natural shellfish waste. All other organic flocculants are commercial industrial products made in Japan or France.

Synthesis

To synthesize the inorganic and organic composite polymers, organic polymer (OP, 0.1% solution) was injected into the finished PAC product (0.1 mol/L Al solution) at a rate of 0.1 mL/min under strong stirring. The amount of each reagent was quantified according to the designed OP/Al ratios (mg/L). All the composite samples were subjected to an ageing time of 24 hours before chemical speciation. The composites were abbreviated as PACPc, a, or n to represent combinations with cationic, anionic and nonionic polymers, respectively. The appearances of the resulting samples were listed in Table 2, where the mark "—-" denoted that no precipitation occurred after processing and the marks "+" or "+ +" denoted that some flocs occurred or notably occurred, respectively. When anionic or nonionic organic polymers were mixed with PAC of higher basicities like 1.0, some precipitates could be observed initially. However, if the PAC was preheated to 40-60°C during injection, the problem of precipitation could be decreased significantly. The basicity without precipitation remained at 1.5 to 2.0. The marks in brackets denote the appearance at higher temperatures (60°C)

In an alternative synthesis, the organic polymer was first mixed with AlCl₃ solution (0.1 mol/L) and then the NaOH solution was micro-titrated into the mixture for basification of aluminum. Mixing the organic polymer and aluminum polymer with low or progressively increasing basicities would slow down the precipitation process and the tenuous flocs could be observed only above a basicity of 1.8.

Table 2. Appearance of the composites to precipitation

Basicity (OH/Al)	OP/Al (g/L/g/L)	Cationic		Anionic		Nonionic
		C109P	Chitosan	AH200P	AN910SH	N505P
		PACPc-1	PACPc-2	PACPa-1	PACPa-2	PACPn
0.5	0.02	---	---	---	---	---
	0.05	---	---	---	---	---
	0.10	---	---	---	---	---
	0.20	---	---	---	---	---
1.0	0.02	---	---	---	---	---
	0.05	---	---	---	+ (---)	---
	0.10	---	---	+ (---)	+ (---)	+ (---)
	0.20	---	---	+ (---)	+ (---)	+ (---)
1.5	0.02	---	---	+ (---)	+ (---)	+ (---)
	0.05	---	---	+ (---)	+ (---)	+ (---)
	0.10	---	---	+ (---)	+ (---)	+ (---)
	0.20	---	---	+ (---)	+ (---)	+ (---)
2.0	0.02	---	---	+ + (---)	+ + (---)	+ + (---)
	0.05	---	---	+ + (---)	+ + (---)	+ + (---)
	0.10	---	---	+ + (---)	+ + (+)	+ + (---)
	0.20	---	---	+ + (+)	+ + (+)	+ + (+)

Chemical speciation of composite flocculants

In order to trace the transformation of chemical species in PAC and organic polymers after synthesis, a series of identifications with ^{27}Al NMR, Al-ferron complexation and FTIR spectrometry were carried out. The prepared composite samples listed in Table 2 were examined.

^{27}Al NMR identification

The ^{27}Al nuclear magnetic resonance was performed using a JEOL FX-90Q spectrometer with an interior standard of NaAl(OH)$_4$ for quantitative determination. The NMR spectrograms for PAC, PACPc-1 and PACPa-1 are shown in Figures 1 and 2. Table 3 shows their Al$_T$ concentrations and species distribution (%) as calculated by the spectrograms. Al$_m$ denotes the Al species of monomer and oligomer present at 0 ppm on the spectrogram; Al$_{13}$ denotes the Al species of tridecamer AlO$_4$Al$_{12}$(OH)$_{24}$$^{7+}$ at the chemical shift of 62.5 ppm, which represents the most active coagulating component in polyaluminum flocculants. Al$_n$ denotes the Al species that cannot be observed in the NMR spectrogram. The peak of interior standard is situated at the point 80 ppm.

Table 3. Chemical speciation of the composites by ^{27}Al NMR

PACPc-1				
OP/Al	0	0.05	0.10	0.20
Al_T	0.0403	0.0386	0.0394	0.0408
$Al_m\%$	23.65	21.79	23.36	21.82
$Al_{13}\%$	58.27	55.71	54.12	52.06
$Al_n\%$	18.08	22.50	22.52	26.12

PACPa-1				
OP/Al	0	0.02	0.04	0.08
Al_T	0.0403	0.0395	0.0388	0.0407
$Al_m\%$	23.65	22.20	23.13	21.28
$Al_{13}\%$	58.27	53.36	53.46	47.91
$Al_n\%$	18.08	24.44	23.41	30.81

Table 4. Chemical speciation of composite by Al-ferron time complexation

PACPc-1				
OP/Al	0 (PAC)	0.05	0.10	0.20
Al_T	0.0403	0.0386	0.0394	0.0408
Ala%	27.10	27.17	26.70	26.11
Alb%	49.14	48.24	46.50	45.60
Alc%	23.76	24.59	26.80	28.29

PACPa-1				
OP/Al	0 (PAC)	0.02	0.04	0.08
Al_T	0.0403	0.0401	0.0412	0.0406
Ala%	27.10	26.75	26.88	25.74
Alb%	49.14	45.98	45.54	45.16
Alc%	23.76	27.27	27.58	29.10

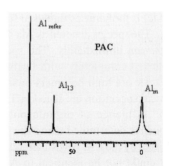

Figure 1. ^{27}Al NMR spectrogram of PAC

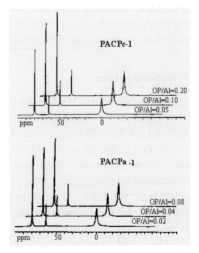

Figure 2. ^{27}Al NMR spectrogram of PACPc-1 and PACPa-1

The monomer and oligomer Al species changed only slightly both in PACPc-1 and PACPa-1, but Al_{13} decreased significantly especially at higher OP/Al ratios (Table 3). On the other hand, the species unobservable by means of NMR increased accordingly. It was assumed that the organic polymers might impair the tetrahedral structure of AlO_4 or move the hydrolysis processes forward to increase the concentration of even higher polymers, sol and precipitates.

Al-ferron time complexation assay

This method can be used to differentiate three kinds of Al species by the kinetics of their complexation with the ferron reagent using the curves of absorption versus time at a wavelength of 370 nm on a spectrometer. The nomenclature (Ala, Alb and Alc) is the same as that used for pure PAC according to the reaction rate with ferron. The species denoted as Ala are the monomers and oligomers that complexed with ferron immediately. The species labeled Alb are medium polymers that complexed completely with ferron in 60 min with our procedure. The species called Alc are high polymers or sol that do not react with ferron. The results of chemical speciation are listed in Table 4.

The influence of organic polymer on aluminum species distribution in composites determined using the Al-ferron assay was very consistent with that identified by the ^{27}Al NMR method. However, the Alb determined using the ferron assay could not be identical to Al_{13} quantitatively and the Alb/Al_{13} ratio was 0.87 ± 0.3. Alb would be identical to Al_{13} if the ratio approached 1, as we (Tang and Luan, 1996) and Parker and Bertsch (1992) found previously for pure PAC. Perhaps the organic polymer additives affected the reaction kinetics of Al with ferron thus decreased the Alb amount.

FTIR spectrogram

To determine the effect of PAC on the groups of organic polymer, IR spectrograms of the organic polymers and 60°C-dried composite samples were examined using a BIO-RAD FTS-20E Fourier transform infrared spectrometer.

In Figure 3, the IR spectrograms of organic polymers C109P and AH200P were compared to those of their mixtures with 10% PAC or $AlCl_3$. When AH200P was combined with Al species, the peak at 1670 cm^{-1} shifted to 1640 cm^{-1} for PAC and 1653 cm^{-1} for $AlCl_3$. At the same time, the spectrogram of PAC changed much at the fingerprint sections after mixing with AH200P. These transfers demonstrated that the anionic organic and cationic inorganic polymers were joined not only with charge interactions but also with chemical bonds. The spectrograms of C109P and PAC changed less after mixing. Accordingly, the interaction between cationic polymers would be weaker because their positive charges repel each other.

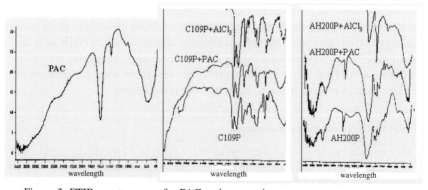

Figure 3. FTIR spectrograms for PAC and composites

The charge intensity of composites

The sign and strength of possessive charge should be the important property of any composite flocculants. It can be indicated by zeta potential (ζ) or streaming potential. We used a SC-2300 Detector to determine the streaming current (SC) that is related to the zeta potential by the expression $SC = \pi \varepsilon \rho r2\zeta/\eta l$, where ε and η are the dielectric constant and viscosity of water, respectively, and ρ, r, and l are the parameters of the capillary in the detector.

The strength of charges for cationic composites indicated by SC values is illustrated in Figure 4. When combined with cationic organic polymers, composites have a higher SC (stronger charge); SC also increased with the OP/Al ratios. The isoelectric points of the composites moved to the left accordingly. When combined with anionic organic polymers, however, the SC curves were close to each other. These results indicate that the charges of PAC did not change significantly after

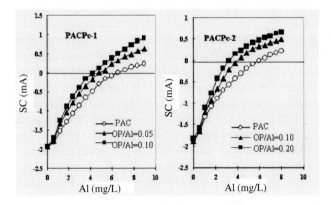

Figure 4. Streaming current for cationic composites

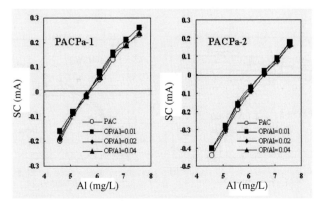

Figure 5. Streaming current for anionic composites

synthesis. Although the SC values of the composites themselves decreased reasonably with increasing OP/Al ratios, they were all a little higher than the SC of PAC. The composites of organic polymers with long chains and stronger hydrophobicity may have increased the SC by adsorbing onto the piston in the detector and giving a stronger response than the original PAC as described in the next section.

In addition, determination of electrophoresis mobility with a BIC ZETA PLUS meter also indicated that the flocculants AH200P and AN910SH are weakly anionic. Their behaviors were similar to nonionic N505P. When they were added to a suspension in which the particles already adsorbed PAC, then they could not change the electrophoretic mobility of the particles very much. It was also suggested that the charge intensity of these anionic polymers was much weaker than that of cationic polymers like PAC.

Surface adsorption of composites

Adsorption of coagulant or flocculant onto particulates is one of the key steps in both charge neutralization and aggregation processes. In general, the adsorption affinity of organic polymer is much stronger than that of inorganic polymer, and that is the purpose of designing the composite. The adsorption of various flocculants in kaolinite suspension was compared and their isotherms are shown in Figure 6.

All the isotherms of cationic polymer composites and PAC are presented as the Langmuir type with a plane line of maximum adsorption. It was assumed that they were adsorbed by single layer on the surface of kaolinite particles and saturated at higher dosing concentrations. Due to the strong positive charges, the composites might repel each other. As a result, the adsorption of cationic composites for kaolinite particles was less than that of PAC and adsorption decreased as OP/Al ratio increased. Charge exclusion was more important than adsorption affinity perhaps because of the lower molecular weight and shorter chain length of cationic polymers in general.

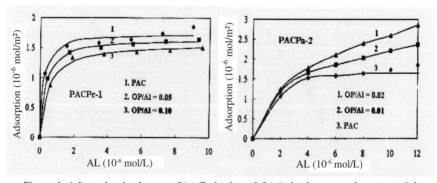

Figure 6. Adsorption isotherms of (a) Cationic and (b) Anionic composites on particles

The isotherms of anionic organic polymer composites were different from those of inorganic polymer composites. They showed the Freundlich type without any saturated maximum value as concentration increased. In contrast to cationics, the adsorption of anionics increased with OP/Al ratio and all of them were much larger than PAC. The behaviors of anionic composites might be the result of their higher molecular weight and longer chain length, making them more hydrophobic. These characteristics made anionic composites highly effective for adsorbing particulates.

Configurations of composites

The configurations of PAC and the composites were observed using transmission electron microscopy (TEM, JEM-100CXII) with a magnification of 2.9×10^3 for all micrographs. A few selected examples are shown in Figure 7. The Figures observed were all aggregates or fragments of flocs formed in water.

PAC aggregates (N1) were a few μm in size, with their Al_{13} units 2~3 nm in

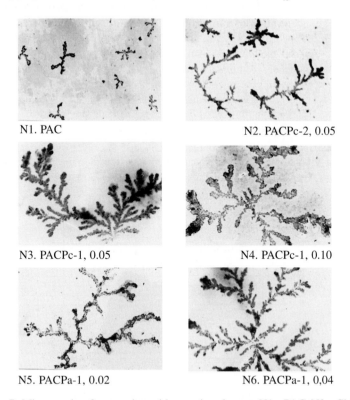

N1. PAC

N2. PACPc-2, 0.05

N3. PACPc-1, 0.05

N4. PACPc-1, 0.10

N5. PACPa-1, 0.02

N6. PACPa-1, 0,04

Figure 7. Micrographs of composites with organic polymers. N1 = PAC, N2 = Chitosan, N3 and N4 = Cationic C109P, N5 and N6 = Anionic AN200P

dimension (Dongsheng *et al.*, 2000).The fractal dimension D_f of polyaluminum aggregates was about 1.8 determined by Bottero and in our laboratory using SAXS (Bottero *et al.*, 1987, Gu, 2000) Chitosan composites (N2) were larger and had more branches than the original PAC.

Composites with cationic C109P (N3 and N4) and composites with anionic AH200P (N5 and N6) were much bulkier and had more multiple branches especially at higher OP/Al ratios. The basicity of PAC could make the composites even bulkier. Based on their bulk and branching, organic polymer composites would be expected to enhance the aggregating power of inorganic polymer flocculants.

Efficiency of composite flocculants

Coagulation and flocculation tests were carried out with a kaolinite suspension, which was prepared in laboratory with conventional procedures. The average size of the kaolinite particles was 1.28 μm (Coulter Counter) and their specific surface area was 16.1 m^2/g (BET). Humic acid (10 mg/L) was also added to the suspension. The initial concentration of kaolinite was 100 mg/L with a pH of 7.50. The jar

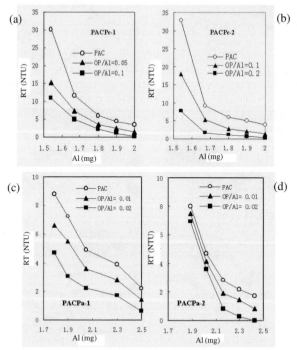

Figure 8. Residual turbidities (RT) after coagulation with various composites. (a) and (b) Cationic composites, (c) and (d) Anionic composites

tests were performed in a PB700 Tester (PHILIPS & BIRD) and turbidity (NTU) was determined using a light scattering turbidometer (NDH-20D).

The jar test results (Figure 8) showed that the residual turbidities after dosing organic composites were much lower than the turbidities after dosing PAC with the same concentration of Al (mg/L). The cationic composites were even better than the anionics for removing turbidity though the adsorption ability of the latter was stronger as the experiments demonstrated in Figure 6. It might be the presentation of humic acid in the suspension that possessed more negative charges to be neutralized but the anionic composites could not contribute the positive charge to overcome them. Furthermore, the OP/Al ratio for anionic composites was much lower than that of cationics.

Nevertheless, it cannot be considered that the additive of cationic polymer was superior to anionic at all conditions. Which sort of composites takes the priority in use should be determined in terms of the composition of specific raw water. For example, the raw water with less particles and low turbidity but weakly negatively charged would be better to use the anionic composites that possess stronger bridging effects. If the raw water is polluted by more organics with negative charges then to use cationic composites would be better for charge neutralization.

Discussion and conclusions

An overview of the properties of cationic and anionic composites is shown in Table 5. Through the study it may be concluded that the organic polymer additives can improve the characters of inorganic polymer flocculants by modifying their weakness of less molecular weight and lower aggregate ability. The cationic composite can enhace their charge neutralization effects at the same time. Both cationic and anionic polymers contributed, though in different aspects, some new properties that enhanced the efficiency in coagulation and flocculation processes.

Table 5. Comparison of the organic polymer enhanced composite flocculants

Flocculants	Inorganic Polymer PAC	Cationic Composites PACPc	Anionic Composites PACPa
Chemical Speciation	$Al_{13} \approx Alb > 50 \sim 80\%$	Al_{13} decreased appreciably, but still predominated	Al_{13} decreased appreciably, but still predominated
Charge Intensity	higher than that of metal salt coagulants	charge-neutralization ability enhanced significantly	charge transformation insignificant
Configuration of Aggregates	MW and chain length unsatisfactory	much bulkier, multiple branches	much bulkier, multiple branches
Surface Adsorption	single layer or local coverage	single layer, adsorption decreased	multi-layer, adsorption much enhanced
Applicability to Raw Water	superior to metal salt coagulants	suitable for more negatively-charged raw water	suitable for low turbidity, low temperature raw water

Weak anionic polymer additives would be more suitable to the requisition of PAC to improve its aggregation power significantly with lower ratios of OP/Al and thereby decrease the PAC's positive charge not much. From this point, weak anionic polymer additives are better than polysilicate additives that need a high Si/Al ratio and thus decrease the charge intensity of PAC too much (Wang and Tang, 2001).

Cationic polymer additives can improve both the charge neutralizing and aggregating power of PAC but owing to their lower molecular weight than anionic polymers commonly the OP/Al ratio should be higher and thus increase the production cost. However, the cationic polymer composites may be more powerful to the raw water polluted by various negatively charged organics.

While laboratory-scale experiments on composites have been promising, the industrial technology for production of organic polymer composites and their economical profitability should be studied further.

References

Bertsch, P.M., Parker, D.R. (1995) Aqueous Polynuclear Aluminum Species. In: The Environmental Chemistry of Aluminum, 2nd Edition, 139-140, Sposito, G (Ed.). Lewis Publishers, pp.464

Bottero, J.Y. (1987.) Mechanism of formation of aluminum trihydroxide from Keggin Al_{13} polymers. *J. Colloid Interface Sci.* **117**, 47

Dongsheng, W., Tang, H.X, Cao, F. (2000) Particle Speciation Analysis of Inorganic Polymer Flocculants: an Examination by Photon Correlation Spectroscopy. *Colloids and Surfaces* **166**, 27-32

Gu, J. (2000) Identification of Inorganic Polymer Flocculants with Light Scattering Spectroscopy and Small Angle X-ray Spectroscopy, Postdoctorial Thesis

Parker, D.R., Bertsch, P.M. (1992) Identification and Quantification of the "Al_{13}" Tridecameric Polycation Using Ferron. *Environ. Sci. Technol.* **26**, 908-914

Shi, B., Tang, H.X. (1999) Study on the Electrical Charge and Coagulation Behaviors of Polyaluminum-Organic Polymer Composite Flocculants. *Environmental Chemistry* (in Chinese) **18,** 302-308

Shi, B., Tang, H.X. (2000) The Speciation of Polyaluminum-Organic Polymer Composite Flocculants. *Acta Scientiae Circumstantiae* (in Chinese) **20**, 391-396

Tang, H.X. Luan, Z.K.(1996) The Differences of Behaviour and Coagulating Mechanism between Inorganic Polymer Flocculants and Traditional Coagulants. In: Chemical Water and Wastewater Treatment IV, H.H. Hahn, E. Hoffman, H. Ødegaard (Eds.) Springer Verlag, pp. 83-93

Tang, H.X., Luan, Z.K., Wang, D.S., Gao, B.Y. (1998) Composite Inorganic Polymer Flocculants. In: Chemical Water and Wastewater Treatment V, H.H. Hahn, E. Hoffman, H. Ødegaard (Eds.). Springer Verlag, pp. 25-34

Wang, D.S., Tang, H.X. (2001) Modified Inorganic Polymer Flocculant-PFSi: Its Preparation, Characterization and Coagulation Behavior. Water Research 35, 3418-3428

H.H. Hahn, E. Hoffmann, H. Ødegaard (Eds.)
Chemical Water and Wastewater Treatment VII, pp. 29-38
© IWA Publishing, London
ISBN: 1 84339 009 4

Break-up and Re-formation of Flocs Formed by Hydrolyzing Coagulants and Polymeric Flocculants

J. Gregory and *M. A. Yukselen*

*University College London, Dept of Civil & Environmental Engineering
email: j.gregory@ucl.ac.uk

Introduction

The properties of flocs formed in water and wastewater treatment are of great importance in solid-liquid separation. The size, density and strength of flocs are affected by many process variables, including the nature and concentration of the coagulant, the initial mixing conditions and the shear regime during floc formation. These properties are closely related (Gregory, 1997), since floc size in a sheared suspension is limited by floc strength, and increased floc density leads to greater floc strength. The factors that determine these properties under practical conditions are not well understood, although some progress has been made in model systems (e.g. Spicer *et al.*, 1998).

Hydrolyzing metal coagulants, based on aluminium or iron are widely used in water treatment and their mode of action is broadly understood in terms of charge neutralization and "sweep flocculation" in which hydroxide precipitation plays a major role (Gregory and Duan, 2001). The relative importance of these mechanisms in the case of pre-hydrolyzed products, such as polyaluminium chloride, is not so well established. Polymeric flocculants are also commonly used and their behaviour can usually be explained by either charge neutralization (including "electrostatic patch" effects) or polymer bridging. There is a superficial similarity between these and the corresponding mechanisms for hydrolyzing metal coagulants and it is of some interest to compare the properties of flocs produced by these agents.

The present work is part of a comprehensive study (Gregory *et al.*, 2000) of dynamic aspects of flocculation, including the formation, break-up and re-formation of flocs under controlled conditions. Here, we compare the behaviour of a number of hydrolyzing coagulants, including pre-hydrolyzed forms, and some commercial polymeric flocculants. Dynamic monitoring, using a flow-through optical technique, and conventional turbidity measurements have been used, with particular attention to the effects of floc breakage and re-formation.

Materials and methods

Suspension

A suspension of kaolin clay (Imerys, St Austell, Cornwall, UK) was prepared as previously described (Gregory *et al*, 2000) and diluted to give a stock suspension with 50 g/L kaolin. The particles were mostly below about 5 μm in size, with a mean size of about 2 μm, determined by an Elzone particle counter.

For the flocculation tests, the stock suspension was diluted in London tap water to give a clay concentration of 50 mg/L. London tap water has a high alkalinity and a pH value of around 7.4, and so it is a quite convenient medium for studies of hydrolyzing coagulants. However, its high calcium content (around 2 mM Ca^{2+}) causes destabilization of the kaolin particles and slow coagulation. To avoid this difficulty, a small amount of commercial humic acid (Aldrich) was added to the stock kaolin suspension. Humic acid adsorbs on the clay particles and gives enhanced stability against divalent metal ions. Humic acid solution was included in the stock 50 g/L kaolin to give a concentration of 0.5 g/L, so that the suspensions used in the flocculation tests had 50 mg/L kaolin and 0.5 mg/L humic acid. Under these conditions it is very likely that almost all of the humic acid is adsorbed on the clay particles, with very little in solution.

Coagulants

Aluminium sulphate hydrate ($Al_2(SO_4)_3 \cdot 16H_2O$; Fisons) '**alum**', and **ferric sulphate** hydrate ($Fe_2(SO_4)_3 \cdot 6H_2O$; Sigma) were used. M/10 stock solutions were prepared, kept in refrigerator at 5°C and renewed every two weeks.

Two commercial products from Kemira Kemi AB, Helsingborg, Sweden were used. A polyaluminium chloride (PACl) product, **PAX-XL9** that had a degree of neutralization, r (=OH/Al), of 2.1, supplied as a 4.6 wt % Al solution and an iron-based product, **PIX-115** supplied as a 11.4 wt % Fe solution. These solutions were used directly, without prior dilution.

Two cationic polyelectrolytes and one anionic polyelectrolyte from Allied Colloids Ltd (now Ciba Specialty Chemicals) were used:

Magnafloc 1697 (poly(diallyldimethylammonium) chloride, polyDADMAC), with an intrinsic viscosity (IV) of about 0.2, or a molecular weight of around 50000, and a high cationic charge density (about 6 meq/g). This was supplied as a 40 % aqueous solution and diluted to give a stock solution of 0.1% or 1 g/L.

Zetag 64, a copolymer of acrylamide and a cationic monomer (about 40 mole % cationic), with IV of about 12 and a molecular weight of several millions. This was made up to a 0.1 % solution by first wetting 0.1g of the solid polymer with 2 mL acetone and then adding 98 mL of deionized water. The mixture was rotated for several hours to ensure complete dissolution of the polymer.

Magnafloc 156, about 40 mole % anionic, with IV of about 13. This was made up to a 0.1 % solution in the same way as for Zetag 64.

Apparatus

A continuous optical flocculation monitor (PDA 2000, Rank Brothers Ltd., Cambridge, UK) was used in a modified jar test procedure. The test suspension was contained in 1-L beakers with stirrer units from a Flocculator 90, semi-automatic jar test device (Kemira Kemwater, Helsingborg, Sweden). This enables the rapid mixing and slow stirring speeds and times to be pre-set. For dynamic monitoring, sample from one beaker was circulated through transparent plastic tubing (3 mm i.d.) by means of a peristaltic pump. The pump was located after the PDA instrument to avoid effects of possible floc breakage in the pinch portion of pump. The tubing was clamped in the PDA instrument so that the flowing sample was illuminated by a narrow light beam (850 nm wavelength). The PDA 2000 measures the average transmitted light intensity (dc value) and the rms value of the fluctuating component. The ratio (rms/dc) provides a sensitive measure of particle aggregation (Gregory and Nelson, 1986). In this work, the ratio value is called the *Flocculation Index* (FI).

Procedure

Using standard jar tests with 800 mL of test suspension (50 mg/L kaolin and 0.5 mg/L humic acid in London tap water) the optimum dosages of coagulants were determined. After rapid mix at 400 rpm (G = 518 sec^{-1}) for 10 seconds, slow stirring at 50 rpm (G = 23 sec^{-1}) for 10 to 60 minutes and settling for 30 minutes, supernatant samples were withdrawn for turbidity measurements. The G values were calculated following Mejia and Cisneros (2000), who used a similar jar test device. The 10s rapid mix period has been found to be about optimum (Gregory and Yukselen, 2002). Longer rapid mix times give significantly smaller flocs.

For dynamic tests, sample was pumped from a stirred beaker at about 25 ml/min through the tubing and the average (dc) and fluctuating (rms) components of the transmitted light intensity were monitored by the PDA instrument. Readings were taken every two seconds and the results were stored in a computer for subsequent spreadsheet analysis. After allowing 1 minute for steady state readings to be established, coagulant was dosed and the suspension was stirred at 400 rpm for 10 seconds. The stirring speed was reduced to 50 rpm and held at this value for the required time (10 to 60 min). In order to investigate the floc breakage and re-formation, the stirring speed was increased to 400 rpm for times ranging from 5 to 300 seconds and then reduced back to 50 rpm. All experiments were carried out 2-3 times and very little variation was observed.

Results

Metal-based coagulants

The optimum dosage for alum was found to be about 3.4 mg/L as Al. For comparison, the dosage of PAX-XL9 was chosen to give the equivalent concentration as Al. In the case of ferric sulphate, the dosage was chosen to give about the same molar concentration as Al (i.e. 7 mg/L as Fe) with a corresponding dosage as Fe for PIX-115.

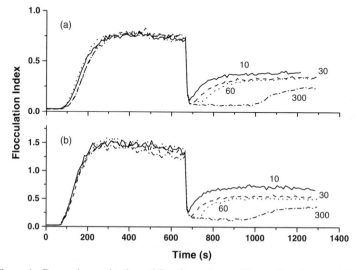

Figure 1. Dynamic monitoring of floc formation at 50 rpm, breakage at 400 rpm and re-formation at 50 rpm. (a) alum and (b) PAX XL9

Floc formation, breakage and re-formation were investigated systematically. For the metal-based coagulants, no significant changes were observed in the Flocculation Index (FI) values after 10 minutes of slow stirring. So, for coagulation with these additives, 10 minutes at 50 rpm was adopted as the standard floc formation period.

Figure 1 shows the results of dynamic monitoring for alum and PAX XL9. Floc formation conditions are the same in all cases and the only variable is the time of floc breakage at 400 rpm (from 10 to 300 s). The initial parts of the flocculation curves show the reproducibility of the procedure. The FI value rapidly reaches a maximum value, corresponding to a limiting floc size. This behaviour is generally explained in terms of a balance between floc growth and breakage. For the PACl product the maximum value is about twice that for alum and is reached more quickly, indicating larger and hence stronger flocs.

When the stirring speed is increased to 400 rpm, floc breakage occurs immediately and is nearly complete within a short period. The 30 and 60 second breakage periods give nearly the same decrease in FI. Also, in these cases, there is a similar degree of recovery after the stirring speed is reduced to 50 rpm. With only 10 seconds at 400 rpm the FI value shows a slightly smaller decrease, presumably because there is insufficient time to complete the initial breakage, and there is rather more recovery at 50 rpm. With 300 seconds at 400 rpm there is evidence of a much more gradual decrease in FI with a more limited recovery at 50 rpm. The new plateau was always lower than the first value and was reached

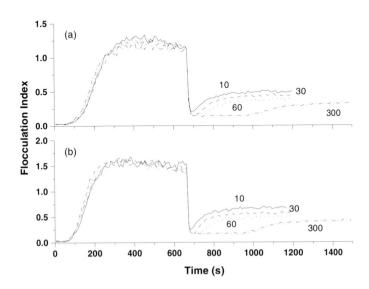

Figure 2. As Figure 1, but with (a) ferric sulphate and (b) PIX 115

rather more slowly indicating that the breakage of alum flocs under the stated conditions is not reversible. The floc formation/breakage/re-formation cycle can be repeated several times, giving a steadily decreasing size of the re-formed flocs (Yukselen and Gregory, 2002). Although the PACl coagulant gives larger flocs at 50 rpm, there is about the same relative size reduction at 400 rpm and about the same degree of recovery at 50 rpm.

Results for the two iron-based coagulants are shown in Figure 2. The curves are of similar general form to those in Figure 1, but the initial value of FI is significantly greater for ferric sulphate than for alum. Also, PIX-115 gives a slightly higher FI value than PAX-XL9. Again, floc breakage at 400 rpm was very rapid and re-growth at 50 rpm occurred to only a limited extent.

Polymeric flocculants

With the polymeric additives, floc formation is considerably slower than for metal-based coagulants, especially for the higher molecular weight materials. For this reason, longer slow stirring times were used (30 minutes for polyDADMAC and 60 minutes for the other two polymers). The dosage for all polymers was chosen as 125 µg/L, which is within the optimum range for each.

The results are shown in Figure 3, which also has the results from Figures 1 and 2 for comparison. Apart from the longer slow stirring times with the polyelectrolytes, the conditions are the same in all cases. The time of floc breakage at 400 rpm is chosen as 300 seconds. The results show dramatic differences in behaviour between hydrolyzing coagulants and polyelectrolytes. In the latter case

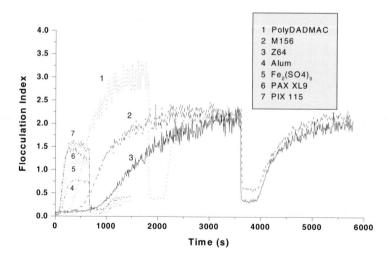

Figure 3. Monitoring of Floc formation with different additives

significantly larger flocs are formed (higher FI values) and the recovery after floc breakage is much greater. However the onset of flocculation is significantly delayed with these additives, especially for Zetag 64 and Magnafloc 156, where significant floc growth was apparent only after about 10 minutes of slow stirring. It is noteworthy that the largest flocs are formed with polyDADMAC, which has the lowest molecular weight and the highest charge density. This also shows considerably less recovery after breakage than the higher molecular weight polymers. The latter show fairly similar patterns of floc formation and breakage, despite their different charge.

Jar test results

For each coagulant the residual turbidities before floc breakage (Peak I) and for re-formed flocs (Peak II) after breakage at 400 rpm for 300 seconds are shown in Figure 4. It is apparent that the residual turbidity with re-formed flocs is significantly higher than the previous value for hydrolysing coagulants. PolyDADMAC shows slightly higher turbidities after floc breakage and re-formation, but Zetag 64 and M156 give similar values. These are in line with the FI results in Figure 3. However, the residual turbidity values before floc breakage show less difference between the various additives than expected from the FI values. For instance, Zetag 64 gives a much higher FI value than PAX XL9, but the residual turbidities before floc breakage are about the same.

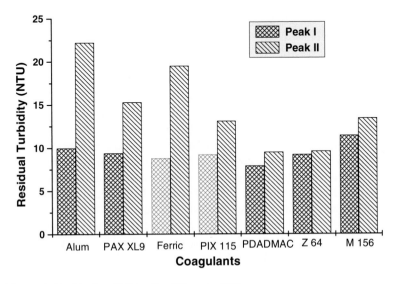

Figure 4. Residual turbidities for different additives

Discussion and Conclusions

The results presented here highlight very significant aspects of floc formation, breakage and re-growth. Under the conditions used, the metal-based coagulants showed a more rapid onset of flocculation, but significantly smaller flocs than with the polyelectrolytes. The lag time in the latter case is due to the relatively slow adsorption of these materials on the clay particles. With the dilute suspensions used and the low polymer concentrations it is expected that adsorption of sufficient polymer to give particle destabilization can be quite long compared to the particle collision rate (Gregory, 1988). The low molecular weight polyDADMAC shows a shorter lag time, which may be a result of its more rapid diffusion than the much larger molecules of the other two polymers. For the hydrolyzing coagulants, hydroxide precipitation occurs quite rapidly and so flocculation begins soon after coagulant dosing. It is noticeable that the pre-hydrolyzed products show an earlier onset of flocculation than with alum and ferric sulphate, most likely because of a more rapid precipitation.

The experiments revealed that iron-based coagulants give larger flocs than those based on aluminium. The behaviour of iron (III) is similar to that of aluminium in that sweep flocculation by the hydroxide is fast and dominates at around neutral pH. Because of the much lower solubility of ferric hydroxide, rather more precipitate would be expected than with alum for equimolar dosages (as in the present case). For both aluminium and iron, the pre-hydrolyzed product gave significantly increased floc size, but only marginal differences in residual turbidity before floc breakage. However, the re-formed flocs showed better removal by sedimentation. The similar form of the FI curves in Figures 1 and 2 strongly suggests that flocs are of broadly similar nature for all the metal-based coagulants, but are significantly stronger with the pre-hydrolyzed products. The underlying mechanisms are not well understood and it is unlikely that simple charge effects can explain the results.

The polymeric flocculants gave significant differences in floc size, with polyDADMAC giving larger FI values. This finding is difficult to explain in terms of the different molecular weights, since it is expected that higher molecular weight polymers should give stronger flocs, either as a result of polymer bridging or 'electrostatic patch' effects (Gregory, 1996). For the dilute suspensions used here and the relatively low particle collision rates, it is expected that adsorbed polymers would adopt a rather flat configuration, which makes bridging unlikely. It has been shown (Kam and Gregory, 2001) that high charge density polyelectrolytes are more effective in removing humic substances from water. Since the suspensions used are stabilized by humic acid and polyDADMAC has a much higher charge density than Zetag 64, charge effects may partly explain the observed behaviour. The fact that an anionic polyelectrolyte (Magnafloc 156) gives similar results is surprising. The high level of calcium in the tap water used may play a significant part.

Although there have been previous reports of the irreversible nature of floc breakage with hydrolysing coagulants, the reasons remain unclear. Irreversible breakage suggests that chemical bonds are broken. When particle interaction is of a physical nature (such as van der Waals or electrostatic attraction) there is no obvious reason why aggregates should not re-form after breakage.

With polymeric flocculants, irreversible floc breakage is well known, but it is usually associated with bridging interactions (Ditter *et al.*, 1982). High shear rates (especially with turbulence) may cause scission of polymer chains (Horn and Merrill, 1984) and adsorbed polymer could adopt a more flat configuration during the breakage phase. However, these considerations do not apply when charge neutralization or 'electrostatic patch' effects are responsible for flocculation, so that reversible floc breakage might be expected in such cases. The lack of complete reversibility in the case of polyDADMAC is thus quite difficult to explain. A more detailed investigation of these effects is needed.

Dynamic monitoring of floc formation, breakage and re-formation gives much more detailed information on these processes than a simple jar test procedure. The findings cannot be adequately explained on the basis of current models, but further studies on other systems should be fruitful.

Acknowledgements

The work was funded by an EPSRC Visiting Fellowship to M.A. Yukselen and was also partly supported by INTAS grants 99-0510 and 00-0493.

References

Ditter, W., Eisenlauer, J. and Horn, D. (1982) Laser optical method for dynamic flocculation testing in flowing dispersions. In *The Effect of Polymers on Dispersion Properties* (ed. T.F. Tadros), Academic Press, London, pp. 323-342

Gregory, J. (1988) Polymer adsorption and flocculation in sheared suspensions. *Colloids and Surfaces* **31**, 231-253

Gregory, J. (1996) Polymer adsorption and flocculation. In Finch, C.A. (Ed.) Industrial Water Soluble Polymers , Royal Society of Chemistry, Cambridge, UK, pp. 62-75

Gregory, J. (1997) The density of particle aggregates, *Wat.Sci.Tech.* **36** (4), 1-13

Gregory, J. and Duan, J. (2001) Hydrolysing metal salts as coagulants. *Pure and Applied Chemistry* **73**(11), 1-10

Gregory, J. and Nelson, D.W. (1986) Monitoring of aggregates in flowing suspensions. *Colloids and Surfaces* **18**, 175-188

Gregory, J. and Yukselen, M.A. (2002) Monitoring the effect of rapid mixing on the break-up and re-formation of flocs. 4th World Congress on Particle Technology, Sydney, 21-25 July

Gregory, J., Rossi, L. and Bonechi, L. (2000) Monitoring flocs produced by water treatment coagulants, In: *Chemical Water and Wastewater Treatment VI* (H.H.Hahn, E.Hoffmann and H.Ødegaard, Eds.), Springer, Berlin, pp 57-65

Horn, A.F. and Merrill, E.W. (1984) Midpoint scission of macromolecules in dilute solution in turbulent flow. *Nature* **312**, 140-141

Kam, S.K. and Gregory, J. (2001) Interaction of humic substances with cationic polyelectrolytes. *Water Research* **35**(15), 3557-3566

Mejia, A.C. and Cisneros, B.J. (2000) Particle size distribution (PSD) obtained in effluents from an advanced primary treatment process using different coagulants. In: *Chemical Water and Wastewater Treatment VI* (H. Hahn, E. Hoffmann and H. Ødegaard, Eds.), Springer, Berlin, pp. 257-268

Spicer, P.T., Pratsinis, S.E, Raper, J., Amal, R., Bushell, G., Meesters, G. (1998) Effect of shear schedule on particle size, density, and structure during flocculation in stirred tanks. *Powder Technology* **97**, 26-34

Yukselen, M.A. and Gregory, J. (2002) Breakage and reformation of alum flocs. *Environmental Engineering Science* (Submitted)

Flocculation and Floc Separation

H.H. Hahn, E. Hoffmann, H. Ødegaard (Eds.)
Chemical Water and Wastewater Treatment VII, pp. 41-50
© IWA Publishing, London
ISBN: 1 84339 009 4

Optimisation of the Flocculation Process Using Computational Fluid Dynamics

K. Essemiani and C. de Traversay*

*Anjou-Recherche Vivendi Water, Chemin de la Digue, Maisons Laffitte, France
email: karim.essemiani@generale-des-eaux.net

Introduction

Flocculation plays a dominant role in water treatment processes. Because there is a complex interdependence of numerous factors inherent to the coagulation and flocculation processes, a thorough understanding of the phenomena involved is essential.

The coagulation-flocculation steps aim to agglomerate the suspended particles to form larger flocs that are easier to settle. These flocs are formed in a stirred tank vessel. The performance of the flocculation process depends on the mixing rate (turbulence generated by the impeller), impeller type, reactor geometry, efficiency of the coagulation step (type and dose of coagulant), nature of the water and the hydraulic residence time (inlet flow rate). All these parameters can have various degrees of impact on the flocculation efficiency. The effect of some of these parameters can be evaluated using Computational Fluid Dynamics (CFD) modelling tools for an optimal design and scale up of the process. Understanding the hydraulic behavior of unit processes is a frequent interrogation, and computational fluid dynamics can provide the information necessary for designing, optimizing or retrofitting of various treatment processes (Levecq *et al.,* 2001).

In this study, FLUENT CFD software was used to determine the local hydrodynamics induced by the impeller and to evaluate the mixing efficiency needed for flocculation. The optimal configuration could be determined by testing different configurations.

Experimental apparatus

The pilot unit (Figure 1) consisted of a rectangular flocculation stirred vessel equipped with a Mixel TT axial impeller and fed with a central inlet tube (0.05 m diameter). The liquid volume was 1 m³, and the liquid height 1m.

The hydrodynamics of the flocculation vessel were modified by varying the inlet flow rate (from 2 to 6 m³•h⁻¹) and the impeller rotating speed (36 and 73 rpm). All the other parameters (water quality, impeller type, coagulation step, flocculant type) were kept constant.

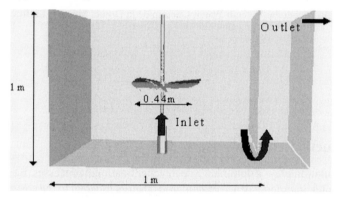

Figure 1. Geometry of the flocculation stirred vessel

CFD modelling

CFD is a computer-based methodology for solving the fundamental, non-linear equations of fluid flow, namely, the Navier-Stokes equations coupled with the transport equations of the turbulence. The Fluent software is a CFD commercial tool. It allows the hydrodynamics of processes to be simulated by taking into account the flow patterns and local variables such as velocity, turbulent kinetic energy and local velocity gradient G.

The methodology of the CFD procedure involves three steps:

1   Construction of the mesh which consists of dividing the geometry into small cells.
2   Resolving the equations of conservation of mass, momentum and energy for each cell.
3   Interactive analysis of the results, which consists of the visualisation of different variables in each cell (velocity components, concentration, turbulent kinetic energy, pressure, and so on).

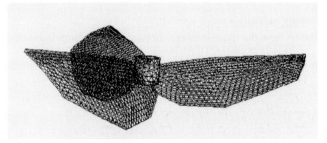

Figure 2. Mesh of the Mixel TT axial impeller

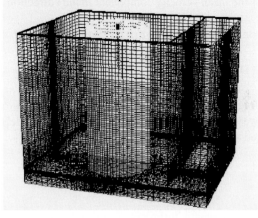

Figure 3. Rotating reference region and the mixing tank mesh

A large body of knowledge has been accumulated on the flow in stirred tanks using the numerical approach (Kresta, 1991; Bakker, 1992; Harvey III *et al.* 1997; Lee and Yianneskis, 1998; Essemiani, 2000).

Two approaches in computational fluid dynamics can be applied to mixing: (1) one can assign to the impeller some boundary conditions that are equivalent to the impeller in terms of mean flow and turbulence or (2) one can take into account the real geometry of the impeller (Figure 2). The second approach, called the Multiple Reference Frames approach (MRF), has the main advantage of being independent of any experimental data for the impeller's boundary conditions.

The resolution with the MRF approach is based on the use of two references, one related to the lateral walls and the baffles and the other to the movement of the rotating shaft and the impeller. The fluid zone is divided into two regions, one related to the fixed reference and one close to the impeller related to the moving reference (Figure 3). We solve the Navier-Stokes equations and other closure models in the two domains. Two new terms appear in the equations written in the moving reference: the coriolis and the centrifugal forces.

The following equations were solved:
For the mean flow the mass conservation and momentum balance equations

$$\sum_i \frac{\partial \overline{u}_i}{\partial x_i} = 0 \tag{1}$$

$$\frac{\partial \overline{u}_i}{\partial t} + \sum_j \overline{u}_j \frac{\partial \overline{u}_i}{\partial x_j} = -\frac{1}{\rho} \frac{\partial \overline{p}}{\partial x_i} + v \sum_j \frac{\partial^2 \overline{u}_i}{\partial x_j^2} - \sum_j \frac{\partial \overline{u'_i u'_j}}{\partial x_j} \tag{2}$$

Where $\overline{u}_i$ is the mean velocity component in the i direction and $\overline{P}$ is the mean pressure.

This equation contains one term that needs to be modelled: the Reynolds stress

$$\tau^t_{ij} = -\rho \overline{u'_i u'_j} \tag{3}$$

The modelling of this term is considered the problem of turbulence modelling. In this paper, we restrict our analysis to a classical (k-ε) model of turbulence. The equation related to turbulence modelling is:

$$v_t = C_\mu \frac{k^2}{\varepsilon} \tag{4}$$

Where v_t is the kinematic turbulent viscosity, k is the turbulent kinetic energy (TKE), and ε is its dissipation rate.

The transport equations of the turbulent kinetic energy and its dissipation rate are:

$$\frac{\partial \varepsilon}{\partial t} + \frac{\partial \varepsilon \overline{U}_j}{\partial x_j} = \frac{\partial}{\partial x_j}\left(\frac{v_t}{\sigma_\varepsilon} \frac{\partial \varepsilon}{\partial x_j}\right) - C_{\varepsilon 1} \frac{\varepsilon}{k} \overline{u'_i u'_j} \frac{\partial \overline{u}_i}{\partial x_j} - C_{\varepsilon 2} \frac{\varepsilon^2}{k} \tag{5}$$

$$\frac{\partial \varepsilon}{\partial t} + \sum_j \frac{\partial \varepsilon \overline{u}_j}{\partial x_j} = \sum_j \frac{\partial}{\partial x_j}\left(\frac{v_t}{\sigma_\varepsilon} \frac{\partial \varepsilon}{\partial x_j}\right) - \sum_j \sum_i C_{\varepsilon 1} \frac{\varepsilon}{k} \overline{u'_i u'_j} \frac{\partial \overline{u}_i}{\partial x_j} - C_{\varepsilon 2} \frac{\varepsilon^2}{k} \tag{6}$$

σ_k, C_μ, σ_ε, $C_{\varepsilon 1}$ and $C_{\varepsilon 2}$ are empirical constants with values published in literature (Bakker, 1992).

Results and discussion

The pilot unit was simulated for the same operating conditions and by varying the same parameters (inlet flow rate and rotating velocity). In this paper we present the results obtained for the 2m³•h⁻¹ inlet flow rate and the 73 rpm impeller rotating velocity.

The velocity field induced by the axial impeller is presented in Figure 4. It shows how the central inlet affects the down pumping flow. Normally the Mixel TT axial impeller induces two large recirculation loops in the vessel, but the incoming jet from the central inlet and the down pumping flow from the impeller induce four recirculation loops in the vessel. The behaviour is close to that of radial impellers. The highest velocities are thus located in the impeller region.

Figure 5 shows the particle pathlines injected from the inlet tube. The pathlines indicate a strong 3D rotational/axial down pumping flow with important recirculation in the upper and bottom part of the vessel.

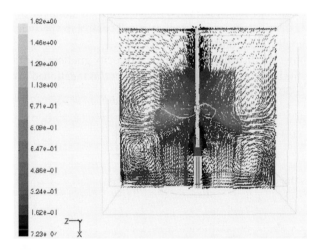

Figure 4. Velocity vectors in a vertical plane (m.s^{-1})

Figure 5. Particle pathlines in the flocculation vessel

In literature most of the time the performance of a flocculation vessel is described by the global hydraulic gradient G (s^{-1}). This parameter is defined as

$$G = \sqrt{\frac{P/Volume}{\mu}} \qquad (7)$$

Where P/Volume is the power dissipated per unit volume and μ is the dynamic viscosity (Pa•s). This approach gives an average value, however it does not tell anything about the local shear rate distribution in the tank. The rate of floc formation is directly proportional to the local shear rate distribution in the vessel. Therefore, we can expect different performances from different mixing rates and from the different types of impellers at the same global hydraulic gradient G.

The local G distribution is important as it will control the particle suspension, distribution, coalescence, and break up efficiency.

Figure 6 shows the distribution of the local velocity gradient G. This variable is calculated on the basis of a turbulent viscosity effect, as mentioned in the literature (Korpijärvi, 2000) as follows:

$$G = \left(\mu_t/\mu + 1\right)\sqrt{2\sum_i \left(\frac{\partial \overline{u_i}}{\partial x_i}\right)^2 + \sum_{i,j}\left[\frac{\partial \overline{u_i}}{\partial x_j} + \frac{\partial \overline{u_j}}{\partial x_i}\right]^2} \qquad (8)$$

where μ_t is the dynamic turbulent viscosity (Pa•s).

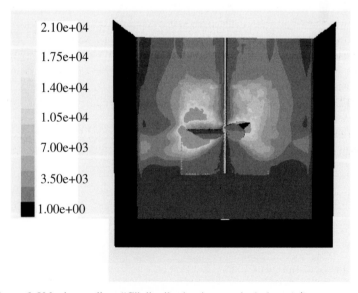

Figure 6. Velocity gradient "G" distribution in a vertical plane (s^{-1})

The maximum G values are located in the impeller region, with the highest value being 38000 s^{-1} and the mean value being 5549 s^{-1}. This local distribution makes it possible to identify the region where the highest (regions of floc breakup) and lowest (flocs coalescence) shear stress occurs.

The experimental and the numerical values were compared by means of three parameters: a Residence Time Distribution (RTD) and a comparison to published correlations (Roustan, 1999) of the power and pumping numbers (these two non-dimensional parameters are constants in turbulent regime flow).

The power number is defined as follows:

$$N_P = \frac{P}{\rho N^3 D^5} \tag{9}$$

where N is the rotating velocity, D the impeller diameter and ρ the fluid density. The power number is equal to 1 for the Mixel TT.

The pumping number is defined as follows:

$$N_{Q_P} = \frac{Q_P}{ND^3} \tag{10}$$

where Q_p is the impeller pumping flow rate. The pumping number is equal to 0.56 for the Mixel TT.

Numerically the power dissipated is calculated on the basis of the torque applied on the drive shaft and the impeller. The global G gradient is calculated from the power number using Equation 7.

Table 1 summarises the comparative values of the power number, pumping number, power dissipated, and the global hydraulic gradient G. All the experimental/numerical values are close except the pumping number, where the percent error is higher. This is due to the central inlet tube that modifies the pumping behaviour of the impeller; this specificity is not taken into account in the correlation (batch reactor).

The CFD global G value obtained using Equation 7 is close to the value obtained by the correlation, but the mean value obtained by the local G distribution (considered as a variable) is much higher (5549 s^{-1}, factor 34). This is due to the integration of the turbulence effects through the turbulent viscosity term in the G

Table 1. Comparison between correlations and CFD

	Correlation	CFD	Error %
Power number	1	0.9	10
Pumping number	0.56	0.74	24
Power dissipated (W)	30	27	10
Global G (s^{-1})	174	163	6

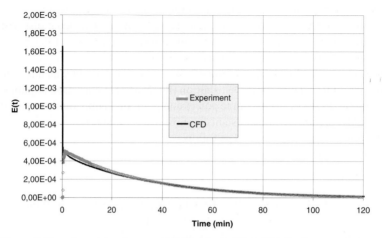

Figure 7. Experimental and numerical Residence Time Distribution

calculation (Equation 2.8). Thus it is really interesting to integrate this "local" definition of the velocity gradient G for a detailed description of the hydrodynamics induced by the impeller and relate it to the flocculation efficiency through a description of the spatial floc size distribution in the vessel.

Figure 7 presents the Residence Time Distribution curves obtained by CFD and by a tracer experiment in the pilot unit for the fixed operating conditions. The curves are very close. Thus the CFD results are validated by the experimental results.

Conclusion

A CFD tool was used to simulate the hydrodynamics induced by an axial impeller in a flocculation vessel. The CFD offers all the advantages of an experimental platform for batch through to full-scale applications (batch and continuous reactors), once validated by experimental measurements. This step was performed by comparing the pumping number, the power number (and thus the power dissipated and the global hydraulic gradient) and a tracer test (RTD) for fixed operating conditions. The local distribution of the G gradient indicates a high level of G intensity in the region close to the impeller, with lower levels far from this region. The global G definition does not take into account the eventual modifications of the process (geometry, impeller position, continuous operating mode, residence time, and so on) that directly affect the velocity gradient distribution and thus the turbulence production and floc size distribution spatially in the vessel. CFD showed a difference of a factor of 34 between the global G value obtained by classical correlation and the volume integral of the G distribution in the vessel. This significant difference can be explained by the fact that the new approach takes into account the local hydrodynamics induced by the impeller under real operating conditions (interaction with the inlet flow, non-conventional stirred vessel geometry, and so on), which can be very different from the ideal batch stirred contactor.

The next step of the study will be to test different configurations, and compare them on the basis of new hydrodynamic parameters (volume integral of local G distribution, mixing efficiency, by-pass, mixing time/residence time ratio, and so on) and the coupling of a population model that describes the break-up and coalescence of flocs with the local hydrodynamic conditions for a defined floc population.

References

Bakker, A. (1992) Hydrodynamics of stirred gas-liquid dispersions. *PhD Thesis, Delft University of Technology.* The Netherlands

Essemiani, K. (2000) Potentialities of computational fluid dynamics in stirred vessels. *PhD Thesis. University of Toulouse.* France

Harvey III, A.D., Wood, S.P., Leng, D.E. (1997) Experimental and computational study of multiple impeller flows. *Chem. Eng. Sci.* **52**(9), 1479-1491

Korpijärvi, J., Laine, E., Ahlstedt H. (2000) Using CFD in the study of mixing in coagulation and flocculation. In Hahn, H.H., Hoffmann, E., Ødegaard, H. (Eds.) Chemical water and wastewater treatment VI, Springer Berlin, 89-99

Kresta, S M. (1991) Characterization, measurement and prediction of the turbulent flow in stirred tanks. *PhD Thesis, MacMaster University.* Canada

Lee, K. C., Yianneskis M. (1998) Turbulence properties of the impeller stream of a Rushton turbine. *AIChE. J.* **44**(1) 13-24

Levecq, C., Essemiani K., De Traversay C. (2001) Hydraulic study and optimisation of water treatment processes using numerical simulation. In *Proc of the IWA 2001 congress.* Berlin. 15-19 October

Lunden, M., Stenberg, O., Andersson, B. (1994) Evaluation of a method for measuring mixing time using numerical simulation and experimental data. *Chem. Eng. Commun.* **139.**, pp 115

Roustan, M. (1999) Agitation, Mélanges, Caractéristiques des mobiles d'agitation. *Techniques de l'Ingénieur, Form.* J3 802, 1-10

H.H. Hahn, E. Hoffmann, H. Ødegaard (Eds.)
Chemical Water and Wastewater Treatment VII, pp. 51-58
© IWA Publishing, London
ISBN: 1 84339 009 4

Effect of Pre-Coagulation/Sedimentation on the Ultrafiltration Membrane Process

N-Y. Jang*, Y. Watanabe, G. Ozawa and M. Hosoya

*Hokkaido University, Dept. of Urban Environmental Engineering, Sapporo, Japan
email: nyjang@eng.hokudai.ac.jp

Abstract

Membrane fouling and low effectiveness in removing dissolved organics are the main disadvantages of the ultrafiltration (UF) and microfiltration (MF) membrane processes. To control membrane fouling and to enhance the removal efficiency of dissolved organics in UF membrane processes, we combined the pre-coagulation/sedimentation process using a jet mixed separator (JMS) having inclined tube settlers with a UF membrane. A pilot plant study was conducted to compare the effect of three coagulants, namely, aluminum sulfate (AS), polyaluminum chloride (PACl) and polysilicato-iron (PSI) on the performance of the UF membrane.

The experimental results showed that membrane filterability and removal efficiency of dissolved organics were significantly improved by applying pre-coagulation/sedimentation. In a longterm operation, PSI was found to be the most effective coagulant in terms of dissolved organics removal and control of membrane fouling.

Introduction

Ultrafiltration (UF) and microfiltration (MF) membranes have been used successfully in water purification since the mid-1980s to remove suspended particles and microorganisms from surface waters or ground waters (Jacangelo *et al.*, 1995a; Jacangelo *et al.*, 1995b).

[51]

However, these low-pressure membrane processes are much less effective in removing dissolved molecules than reverse osmosis (RO) or nanofiltration (NF) membranes (*AWWA* Membrane Technology Research Committee, 1998; Bian *et al.*, 1999). Furthermore, membrane fouling, which is mainly caused by natural organic matter (NOM) and suspended particles, is also another important obstacle that limits the use of UF and MF membranes (Connell *et al.*, 1999; Fane *et al.*, 1987; Fu and Dempsey, 1998; Jones and O'Melia, 2001; Wei and Andrew, 1999).

To overcome these two obstacles, other processes, such as coagulation/ sedimentation, activated carbon adsorption, biological oxidation, and ozonation have been combined with UF/MF membranes in water purification technology (*AWWA* Membrane Technology Research Committee, 1998; Chang *et al.*, 1998; Jang *et al.*, 2001; Jang *et al.*, 2002).

Jang *et al.* (2001) used the jet mixed separator (JMS) as a pre-coagulation/ sedimentation unit for the UF/MF membranes. The JMS has several porous plates inserted vertically in the channel perpendicular to the flow; inclined tube settlers are located in another section of the JMS. Simultaneous flocculation and sedimentation occur in the porous plate part, and residual flocs are removed in the inclined tube settlers (Watanabe *et al.*, 1990; Watanabe *et al.*, 1998).

Aluminum-based coagulants, such as aluminum sulfate (AS) and polyaluminum chloride (PACl), are commonly used for water purification. When ingested, however, aluminum can produce severe encephalopathy, leading to dementia (Steve *et al.* 1995). Aluminum inhibits plant growth when the sludge containing aluminum is introduced to the agricultural land. Therefore, it is important to develop effective iron-based coagulants. Hasegawa *et al.* (1991) developed a new inorganic polymerized iron coagulant, called polysilicato-iron (PSI), which has strong bridging properties derived from polysilicate acid particles. Watanabe *et al.* (2000) used PSI in the pre-coagulation of municipal wastewater, with the goal of recycling phosphorous from pre-coagulated sludge.

This paper deals with the experimental results obtained in a pilot plant to test the performance of UF membranes when filtration was combined with pre-coagulation/sedimentation. The effects of three coagulants, namely, aluminum slufate (AS), polyaluminum chloride (PACl) and polysilicato-iron (PSI) were also compared.

Experimental methods

Figure 1 shows the schematic flow diagram of the pilot plant, which was operated at the Kami Ebetsu water purification plant. The pilot plant consists of a rapid mixing tank, JMS, and three UF membrane filters. The effective volume of JMS is 5.95 m^3 and its hydraulic retention time was around 60 min. Chitose River water was used as raw water.

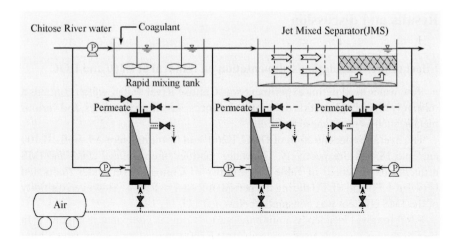

Figure 1. Schematic flow diagram of pilot plant

The operating conditions of the pilot plant are summarized in Table 1. The various runs were designed to test the effects of different coagulants types and dosage. PSI, an inorganic polymeric iron coagulant, has a molar ratio of Fe to Si of 1:1 to 1:5 and a molecular weight of 200 kDa to 500 kDa. PSI with molar ratio of Fe to Si of 1:1 was used in the pilot plant experiment.

In the pilot plant study, an UF membrane made of specially polymerized polyacrylonitrile (PAN) was used, which is made of an external pressure type hollow fiber with a nominal average pore size of 0.01μm.

Table 1. Operating conditions of pilot plant experiment

Items		Run-1*	Run-2	Run-3	Run-4
Coagulation/ Sedimenta- tion process	Coagulant	AS	PACl	PACl	PSI
	Dosage (mmol-Al/L or mmol-Fe/L)	0.37	0.19	0.37	0.21
	PH	7.0	6.5	7.0	6.2
Membrane filtration process	Filtration mode (dead-end)	CFR	CFR	TMP	CFR
	TMP (kPa)	–	–	50	–
	CFR (m^3/m^2/day)	0.9	0.9	–	0.9 to 1.5
	Interval of physical cleaning	1 hour	1 hour	30 min.	1 hour

* Activated silicate was used as a coagulant aid

Results and discussion

Effect of pre-coagulation/sedimentation on removal of E260 and DOC

Raw water used in this experiment was Chitose River water, which contains a relatively high concentration of humic substances, suspended particles, and various metals, such as manganese and iron.

The average concentration of DOC, E260, and turbidity were 2.7 mg/L, 0.109 cm^{-1} and 18.4 TU, respectively. The water qualities of coagulated water and JMS effluent are presented in Table 2. Turbidity of Chitose River water fluctuated between 4 TU and 87 TU during the period of Run-1 to Run-4. However, turbidity of the JMS effluent was constantly below 1.0 TU.

E260 removal efficiency among the various runs is compared in Figure 2. In Run 4, where PSI was used as coagulant, E260 removal efficiency was a little higher than that in the other runs where AS or PACl was used as coagulant.

DOC removal efficiency among the various runs is compared in Figure 3. The highest DOC removal efficiency was obtained in Run 4. About 40 % of the DOC was coagulated to form flocs that were larger than the UF membrane pore size.

Table 2. Average turbidity, E260, and DOC of Chitose River water (CR), coagulated water (Coag), and JMS effluent (Coag/Sed)

	Run-1			Run-2		
	CR	Coag	Coag/Sed	CR	Coag	Coag/Sed
Turb. (TU)	7.8	11.0	0.9	20.2	22.0	0.9
E260 (cm^{-1})	0.083	0.025	0.024	0.098	0.028	0.026
DOC (mg/L)	2.37	1.78	1.67	3.05	2.2	1.96
	Run-3			Run-4		
	CR	Coag	Coag/Sed	CR	Coag	Coag/Sed
Turb. (TU)	19.2	20.8	0.5	20.3	27.1	0.5
E260 (cm^{-1})	0.111	0.044	0.039	0.120	0.027	0.024
DOC (mg/L)	3.0	2.33	2.08	2.31	1.11	0.93

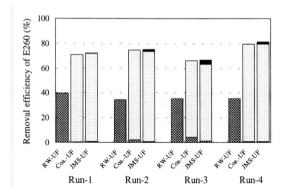

Figure 2. Removal efficiency of humic substances from Chitose River water

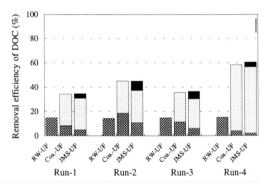

Figure 3. Removal efficiency of DOC from Chitose River water

Effect of pre-coagulation/sedimentation on membrane fouling

Figure 4 shows how membrane resistance increases with increasing filtration time in Run-1 (constant flow rate mode). When Chitose River water was filtered directly, it took about 400 hours filtration time to reach a membrane resistance of $12 \cdot 10^{12} \, m^2/m^3$. With pre-coagulation, the filtration time was about 2.5 times longer, even though the turbidity of the coagulated water was a little higher than that of Chitose River water. This is because the humic substances were coagulated into larger flocs, resulting in the formation of a cake layer with lower specific resistance. This result suggests that humic substances have a greater impact than suspended solids on membrane fouling.

When the JMS effluent was filtered, it took around 2400 hours to reach the same membrane resistance. This result may be due to the removal of humic substances and suspended solids, which form the cake layer on the membrane.

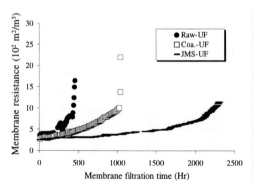

Figure 4. The effect of pre-coagulation and pre-coagulation/sedimentation on the UF membrane system (Run-1)

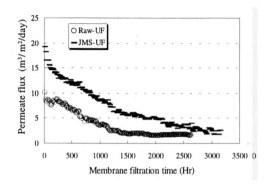

Figure 5. The effect of pre-coagulation/sedimentation on the UF membrane system (Run-3)

Figure 5 shows the effect of pre-coagulation /sedimentation in Run 3 (constant trans-membrane pressure mode). The initial flux of raw water filtration was almost $10 \text{ m}^3/\text{m}^2/\text{d}$, and then it decreased with increasing membrane filtration time. It was hardly changed after 1800 hours of membrane filtration time, resulting to the flux of about $1.8 \text{ m}^3/\text{m}^2/\text{d}$. On the other hand, in case of filtering the JMS effluent, not only the initial flux was about two times higher, also the higher flux was maintained from initial filtration time to about 3000 hours of filtration time.

From the results of Figure 4 and 5, we can conclude that the pre-coagulation/ sedimentation was a very effective process both at constant flow rate and constant trans-membrane pressure mode.

We compared the effect of the coagulant species used in the pre-coagulation/ sedimentation on the membrane fouling. Figure 6 shows the comparison of the filtration resistance among Runs 1, 2 and 4 where AS, PACl and PSI was used as coagulant, respectively. As seen in Figures 2 and 3, there is difference in removal

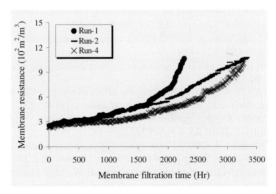

Figure 6. Effect of different coagulants on filtration resistance in the coagulation/ sedimentation - UF membrane system

efficiency of E260 and DOC with different coagulants in the pre-coagulation/ sedimentation. PSI was the most effective, especially in removing DOC. Considering the data shown in Figures 3 and 6, we may conclude that the higher DOC removal in the pre-coagulation/sedimentation gives better performance of UF membrane filtration.

Even though the filtration resistance in Run-2 and Run-4 was almost the same at about 3300 hours of filtration time (the actual TMP reached about 100 kPa. which is the recommended TMF for chemical cleaning), filtration resistance in Run-4 has always been lower than that in Run-2.

As a result, PSI seems to have better effect on both control the membrane fouling and removal of the dissolved organics than AS or PACl in the UF membrane combined with pre-coagulation/sedimentation.

Summary

A longterm pilot plant study was carried out to demonstrate the effect of pre-coagulation/sedimentation on UF membranes. Pre-coagulation/sedimentation improved UF membrane performance in terms of removing soluble organics and controlling membrane fouling. Of the three coagulants tested, polysilicato-iron was superior to aluminium chloride and polyaluminium chloride for removing soluble organics, resulting in the lowest filtration resistance.

References

AWWA Membrane Technology Research Committee (1998) Committee Report; Membrane Processes. *AWWA*. **90**(6), 91–105
Bian R., Watanabe Y., Tambo N., and Ozawa G. (1999) Removal of humic substances by UF and NF membrane systems. *Wat. Sci. Tech.* **40**(9), 121–129

Chang Y-J., Choo K-H., Benjamin M.M., and Reiber S. (1998) Combined adsorption-UF process increases TOC removal. *AWWA.* **90**(5), 90–102

Connell H., Zhu J., and Bassi A. (1999) Effect of particle shape on crossflow filtration flux. *Membr. Sci.* **153**, 121–139

Fane A.G., and Fell C.J.D. (1987) A review of fouling and fouling control in ultration. *Desalination*, **62**, 117–136

Fu L.F., and Dempsey B. A. (1998) Modeling the effect of particle size and charge on the structure of the filter cake in ultrafiltration. *Membr. Sci.* **149**, 221–240

Hasegawa T., Hasimoto K., and Tambo N. (1991) Characteristics of metal-polysilicate coagulants. *Wat.Sci. Tech.* **23**, 1713–1722

Jones K. L., and O'Melia C. R. (2001) Ultrafiltration of protein and humic substances;effect of solution chemistry on fouling and flux decline. *Membr. Sci.* **193**, 163–173

Jacangelo J.G., Aieta E.M., Carns K.E., Cummings E.W., and Mallevialle J. (1995a) Assessing Hollow-Fiber Ultrafiltration for Particulate Removal. *AWWA.* **87**(11), 68–75

Jacangelo J.G., Adham S.S., and Laîné J.-M. (1995b) Mechanism of Cryptosporidium, Giardia, and MS2 virus removal by MF and UF. *AWWA.* **87**(9), 107–121

Jang N-Y., Watanabe Y., Minegishi S., and Bian R. (2001) The evaluation of dead-end ultrafiltration membrane process combined with pre-coagulation/sedimentation. *JWWA*, **70**(2), 1–14

Jang N-Y., Watanabe Y., and Ozawa G. (2002) The study on microfiltration membrane process combined with pre-ozonation. *JWWA*, **71**(2), 1–13

Steve R., Walter K., and Perri S.-L. (1995) Drinking water aluminum and bioavailability. *AWWA.* **87**(5), 86–99

Watanabe Y., Fukui M, and Miyanoshita T. (1990) Theory and performance of jet mixed separator, *Water SRT, Aqua*, **39**(6) 387–395

Watanabe Y., Kasahara S., and Iwasaki Y. (1998) Enhanced flocculation/sedimentation process by a jet mixed separator. *Wat.Sci. Tech.* **37**(10), 55–67

Watanabe Y., Tadano T., Hasegawa T., and Shimanuki Y. (2000) Phosphorous Recycling from Pre-Coagulated Wastewater Sludge. Chemical Water and Wastewater Treatment VI. *Proc.* Of the 9th Gothernburg Symposium (Hahn H.H, Hoffmann E. and Ødegaard H, Eds.) Springer. Berlin . 359–371

Wei Y., and Andrew L.Z. (1999) Humic acid fouling during microfiltration. *Membr. Sci.* **157**, 1–12

H.H. Hahn, E. Hoffmann, H. Ødegaard (Eds.)
Chemical Water and Wastewater Treatment VII, pp. 59-68
© IWA Publishing, London
ISBN: 1 84339 009 4

Hybrid Membrane Processes for Drinking Water Treatment

N.A. Booker, T. Carroll, S.R. Gray, and J. Meier-Haack*

*Berwickshire, UK
email: nic.booker@btopenworld.com

Abstract

The microfiltration of fractionated NOM identified that the main contributor to hydrophobic microfiltration membrane fouling was the low molecular weight neutral hydrophilic fraction. Attempts to identify a suitable process for selectively removing this fraction of NOM from water before microfiltration were unsuccessful, and led to the idea that the membrane surface should be modified to enhance rejection of this fraction of NOM. Polyelectrolytes were grafted onto the surface of polypropylene membranes using expertise developed at the Institute for Polymer Research (IPF) in Germany. Experiments with membranes grafted with anionic polyacrylic acid hydrophilic polymers demonstrated that fouling rates were 50 % lower than for the non-grafted membranes and rejection of NOM was not compromised.

Introduction

Both microfiltration (MF) and ultrafiltration (UF) membrane processes have been used for potable water treatment, where they have been particularly suitable for the removal of finely divided suspended solids, especially bacteria, algae and protozoa such as *Giardia* and *Cryptosporidium.* They have been less successful in cost terms for the removal of dissolved contaminants such as colour and chemical pollutants. The use of tighter membrane filtration processes such as nanofiltration (NF) and reverse osmosis (RO) for the removal of soluble contaminants from drinking water results in higher energy consumption and lower rates of water recovery; these are issues which may limit their application in remote areas and where water supplies are scarce. There is potential for combining conventional chemical coagulation processes with the relatively low operating costs and high water recovery rates of MF systems to achieve cost effective soluble contaminant removal from drinking water.

The role of natural organic matter in the fouling of microfiltration membranes was evaluated in a series of laboratory tests performed at CSIRO Molecular Science, Clayton, Victoria (Carroll *et al.*, 2000a) and also formed the basis of postgraduate student research at both the RMIT University, Melbourne (RMIT) (Fan *et al.*, 2000) and the University of New South Wales (UNSW) (Schaefer *et al.*, 1998a and 1998b). This was followed by research at both CSIRO (Carroll *et al.*, 2000b) and RMIT (Fan *et al.*, 2001) aimed at the development of hybrid MF processes based on the use of coagulation and/or adsorption processes.

Polyelectrolyte-grafted MF membranes were developed at the Institute for Polymer Research (Institut fur Polymerforschung (IPF)) in Dresden, Germany in an attempt to control fouling of hydrophobic hollow fibre membranes by NOM and these membranes were evaluated by CSIRO Molecular Science staff, both in Dresden and in Melbourne (Carroll *et al.*, 2001 and Meier-Haack *et al.*, 2001a, 2001b).

Procedures

The evaluation of membrane performance was based on the use of polypropylene hollow fibre microfiltration membranes similar to those used commercially in the Australian water industry. New hollow fibre membranes were donated to the project by US Filter Memcor, Windsor, Sydney. The MF membrane fibres had a nominal pore size of 0.2 µm, an internal diameter of 250 µm and an outer diameter of 550 µm. Without any surface modification, the membrane fibres were highly hydrophobic. A membrane test rig was designed and constructed at CSIRO and used a single 0.9 m long hollow fibre membrane with a nominal surface area of 7 cm². Each test used a new hollow fibre that was wetted and degreased with ethanol and flushed thoroughly with ultra-pure water before use. The feed water was pumped onto the membrane from a stirred tank using a peristaltic pump and forced through the fibre wall under pressure, to emerge as permeate from the open ends. MF permeate was collected in a vessel mounted on an analytical balance and the transmembrane pressure determined using pressure transducers. The signals from the analytical balance and pressure transducers were processed to calculate the permeate flow rate as a function of permeate throughput, at a constant transmembrane pressure of 50 kPa. Based on these measurements, the membrane performance under a range of hybrid treatment systems could be observed.

Water Sources and Fractionation

Ultra-pure water, used for control purposes, dilution and making up of standard solutions, was produced by a laboratory Milli-Q system, marketed by Millipore.

The raw water used in the laboratory tests was harvested from two regions of Victoria, Australia and stored in a refrigerator at 4°C until needed. The two raw water sources used were the East Moorabool River near Anakie and the Maroondah Aqueduct near Yering Gorge. The East Moorabool River water had a relatively high dissolved organic carbon (DOC) content (8-12 mg/L) and relatively low turbidity (3-7 NTU). The Maroondah aqueduct was fed from Melbourne Water's Yarra Ranges reservoir system where it had been stored for a number of years; as such the water had a moderate dissolved organic carbon concentration (2.5 - 3.5 mg/L), and a very low turbidity (0.6-2.6 NTU).

In the NOM characterisation and fractionation tests, water from these sources was concentrated using RO to a DOC concentration of approximately 500 mg/L. The raw water concentrate was then fractionated according to an established ion-exchange procedure (Carroll *et al.*, 2000a). The concentrate was filtered through a 0.45 μm membrane, adjusted to pH 2, and fed onto a Supelite DAX-8 resin, which retains strongly hydrophobic substances (e.g. humic and fulvic acids). This fraction was eluted with NaOH and acidified on an IR-120 resin. The unadsorbed concentrate was fed onto an Amberlite XAD-4 resin, which retains weakly hydrophobic substances. This fraction was eluted with NaOH and separately acidified on an IR-120 resin. The unadsorbed hydrophilic (non-humic) concentrate, attributed to proteins, amino acids and carbohydrates, was fed onto an Amberlite IRA-958 resin, which retains anionic charged material. This fraction was eluted with a NaOH/NaCl mixture. The remaining neutral material was not retained by any of the resins. The TOC concentration and relative proportion of each of the NOM fractions in East Moorabool water are shown in Table 1.

Before they were used in the membrane tests, the four East Moorabool NOM fractions were separately reconstituted to the same respective TOC concentrations that they had in the raw water. Calcium and sodium concentrations of each fraction were corrected to the raw water values to eliminate electrostatic-mediated contributions to fouling. The four different fractions were then passed through the membrane fibres and the flux versus throughput determined for each fraction individually.

Table 1. TOC concentrations of East Moorabool NOM fractions

Fraction	TOC (mg/L)	Percent of Total
Strongly hydrophobic	3.1	34
Weakly hydrophobic	1.6	18
Charged hydrophilic	3.0	33
Neutral hydrophilic	1.4	15
Unfractionated	9.0	100

Coagulation and Adsorption

Coagulation and/or adsorption processes were tested to determine their ability to remove each of the NOM fractions and to determine whether they could reduce the fouling of the MF membranes by the NOM fractions.

The water was treated in standard jar tests with several different coagulants including alum, polyaluminium chloride, polyaluminium chlorohydrate, ferric chloride, polyferric sulphate and a 50/50 alum/polyferric sulphate mixture. The first step in the jar test procedure was to adjust the pH of one litre of raw water from 7.5 to 6.0 with sulphuric acid. The water was then dosed with coagulant at a dose determined in previous jar tests to yield optimum TOC removal. After each chemical addition the water was stirred for 60 s at 250 rpm. The pH was maintained at 6.0 throughout the procedure by adding sulphuric acid or sodium hydroxide. Once dosing was complete the mixture was stirred for 90 seconds at 100 rpm. The treated water and chemical floc mixture was then filtered through the hollow fibre MF membrane unit as discussed previously.

The adsorption of the hydrophilic neutral NOM fraction by various adsorbents was determined in a number of laboratory scale jar tests. This was attempted after it became apparent that the NOM remaining after alum coagulation and floc separation still led to membrane fouling. Powdered polypropylene and finely chopped hollow fibre polypropylene membranes were both used in jar tests to determine whether there was significant adsorption of the neutral hydrophilic NOM fraction by these materials. A range of powdered activated carbons, magnetic ion exchange resin (MIEX®, Orica p/l), magnetite and clay were also used in jar tests to determine whether this specific NOM fraction could be selectively adsorbed from the water.

Modified Membranes

The polypropylene hollow fibre MF membranes, described earlier, were modified by grafting with polyacrylic acid at 7.2 % w/w to form a negatively charged surface layer using a solution phase grafting technique developed by IPF. Microfiltration experiments were carried out on the original and graft-modified single polypropylene hollow-fibre membranes for comparison.

The raw water source used in these tests was the East Moorabool River that, at the time of harvesting, had a DOC concentration of 9.0 ppm, a turbidity of 4.3 NTU, a pH of 7.4 and a calcium concentration of 10 mg/L as $CaCO_3$ equivalents. The raw water was tested in the MF filtration rig both untreated and after coagulation pre-treatment with alum. In the former case, the pH was adjusted to 6.5 before filtration. In the latter case, alum was dosed at 5.0 mg/L (as Al^{3+}), the pH was adjusted to 6.5, and the water was filtered after a 5-minute delay. The effect of water hardness was investigated by adding calcium chloride to the raw water (adjusting to pH 6.5) at concentrations of 100 and 350 mg/L as $CaCO_3$ equivalents.

Results

Membrane fouling by NOM

The decline in permeate flow rate with permeate throughput during microfiltration of the four reconstituted East Moorabool NOM fractions is shown in Figure 1. The decline for the raw water pre-filtered through a 0.2 µm membrane is also shown for comparison. The rates of fouling were relatively low for the strong hydrophobic, weak hydrophobic, and charged hydrophilic fractions (decline in permeate flux rates of approximately 16 % after 1000 L/m² was filtered). The rates of fouling were significantly faster for the neutral hydrophilic fraction and the pre-filtered raw water (40 % and 50 %, respectively, after 1000 L/m² was filtered). A parallel study at RMIT (Fan *et al.*, 2001) confirmed these findings and also demonstrated that this neutral fraction of the NOM had a low molecular weight distribution with an average size below 1000 Daltons.

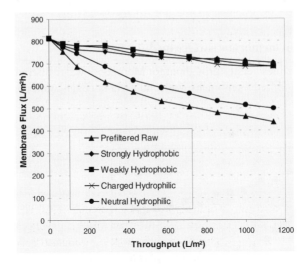

Figure 1. The decline in permeate flux caused by fouling of polypropylene microfiltration membranes by the four NOM fractions (pH = 6.0)

Coagulation and Adsorption

The microfiltration membrane permeate flux, as a function of permeate throughput, for East Moorabool water, pre-treated by the aluminium and iron-based coagulants was determined. In all cases, the rate of fouling after coagulation pre-treatment was lower than for the raw water. For the aluminium-based coagulants, the rates of fouling were similar. The declines in permeate flow rate after 1000 L/m² of permeate was filtered were 51%, 52% and 51% for alum, polyaluminium chloride, and polyaluminium chlorohydrate, respectively, compared to a 78% flux decline for the raw water. For the iron-based coagulants, the rates of fouling by ferric chloride and polymerised ferric sulphate were slightly higher than for alum, but the rate for alum/polymerised ferric sulphate was similar to alum. The declines in permeate flow rate after 1000 L/m² of permeate was filtered were 61 %, 62 %, 67 %, 71 % for alum, alum/polyferric sulphate, ferric chloride, and polyferric sulphate, respectively, compared to a 88 % flux decline for the raw water used in these tests. Coagulation was unsuccessful at removing the neutral hydrophilic NOM fraction, no matter what coagulant was used.

None of the adsorbent materials tested showed any appreciable capacity for the neutral hydrophilic NOM fraction. Only powdered activated carbon at doses greater than 1 g/L showed any capacity for this specific NOM fraction, removing 25 % of these organics, after 24 hours of contact. MIEX resin, montmorillonite clay, magnetite, powdered polypropylene and chopped polypropylene membrane fibres removed none of the organics in this specific fraction of the NOM.

Microfiltration membrane surface modification

The flux declines versus throughput for Moorabool water filtered through both original and modified membranes, with and without coagulation pretreatment are

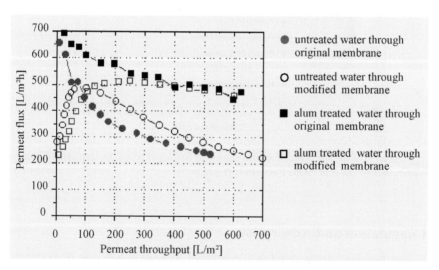

Figure 2. Flux decline with throughput for untreated water through original and polyacrylic acid modified membranes and alum treated water through original and polyacrylic acid modified membranes

shown in Figure 2. The flux decline was most pronounced for the untreated water filtered through the original membrane, dropping by 60 % after 500 L/m² throughput. Pretreatment with alum reduced the flux decline for the original membrane to 30 % after 500 L/m² throughput. This reduction was most likely due to the effect of flocculation on particle size and cake resistance (Meier-Haack *et al.*, 2001a).

The initial fluxes for the modified membranes (7.2 % w/w graft yield) were approximately 60 % lower than for the original membranes. However, the flux increased upon filtration, by 75 % and 125 % for the untreated and alum-treated water, respectively. In the former case, the flux reached a maximum at 100 L/m² throughput, while in the latter case, the maximum flux was at 250 L/m² throughput. In both cases, the flux declined at higher throughputs. In the untreated water case, the flux was consistently higher for the modified membrane than for the original membrane after the maximum flux of the former was reached. This enhancement corresponded to an increase in throughput of approximately 100% for a given flux over the range investigated. However in the alum-treated water case, the fluxes through the original and modified membranes were similar after the maximum flux was reached.

The flux declines as a function of throughput for raw and artificially-hardened Moorabool water filtered through modified membranes are shown in Figure 3. In all cases, the initial flux was lower than for the original membranes and the flux increased initially but declined at higher throughputs, as described earlier. As the calcium concentration increased the flux enhancement was reduced, and the throughput at maximum flux was also lowered. The flux at a given throughput was consistently highest for the softest water.

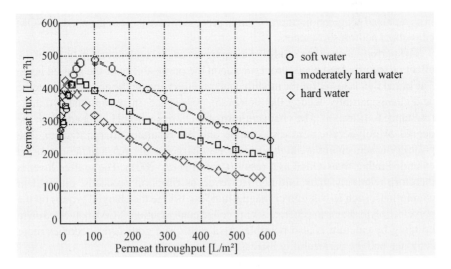

Figure 3. Flux decline with throughput for soft, moderately hard and hard water

Discussion

The neutral hydrophilic component of NOM caused more severe fouling of the polypropylene microfiltration membrane than hydrophobic or charged components. The fouling of microfiltration membranes by neutral hydrophilic NOM has several implications for microfiltration performance. The concentration of neutral hydrophilic NOM in the feed water may be a better indicator of potential fouling than the total NOM concentration. Membrane chemical cleaning may be more effective if the chemical agent selected specifically targets these neutral hydrophilic substances.

A combined coagulation-microfiltration process will not eliminate membrane fouling by NOM since aluminium or iron salts preferentially coagulate the large, charged, hydrophobic NOM fractions (Carroll *et al.*, 2000a). Coagulation in combination with an adsorption step specific to the small neutral substances would be an ideal pre-treatment before microfiltration. However, of the adsorbents tested, none showed any appreciable capacity for this fraction of the NOM apart from activated carbon used at uneconomically high doses. The ability to remove this fraction of the NOM from water, apart from minimising membrane fouling, would also be expected to be advantageous in reducing the Trihalomethane (THM).

The small, neutral, hydrophilic NOM should be readily susceptible to bio-degradation. Why this has not occurred within the reservoir during the long periods that the water has been stored is not obvious. One possible reason could be that these specific organics are constantly being produced during the biodegradation of other organic material within the reservoir.

A membrane material that reduces fouling by small, neutral, hydrophilic NOM may give improved microfiltration performance; this would be particularly likely in a combined coagulation-microfiltration process, where most of the residual NOM consists of neutral substances.

The performance of the modified membranes can be understood in terms of the effect of specific feed water properties upon the permeability of the grafted layer. The initial flux is lower for the modified membranes because the grafted polyacrylic acid chains partially occupy the membrane pore space, increasing the hydraulic resistance to filtration. The chain conformation varies with pH, depending on the degree of polyacrylic acid dissociation and the resultant charge density. The polyacrylic acid chains are highly negatively charged at pH 6.5, and the chains are extended rather than coiled as a result (Carroll *et al.*, 2001). The grafted layer is therefore relatively thick initially. However, as filtration proceeds, multivalent counterions, such as calcium or magnesium, can bridge the charged groups on the polyelectrolyte chains to generate a more coiled conformation. NOM-polyelectrolyte bridging by calcium is also possible. As a result, the graft layer becomes more compact, and the permeability increases.

Conclusions

Fouling of polypropylene microfiltration membranes by NOM is more severe for the hydrophilic, neutral fraction of NOM than for hydrophobic substances such as humic or fulvic acids. The hydrophilic neutral fraction of NOM may contain molecules such as carbohydrates and polysaccharides, but the exact nature of the molecules responsible for fouling has not been identified. This fraction of NOM was not significantly removed by conventional treatment approaches such as coagulation and sedimentation. Removal of this fraction using common water treatment adsorbent materials at economically viable doses was also ineffective.

It is possible to minimise the flux decline caused by NOM fouling of a microfiltration membrane either by selecting membrane materials with a hydrophilic surface or by surface modification of hydrophobic membranes. Anionic hydrophilic polymers grafted as a flexible layer onto a microporous polypropylene hollow fibre resulted in a membrane whose flux depended on polymer conformation, as controlled by properties of the feed stream such as pH and multivalent counterions.

Acknowledgements

The authors thank the CRC for Water Quality and Treatment for their support of this project. We also thank the staff of both Barwon Water and Melbourne Water, especially for their assistance during water harvesting and pilot trials.

References

Carroll, T., and Booker, N.A. (2000) Improved performance of water microfiltration with hybrid particle pre-treatment. In *Proc. of the Int'l. Conf. on Membrane Technology in Water and Wastewater Treatment*, Lancaster, UK, March 2000

Carroll, T., King, S., Gray, S., Bolto, B. ,and Booker, N. (2000a) The fouling of microfiltration membranes by NOM after coagulation treatment. *Wat. Res.* **34**(11), 2861-2868

Carroll, T., Vogel, D., Rodig, A., Simbeck, K., and Booker, N. (2000b) Coagulation-Microfiltration Processes for NOM Removal from Drinking Water. In Hahn, H.H., Hoffmann, E., Ødegaard, H. (Eds.) *Chemical Water and Wastewater Treatment VI.* Springer, Berlin, 171-180

Carroll, T., Booker, N.A. ,and Meier-Haack, J. (2001) Polyelectrolyte-grafted microfiltration membranes to control fouling by natural organic matter in drinking water. *Journal of Membrane Science*, In Press

Fan, L., Harris, J.L., Roddick, F.A. ,and Booker, N.A. (2000) Fouling performance during microfiltration of an aquatic dissolved organic matter. Proc. of the 28th Australasian Chemical Engineering Conference - *Chemeca 2000*

Fan, L. Harris, J.L., Roddick, F.A. ,and Booker, N.A. (2001) Influence of NOM characteristics on the fouling of microfiltration membranes. *Wat. Res.* **35**(18), 219-227

Meier-Haack, J., Booker, N.A. ,and Carroll, T. (2001a) A low fouling modified-polypropylene microfiltration membrane for drinking water treatment. Submitted to *Water Science and Technology*

Meier-Haack, J., Booker, N.A., and Carroll, T. (2001b) A Permeability-Controlled Micro-filtration Membrane For Reduced Fouling In Drinking Water Treatment. Submitted to *Water Research*, December 2001

Schaefer, A. I., Fane, A. G., and Waite, T.D. (1998a) Nanofiltration of natural organic matter: Removal, fouling and the influence of multivalent ions. *Proc. Membranes in Drinking and Industrial Water Production*, Amsterdam, September 1998

Schaefer, A. I., Fane, A.G., and Waite, T.D. (1998b) Chemical addition prior to membrane processes for natural organic matter (NOM) removal, In Hahn,H.H., Hoffmann, E., Ødegaard, H. (Eds.) *Chemical Water and Wastewater Treatment V*, Springer, Berlin, 125-137

Chemical Dosing Control

H.H. Hahn, E. Hoffmann, H. Ødegaard (Eds.)
Chemical Water and Wastewater Treatment VII, pp. 71-80
© IWA Publishing, London
ISBN: 1 84339 009 4

Turbidity-Related Dosing of Organic Polymers to Control the Denitrification Potential of Flocculated Municipal Wastewater

A.R. Mels*, A.F. van Nieuwenhuijzen and A. Klapwijk

*Wageningen University, Sub-Department of Environmental Technology, Wageningen, The Netherlands
email: adriaan.mels@algemeen.mt.wau.nl

Introduction

Particles in wastewater represent a large part, 60-70 %, of the organic pollutants in municipal wastewater (Ødegaard, 1999; van Nieuwenhuijzen and Mels, 2002). The removal of these particles in the first treatment step may result in a significant saving of space and energy at wastewater treatment plants, i.e. 50-75 % (Mels and van Nieuwenhuijzen, 2000). Due to particle removal, the load on subsequent unit operations can be reduced thus resulting in a smaller design and lower energy consumption. Moreover, the produced primary sludge can be utilized as a resource for biogas production.

To achieve a high degree of pre-treatment particles can be flocculated by organic polymers and subsequently be removed by either sedimentation or dissolved air flotation. However, if the removal of particles in the pre-treatment is too high, an increased concentration of total-nitrogen (N_{total}) in the effluent of the wastewater treatment plant may result. The increase in the effluent N_{total} is due to the removal of excess biodegradable COD in the pre-treatment leading to insufficient denitrification in a biological post-treatment step (e.g. activated sludge or biofilm system).

This article describes experiments in a feed forward control strategy for the dosage of organic polymer based on continuous measurement of the wastewater turbidity. This dosing control strategy aims to control the removal of particles and of biodegradable COD in the pre-treatment and adapt it to the desired degree of denitrification of the post-treatment. Earlier experiments showed that, due to the fluctuating concentrations in the wastewater, the particle removal of a pre-treatment system can be rather unpredictable when a constant polymer dose per liter of wastewater is applied (Mels *et al.*, 2001). The results also indicated, however, that a predictable removal could be achieved if the polymer dose is adapted to the influent turbidity at a fixed ratio.

The experiments were conducted with a lab-scale pre-treatment system consisting of a polymer dosing unit and a clarifier. The removal efficiency of the pre-treatment system was determined by measurements of COD and turbidity. The removal of biodegradable COD was assessed by so-called *Nitrogen Uptake Rate* (NUR) experiments (Kujawa and Klapwijk, 1999). During NUR experiments the nitrate consumption is determined at anoxic conditions after a pulse dose of wastewater to activated sludge. Based on the nitrate consumption and the biological sludge yield the rapidly and slowly biodegradable COD fractions (S_s and X_s) can be estimated.

Material and methods

Experimental set up and operational conditions. In the set-up (Figure 1), the incoming wastewater entered a buffer vessel where the turbidity of the influent was measured by an on-line turbidity meter. This meter sent a signal to a computer (software *Control EG*) that made a control action to the polymer dosing pump. The polymer was dosed to a stirred vessel (1.5 l) and was mixed intensively (300 rpm) with the wastewater. The formed flocs were settled in a clarifier. With a second turbidity meter the turbidity of the clarifier effluent was measured.

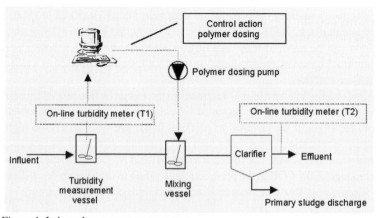

Figure 1. Lab-scale pre-treatment system

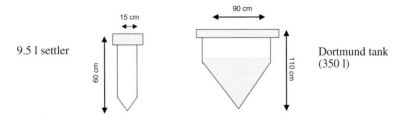

Figure 2. Dimensions of the clarifiers used in the experiments

Two types of settlers were used (Figure 2): a small 9.5-l settler and a large 350 l settler (Dortmund tank). The influent flows that were applied to the clarifiers were 15 l•h⁻¹ for the small settler and 90 l•h⁻¹ for the Dortmund tank, respectively.

Dosage control. In order to achieve a fixed polymer to (influent) turbidity ratio, the flow of the dosing pump was varied from 0 to 100 % proportional to the influent turbidity, according to the following equation. At influent turbidity values > 250 NTU the dosing pump flow was at its maximal rate.

$$Flow(\%) = Turbidity_{influent} \Big/ 250\ NTU \bullet 100\ \%$$

Characteristics of the organic polymer. A linear cationic polyacrylamide-based polymer (*CYTEC*) was used for flocculation. The molecular weight of the polymer was $8 \cdot 10^6$ g•mol⁻¹; the cationic charge density was 24 weight %. The polymer was available as a powder. A concentrated polymer stock solution (10 g•l⁻¹) was prepared in 2 l plastic beakers. After complete dissolution of the polymer (24 h), the stock solution was diluted to the desired concentration and stored for use at 4 °C. To avoid loss of activity a fresh polymer solution was prepared every 3 - 4 days.

Sampling and analyses. The turbidity of the influent and effluent of the clarifier was continuously measured by two Solitax turbidity meters (*Lange Group*). In addition, 24-h and grab samples were taken and analysed for turbidity and $COD_{particulate}$. The turbidity of these samples was measured with a laboratory turbidity meter (*WTW Turb 550*). The laboratory meter and the on-line meters were regularly calibrated and compared. The COD (COD_{total}) was fractionated and determined for two wastewater samples: untreated and filtered through a 0.45 μm membrane filter (*Schleicher&Schuell*). The COD fractions were as follows:

$COD_{particulate}$ = untreated – membrane filtered
$COD_{dissolved}$ = membrane filtered (mg O_2•l⁻¹)

The biodegradable COD (S_s and X_s) of raw and pre-treated wastewater was assessed by *Nitrogen Uptake Rate* experiments as described by Kujawa and Klapwijk (1999).

Results

Relation between COD$_{particulate}$ and turbidity. Earlier results (spring 1999, in Mels, 2001) indicated that turbidity measurements could be used to monitor the COD$_{particulate}$ of the influent and effluent of a pre-treatment system, because a linear relation was found. To confirm these findings additional measurements were done with the same wastewater (both raw and treated with organic polymers) in spring 2000. Figure 3 presents the results of these analyses and compares them with the earlier data. The graph shows that a linear relationship was found in a wide range of COD$_{particulate}$ (100-900 mg O$_2$•l^{-1}) and turbidity (50 - 450 NTU). The small difference in slopes between the data series can be attributed to seasonal variations (indicating that the relationship should be checked regularly).

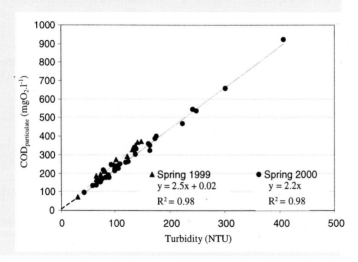

Figure 3. Linear relationship between COD$_{particulate}$ and turbidity as found in spring 1999 and spring 2000 for wastewater from the village of Bennekom

Diurnal particle variations in untreated wastewater. The composition of domestic wastewater varies strongly in the course of a day. Figure 4 presents the influent turbidity (or COD$_{particulate}$) of Bennekom wastewater during two consecutive days of dry weather flow. The graph illustrates that during dry weather conditions the turbidity of this wastewater exhibited a regular, daily pattern. Around 10:00 a.m. a first peak was measured caused by increased activity in the morning. During the day the turbidity was lower, but in the early afternoon and evening there were peaks again.

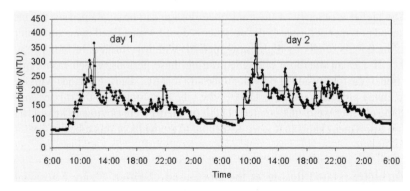

Figure 4. Particle fluctuations in the untreated wastewater of the village of Bennekom during dry weather conditions (results for 10 and 11 August 2000)

Evaluation of fixed dosing and turbidity-related dosing in the 9.5-l settler.
A comparison was made between dosing a fixed concentration of polymer per liter and dosing based on turbidity. The experiments were conducted with the 9.5-liter settler and consisted of two series. The first series compared the two dosing strategies at an average dose of 3.8 mg polymer/100 NTU, the second series at 5.8 mg polymer/100 NTU. The fixed doses in both series were recalculated based on the average influent turbidity. In these experiments fixed doses of 6.0 and 8.6 mg polymer per liter were applied, respectively. Table 1 shows the results of these experiments.

During both series the efficiency of the turbidity-related dosing was higher and resulted in lower effluent values. For the series with 3.8 mg polymer/100 NTU the fixed dosage resulted in 74 NTU, while the controlled dosage exhibited 46 NTU. Turbidity was 53 NTU for the series with 5.8 mg polymer/100 NTU, but only 39 NTU with the controlled dosage.

Table 1. Comparison of fixed polymer dosage and turbidity-related polymer dosage

Strategy	Dose	Influent (24-h av.) (NTU)	Effluent (24-h av.) (NTU)	Particle removal (%)	Number of days
Fixed dosage	3.8 mg pe/100 NTU[a]	157 ± 4	74 ± 13	53	3
Turbidity-related dosage	3.8 mg pe/100 NTU	150 ± 22	46 ± 8	69	3
Fixed dosage	5.8 mg pe/100 NTU[a]	148 ± 8	53 ± 8	64	3
Turbidity-related dosage	5.8 mg pe/100 NTU	152 ± 22	39 ± 8	74	4

[a] average value: recalculated based on dose and average influent turbidity

The higher efficiency of the controlled dosage can be explained by the more balanced polymer addition. If the polymer is dosed with a fixed concentration per liter, the dose is relatively high during periods with low influent turbidity (e.g. at night). In fact an overdose is applied during these periods, which may cause a fraction of the totally added polymer to be still reactive. During periods with fixed dose, many problems with blockage of the tube that connected the mixing vessel with the clarifier were encountered. These problems were obviously linked to the reactive polymer chains that were adsorbing to the walls of this tube.

Turbidity-related dosing in the Dortmund clarifier. Figure 5 shows the turbidity of the influent and effluent of the Dortmund-clarifier for two days without polymer dosing. The average turbidity removal during these two days was 26 %.

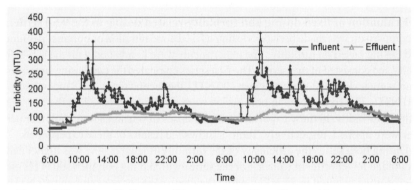

Figure 5. Turbidity of influent and Dortmund-clarifier effluent without polymer dosing (results for 10 and 11 August 2000, dry weather conditions)

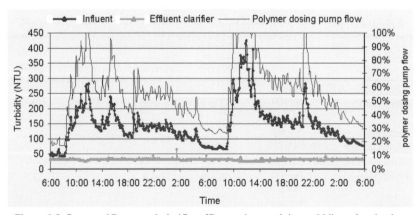

Figure 6. Influent and Dortmund-clarifier effluent when applying turbidity-related polymer dose of 5.5 mg pe/100 NTU

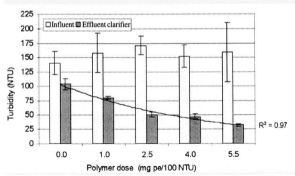

Figure 7. Turbidity (24-h averages) of influent and Dortmund-clarifier effluent as a function of the polymer dose with turbidity-related polymer dosing

Figure 6 shows an example of turbidity-related polymer dosage (5.5 mg polymer per 100 influent NTU). The influent turbidity, the turbidity of the clarifier-effluent as well as the flow of the dosing pump is presented. Figure 6 demonstrates that the control strategy proved highly effective in achieving a constant effluent quality in terms of turbidity (or $COD_{particulate}$). While the influent exhibited an average turbidity of 133 ± 48 NTU, the turbidity of the clarifier effluent had an average value of 31 ± 4 NTU.

Figure 7 shows the results of the turbidity-related dosage at different polymer-turbidity ratios. The figure shows 24-h average turbidity values of the influent and clarifier-effluent as a function of the polymer dosage. Different levels of effluent turbidity could be obtained by applying controlled dosing. The standard deviation of the clarifier-effluent values was fairly low, despite strong fluctuations in the 24-h influent averages.

Table 2 gives the actual values of Figure 7 including the standard deviation. In addition the average COD_{total} value of the 24-h samples is given. The particle removal efficiency of the Dortmund clarifier, shown in the second column, is calculated based on the average influent turbidity (148 ± 35 NTU) during the experimental period.

Table 2. Average clarifier effluent turbidity, particle removal and daily deviation for different levels of polymer dosage

Dosing	Average particle removal efficiency	Effluent clarifier (24-h averages)	COD_{total} (mg O_2 / l)
(mg pe/100 NTU)	(%)	(NTU)	
0	30	104 ± 17	420 ± 24
1.0	46	80 ± 14	n.m.[2]
2.5	66	51 ± 8	n.m.
4.0	68^1	47 ± 8	317 ± 20
5.5	78	32 ± 4	250 ± 38

[1] for these figures the flow was 130 l•h⁻¹ instead of 90 l•h⁻¹; [2] n.m. = not measured

Biodegradable COD at different levels of particle removal. To estimate the effect of different levels of particle removal on the biodegradable COD 24-h wastewater samples were treated with increasing doses of organic polymer. The flocculation and subsequent sedimentation of flocs were carried out in a jar test set-up. The samples were taken during dry weather conditions and had an initial average COD_{total} of 536 ± 80 mg $O_2 \cdot l^{-1}$.

Figures 8a and 8b show the results of the NUR experiments. The left figure (8a) shows the rapidly biodegradable COD (S_s), while the right figure (8b) shows the slowly biodegradable COD (X_s). The biodegradable COD is shown as a function of the COD_{total} of the analysed wastewater samples. The vertical line in the graph represents the average value of dissolved COD. The results show that S_s is not or hardly affected by the pre-treatment with organic polymers. The values that were found vary between 90 and 140 mg $O_2 \cdot l^{-1}$, but show no relationship with particle removal. X_s, however, decreases with increasing particle removal as is shown in

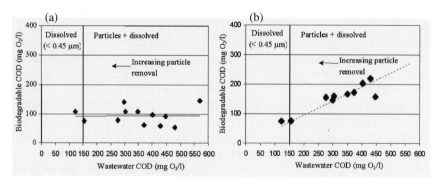

Figure 8. Biodegradable COD of pre-treated wastewater samples (a) Readily biodegradable COD (S_s) (b) Slowly biodegradable COD (X_s)

Conclusions

- The experimental work confirms that, for the investigated wastewater, on-line turbidity measurements can be used to monitor $COD_{particulate}$. Both influent and flocculated clarifier effluent samples showed a linear relationship over a wide range of $COD_{particulate}$ (100-900 mg $O_2 \cdot l^{-1}$) and turbidity (50 to 450 NTU).
- The tested turbidity-related polymer dosing method produced different, constant levels of particle removal, despite large particle concentration variations in the influent. Moreover, it resulted in higher removal efficiencies and a more stable operation compared to the dosing of fixed polymer dose per liter.
- Nitrogen Uptake Rate experiments showed different levels of biodegradable COD at different levels of particle removal. This indicates that turbidity-

related polymer dosing can be used to control the nitrogen concentration in the effluent of an activated sludge plant. Analysis of the biodegradable COD of pre-treated wastewater showed that X_s decreased with increasing particle removal, while S_s remained unaffected. Based on these values and the use of one of the activated sludge models it is possible to assess the nitrogen effluent concentration of an activated sludge plant.

Acknowledgement

This research was funded by the Dutch Foundation for Applied Water Research STOWA.

References

Kujawa, K. ,and A. Klapwijk (1999) A method to estimate denitrification potential for pre-denitrification using batch NUR test. *Water Research* **33** (10): 2291-2300

Mels, A. R., Van Nieuwenhuijzen, A. F. (2000) Cationic organic polymers for flocculation of municipal wastewater—experiments and scenario study. In: *Chemical Water and Wastewater Treatment* VI, Hahn, H.H., Hoffmann, E., Ødegaard, H. (Eds.), Springer-Verlag, Berlin, pp. 23-34

Mels, A.R., van der Meer, A.K., Rulkens, W.H., van Nieuwenhuijzen, A.F., Klapwijk, A. (2001) Flotation with polyelectrolytes as a first step of a more sustainable wastewater treatment system. *Water, Science and Technology* **43** (11) 83-90

Mels, A.R. (2001) *Physical-chemical pre-treatment as an option for increased sustainability of municipal wastewater treatment plants.* Ph.D. Thesis, Wageningen University, Sub-department of Environmental Technology, Wageningen, The Netherlands

van Nieuwenhuijzen, A.F., and Mels, A.R. (2002) Characterisation of particulate matter in municipal wastewater, *Chemical Water and Wastewater Treatment* VII, Hahn, H.H., Hoffmann, E., Ødegaard, H. (Eds.), IWA publishing, London

Ødegaard H. (1999) The influence of wastewater characteristics on choice of wastewater treatment method. *Proc. Nordic Conf. on Nitrogen Removal and Biological Phosphate Removal*, Oslo, Norway 2-4 February 1999

H.H. Hahn, E. Hoffmann, H. Ødegaard (Eds.)
Chemical Water and Wastewater Treatment VII, pp. 81-90
© IWA Publishing, London
ISBN: 1 84339 009 4

Fully Automatic Dose Control of Coagulants by Using Particle Charge Detector in Vienna Sewage Treatment Plant

E. Sailer

Applied Chemicals Handels GesmbH, Wien, Austria
email: erich.sailer@acat.com

Abstract

This report describes a project between EBS/Vienna Treatment Plant (A), Kemira-Kemwater (S), CIBA Specialty Chemicals (UK), Kemifloc (CZ), CDM/ Lasertrim and Applied Chemicals (A) on the use of ferric sulphate in combination with polyelectrolytes in the preprecipitation controlled by particle charge detectors.

The main goal was to achieve fully automatic dose control of coagulants to keep the effluent water quality below the limits set by the government, and, at the same time, to reduce the chemical costs.

Introduction

The construction of the Vienna Sewage Treatment plant started in the late 1960s and it has been in full operation since 1980. Initially, the main goal of sewage water treatment was to eliminate BOD, so a conventional plant with primary sedimentation, aeration, and secondary settlement was built.

The sludges are co-settled and dewatered by centrifuges at the "Fernwärme Wien" incineration plant.

Due to new regulations in Austria as well as in the European Union, it was necessary to improve the reduction of nutrients, such as phosphates, by simultaneous precipitation with ferrous sulphate and later with ferric chloride. The total phosphate concentration from the final effluent was still between 2 and 4 mg per litre. It was expected that the general regulation permitting only a maximum of 1 mg/l total phosphate concentration in the final effluent (average over one year) would take effect no later than 1999. Therefore we started our first extended lab trials in 1994. The target was to remove more particulate substances such as particulate BOD, COD and phosphates in the primary settlement.

Plant Data

The average flow in dry weather conditions is 500000 m³ per day. The flow varies in the course of a day with dry weather conditions between 10000 and 30000 m³ per hour (Figure 1). In rainy conditions the maximum flow can go up to 86400 m³/h. The COD in the influent can vary between 600 and 1,200 mg/l. The residual time in the primary sedimentation varies between 0.5 and 2 hours. The aeration basin volume is 45000 m³.

Only 12 m³/s influent can be treated biologically; if the flow rate is higher, it will pass through to the final effluent only with mechanical treatment.

The load on this typical day changes between 2t COD/h in the morning up to over 8t COD/h in the peak load time. The incoming COD concentration varies between 300 and 800 mg/l.

The total phosphorus level in the final effluent was always close to 2-4 mg/l, wheras the ortho-P was reduced down to 0.1-0.2 mg/l by simultaneous precipitation.

The final effluent had a high turbidity due to very fine particles leaving the final sedimentation step

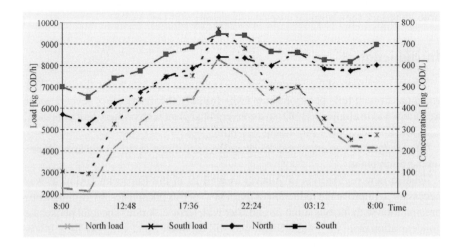

Figure 1. Load variation in the influent during one dry weather day

Principles

Depending on the type of incoming waste water, up to 70 % of the pollutants are in particulate form (Kemira 1991).

These particles usually have an anionic surface charge (Figure 2). By using the streaming current technique, the anionic surface charge can be determined.

The Lasertrim PSA005 particle charge detector titrates the sample online and continuously, usually with a defined solution of Polydadmac in water.

The biggest difference to other on-line PCD measurement systems is, that Lasertrim PSA 005 measures the cationic demand continuously and not by a batch process (Church, 1989).

The amount of Polydadmac solution consumed to reach charge neutralization is recorded, and this value controls the dosage of the coagulant, taking the flow to the plant into account.

This method allows the dose to be load-proportional, based on the incoming charge of the particles.

In our trials we found out that the cationic demand corresponds extremely well with the concentrations of total P and COD.

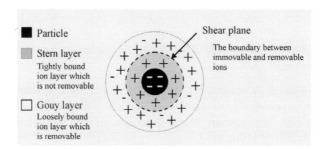

Figure 2. Typical particle charge distribution (adapted from Nixon, 1992)

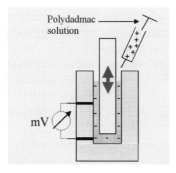

Figure 3. The PSA 005 measuring cell. The mV-signal is measured continuously; Polydadmac solution is added until the mV-signal reaches the setpoint (usually 0 mV)

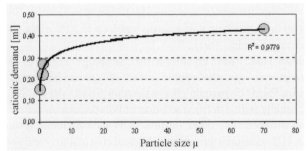

Figure 4. Effect of particle size on cationic demand (Bengtsson,1996)

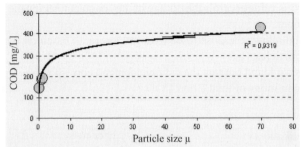

Figure 5. Relationship between COD and particle size (Bengtsson,1996)

Figures 4 and 5 show the correlation between particles, cationic demand and COD. It can be seen that particles below 70 μm show the highest cationic demand as well as COD. The incoming waste water was filtered through different mesh sizes to make this determination.

PSA-005 history

The coagulant dose control system PSA-005 (Particle Stock Analyser), a control system based on continuous cationic demand measurements, was first introduced to the Swedish market. The first experiences were collected at the Borås SWTP. Since 1996 more trials have been conducted and today there are several applications running continuous dosing control in the area of sewage water treatment and potable water production, including Halmstad SWTP, Luleå SWTP, Skellefteå SWPT and Mariestad Waterworks. Vienna SWTP is the first treatment plant using PSA-005 outside Sweden.

The coagulant dose control system has been developed gradually based on process and instrumental experiences. Today the system runs either as an open control system (Vienna application) or as a closed control system (feedback control system) using setpoint for cationic demand (Figure 6).

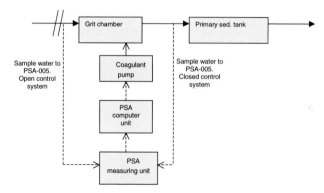

Figure 6. Differences between open control and closed control systems

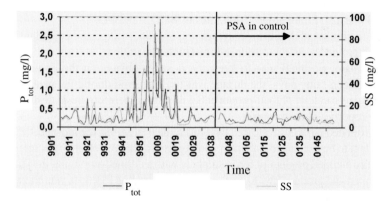

Figure 7. Halmstad SWTP, final treated water that was discharged into a receiving pond

Figure 7 shows some treatment results with the PSA-005 system. Prior to installation of this system, total phosphorus levels (P_{tot}) varied between 0.5 and 3.0 mg/l, particularly from week 41 (1999) to week 6 (2000) and suspended solids (SS) between 20 and 100 mg/l, but after installation, these levels were reduced to below 0.5 mg/l P_{tot} and 20 mg/l SS.

Description of the Vienna process

A sample of the incoming waste water is taken prior to the coagulant dosage point (open control system, Figure 6). This sample is then filtered through a coarse 70-80 µm filter. A flow-controlled pump feeds the sample to the measuring cell. A speed-controlled dosing pump feeds the Polydadmac solution continuously to the sample water. The Polydadmac solution consumed to always reach the setpoint (e.g. 0 mV) generates the cationic demand. That figure is recalculated in the control computer based on the total flow.

This calculation determines how much coagulant would be needed to neutralize all particles.

The computer generates the output signal that controls the dosing pumps for the coagulant.

In Vienna the orthophosphate values from the on-line measurement after the primary sedimantation are used as a correction factor for the dosage as well.

Trials

The first full-scale trials for particle separation started in 1996. For a full comparison, only one of the two separated lines was treated with ferric sulfate and polymer.

In these preliminary trial we succeeded to keep the total phosphorous level below 1 mg/l and also improve reduction rates of BOD and COD.

All facilities were built for polymer preparation, storage and coagulant dosing units to provide a continuous dose for particle separation.

The major problem was to keep the dosages always on the correct levels. Especially on rainy days, the dosage was too high, and during the peak load, the dosage was too low. A load-proportional dose was therefore more desirable than flow-proportional dose control.

In extended lab trials followed by a full scale trial series, the most cost effective coagulant to be used in conjunction with a particle charge detector was selected. A specially blended product of ferric sulphate with a defined amount of Polydadmac, developed and produced by Kemifloc/Kemwater, called PIX 116, was chosen. The ratio of the blended chemicals was specially formulated for the demands on the Vienna Sewage Treatment Works and may vary for other applications.

Since November 1999 PIX 116 is fully automatically dosed under control of a

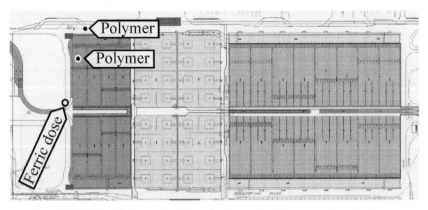

Figure 8. Dosage points at the initial trials

particle charge detector.

Due to the experience of this dose control in several plants in Sweden we started to measure the cationic demand in the outlet of the primary sedimentation.

The dose of the coagulant was controlled with this "feed back" information.

Our experience was at this time that the results were good, but the response was not as quick as we were expecting it, due to the residual time in the primary sedimentation.

Early in year 2000 we changed to measure the cationic demand directly in the influent of the Vienna WWTP (open control system).

Results

End of 1999 the first installation of Lasertrim PSA 005 dose control system was completed, and the graph below shows clearly, that the efficiency of the plant (Figure 8) has increased very much at this time.

Results from PCD dose control showed that the dose variations were much bigger

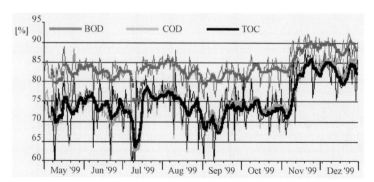

Figure 9. Total reduction of BOD, COD, and TOC as a result of installing lasertrim

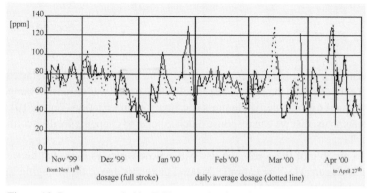

Figure 10. Dosage recorded by PCD control unit and the daily average values measured by the plant

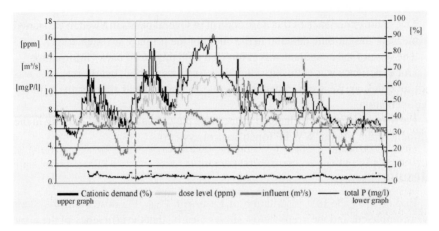

Figure 11. Cationic demand (%), total P (mg/l), influent (m³/s), and dose level (ppm) over one typical week

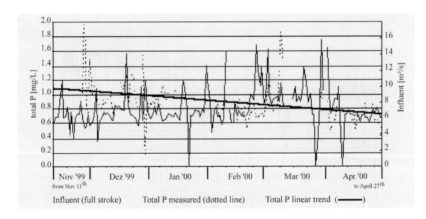

Figure 12. PSA 005: Reduction of total phosphorous

when the samples were taken directly from the influent (this system was started at the end of February 2000) than when "feed back information" was used from samples taken to measure cationic demand after the primary sedimentation (Figure 10).

This meant to us, that we can response much quicker to changing loads incoming to the WWTP.

At the beginning of the week, the dry water flow varies as expected between 3 and 8 m³/s (Figure 11). The cationic demand increases due to the changing load until the middle of the week. At the end of the week it rained, with flow peaks up to 14 m³/s. It can be seen that after the heavy rain, the cationic demand was also dramatically reduced. Total phosphorus was consistently below 1 mg/l.

By continuous improvement of the setup in the control computer, dosages and total phosphorus in the final effluent could be optimised.

It was also seen, that the turbidity of the final effluent improved significantly.

Maintenance of the control unit

To avoid blockage in the pipes and on the filters, the computer can be set to run back washes at different time intervals.

The measuring cell unit was cleaned manually once a month. The cleaning is very simple because the pipework of the unit is made of ¾ inch PVC pipes with quick connects.

The chemical costs for this analytical unit are significantly less than the costs for other methods such as on-line phosphate measurement.

Conclusion

Cationic demand is a useful parameter for controlling coagulant dose. The limit of 1 mg/l total phosphate in the effluent (annual average) was never exceeded. The dosages on average dropped from over 100 ppm to 75 ppm. This process will be used until 2006 when the sewage treatment works is ready to start up the completely rebuilt plant. The plant volume will be expanded from 40000 m^3 in the aeration basin and 60000 m^3 in the secondary clarifiers to additional 171000 m^3 in the aeration basin and 200000 m^3 in the secondary clarifiers.

The capacity of the upgraded plant will allow biological treatment of the total

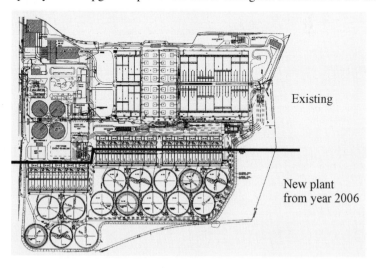

Figure 13. Expansion of the Vienna STW

influent load including nitrogen removal. Future work will be concerned with using the Lasertrim PCD on simultaneous phosphorus removal. It would also be of interest to explore the basic processes that make cationic demand a useful parameter for determining coagulant dosage.

References

Bengtsson, J. (1996) Gothenburg waste water: Particle size vs. Cationic demand, CDM, Lab report

Church, G. (1989) Coagulant control by streaming current technology, Lasertrim Ltd. Water & Waste Treatment

Kemira (1991) Handbuch der Wassertechnologie, Chapter 1.A: Charakter und Zusammensetzung, Kemira Kemwater *(in German)*

Nixon, A.W. (1992) The measurement principles of the particle charge detector, Paper Edge

Oxidation Processes

H.H. Hahn, E. Hoffmann, H. Ødegaard (Eds.)
Chemical Water and Wastewater Treatment VII, pp. 93-108
© IWA Publishing, London
ISBN: 1 84339 009 4

Application of Kinetic Modelling to Optimization of Advanced Oxidation Processes in Wastewater Treatment

T. D. Waite, A. J. Feitz and R. Aplin*

*Centre for Water and Waste Technology, The University of New South Wales, Sydney, Australia
email: d.waite@unsw.edu.au

Introduction

The group of processes known as advanced oxidation processes (AOPs) can be used to treat contaminated water and wastewater. AOPs generate highly reactive oxidants such as hydroxyl radicals (OH$^\bullet$). The hydroxyl radical is a highly reactive and non-selective oxidant and can rapidly oxidise a wide variety of organic compounds (Buxton *et al.*, 1988). Hydroxyl radicals can also be scavenged by natural organic matter (Goldstone *et al.*, 2002) and inorganic ions such as HCO_3^- and CO_3^{2-} (Ollis, 1993; Pelizzetti, 1999). In some AOPs oxidation by other oxidants such as ozone may also occur, and alternative degradation pathways such as direct photolysis and pyrolysis may be possible.

Hydroxyl radicals react with saturated organic compounds by hydrogen abstraction (e.g. Eq. 1), while addition reactions occur between OH$^\bullet$ and aromatic or unsaturated compounds (e.g. Eq. 2) (Huang *et al.*, 1993; Bolton and Cater, 1994). Some organic compounds such as $C_2O_4^{2-}$ react with hydroxyl radicals by electron transfer (Eq. 3) (Getoff *et al.*, 1971).

$$CH_3OH + OH^\bullet \rightarrow {}^\bullet CH_2OH + H_2O \qquad (1)$$

$$C_6H_6 + OH^\bullet \rightarrow {}^\bullet C_6H_6OH \qquad (2)$$

$$C_2O_4^{2-} + OH^\bullet \rightarrow C_2O_4^{\bullet -} + OH^- \qquad (3)$$

Carbon-centred radicals generated by the reaction of hydroxyl radicals with organic compounds may react with O_2 to form organoperoxy radicals (Eq. 4). These radicals can decompose, producing $O_2^{\bullet -}$ and its conjugate acid $HO_2^\bullet$ (Eqs 4 and 5; Pignatello, 1992; Bielski *et al.*, 1985). Carbon-centred radicals may also form dimers (Eq. 6; Walling, 1975).

[93]

$$R^\bullet + O_2 \rightarrow RO_2^\bullet \rightarrow R^+ + O_2^{\bullet-} \tag{4}$$
$$O_2^{\bullet-} + H^+ \leftrightarrow HO_2^\bullet \tag{5}$$
$$R^\bullet + R^\bullet \rightarrow R\text{–}R \tag{6}$$

AOPs can completely mineralise most organic contaminants, converting them to CO_2, H_2O and inorganic ions such as chloride, but this is not always necessary or cost-effective. Partial degradation will usually reduce the toxicity and increase the degradability of the contaminants (Bolton and Cater, 1994), but in some cases may generate intermediates that are toxic to humans or inhibitory to biological treatment processes (Pelizzetti, 1999).

AOPs can be used in combination with other processes. They may be used to degrade nonbiodegradable compounds that remain after biological treatment of contaminated water. Alternatively, they may be used to partially degrade toxic, inhibitory or refractory compounds to more biodegradable compounds before biological treatment (Scott and Ollis, 1995). They may also be used in combination with other physical-chemical processes such as coagulation and sedimentation or membrane separation.

AOPs generate hydroxyl radicals using combinations of oxidants, catalysts and irradiation (Huang et al., 1993; Ruppert et al., 1994) and may involve use of Fenton's reagent ($Fe(II)/H_2O_2$), UV/H_2O_2, ozonation, $h\nu/TiO_2$, wet air oxidation and sonolysis or combinations of two or more of these processes. Individual AOPs typically involve a series of free radical reactions and identification of the rate controlling step and, thence, optimal treatment conditions are often non-trivial and tend to be done empirically. Combining two or more AOPs adds to the complexity of the processes operating.

In this chapter, we examine the potential to mathematically model AOPs and investigate the use of such models in both identification of major transformation pathways and optimisation of process conditions. Examples are drawn from our studies on dark (thermal) Fenton's reactions, modified photo-Fenton's reactions and $TiO_2/h\nu$ mediated contaminant degradation.

Dark Fenton's processes

Fenton's reagent-based processes are finding increasing application for the degradation of organic contaminants in waste streams and subsurface systems. These processes involve the generation of the highly oxidative hydroxyl radical ($OH^\bullet$) by reaction of ferrous iron and hydrogen peroxide. The ferric iron so-produced is re-reduced by hydrogen peroxide or superoxide with the result that, provided excess hydrogen peroxide is present, iron acts to catalyse the production of the key oxidizing intermediate. The hydroxyl radicals so produced degrade organic compounds (RH) principally by hydrogen abstraction. That is:

Hydroxyl radical generation
$$Fe^{2+} + H_2O_2 \rightarrow Fe^{III}OH^{2+} + OH^{\bullet} \tag{7}$$

Reduction of Fe(III)
$$Fe^{III}OH^{2+} + H_2O_2 \rightarrow Fe^{2+} + H_2O + HO_2^{\bullet} \tag{8}$$

$$Fe^{III}OH^{2+} + HO_2^{\bullet} \rightarrow Fe^{2+} + O_2 + H_2O \tag{9}$$

H abstraction
$$RH + OH^{\bullet} \rightarrow R^{\bullet} + H_2O \tag{10}$$

Addition of optimal concentrations of reagents (Fe^{2+} and H_2O_2) requires considerable care since both ferrous ions and hydrogen peroxide also scavenge hydroxyl radicals with the result that their addition may limit rather than enhance reactivity; i.e.

Hydroxyl radical scavenging
$$H_2O_2 + OH^{\bullet} \rightarrow HO_2^{\bullet} + H_2O \tag{11}$$

$$Fe^{2+} + OH^{\bullet} \rightarrow Fe^{III}OH^{2+} \tag{12}$$

While appropriate concentrations of reagents for a given contaminant concentration have normally been determined by "rule of thumb" or empirical "trial and error" studies, the rate constants of most key reactions are known enabling kinetic models of the process to be constructed. As an example, we have investigated the degradation of trichloroethylene by Fenton's reagent over a range of solution conditions and modelled the process using the equation set shown in Appendix 1 with predicted parameter concentrations obtained using the kinetics package ACUCHEM. This program uses implicit, variable order, linear multivalue integration methods with the appropriate method selected by the program to minimise computational effort (Braun *et al.*, 1988). The integration time step is automatically determined by the program based on the integration tolerance specified by the user, which was 0.001 in our case. Optimisation of unknown or poorly defined rate constants was undertaken by visual inspection of the model curve in comparison to the experimental data.

Of the equations for Fenton's reagent degradation of TCE shown in Appendix 1, particular uncertainty exists for the reduction of ferric iron by the hydroperoxy radical $HO_2^{\bullet}$. Pignatello (1992) reported a value of $2.9 \times 10^4 \, M^{-1}s^{-1}$ while Tang and Huang (1997) used a value of $1 \times 10^6 \, M^{-1}s^{-1}$. As shown in Figure 1, a good description of experimental data obtained in our studies is found for a value ($k_6 = 2.7 \bullet 10^5 \, M^{-1}s^{-1}$) between these extremes.

The model describes the biphasic nature of degradation using Fenton's reagent satisfactorily which arises as a result of the initial oxidation of Fe(II) by H_2O_2 with subsequent cycling of iron between the ferric and ferrous states.

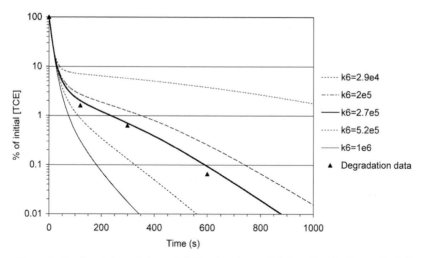

Figure 1. Predicted degradation of TCE using the model described in Appendix 1 for different values of the rate constant k_6

The model can be used to examine the effect of process conditions on rate and extent of TCE removal. Modelling results confirm the results of earlier empirical investigations (e.g. Tang and Huang, 1997) that:

- Optimal contaminant removal is obtained for an initial H_2O_2/Fe(II) concentration ratio ≈ 11, and
- Disproportionately higher concentrations of H_2O_2 are required for degradation of higher concentrations of TCE to similar degrees.

This latter point, coupled with the marked reduction in degradation rate that occurs once the initial Fe(II) oxidation step is concluded, is particularly interesting and suggests that significantly more cost-effective contaminant degradation should be achievable through multistage addition of reagents. An estimation of the savings that might accrue from multistage addition can be deduced from the percentage removal isopleths developed by Tang and Huang (1997). As shown in Table 1, 70 mM H_2O_2 is required for essentially complete degradation of 50 ppb TCE via a single step process while similar degradation can be achieved with less than 2 mM in a three-stage process.

As shown in Figure 2, multi-staged addition of reagents can be described using the model developed above. In this example, a single dose of 100 ppm of H_2O_2 and 9 ppm Fe(II) (a ratio of 11/1) resulted in approximately 70 % removal. A second dosage of the same amounts of H_2O_2 and Fe(II) after 10 minutes results in 99.7 % removal after 12 minutes. Interestingly, a second addition of Fe(II) only yielded a

Table 1. Concentration of H_2O_2 required for degradation of 50 ppb TCE for $[H_2O_2]_{initial}$/ $[Fe(II)]_{initial}$ = 11/1 for one-step compared to three-step process

Step	$[TCE]_{initial}$ ppb	$[TCE]_{initial}$ μM	% Red	$[TCE]_{final}$ μM	$[H_2O_2]$ required mM	$[H_2O_2]$ required ppb
One step process						
1	50	380	100	<10	70	2380
Multi-step process						
1	50	380	70	114	1	34
2	15	114	70	34.2	0.4	14
3	4.5	34.2	100	<10	0.5	17

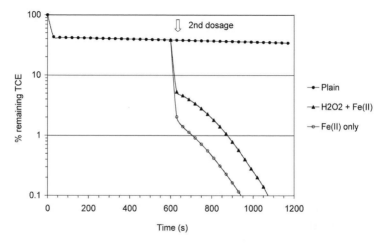

Figure 2. Simulation of degradation of 50 ppb TCE using two stage addition of 100 ppm H_2O_2 and 9 ppm Fe(II) (a ratio of approx. 11/1). The predicted effect of addition of 9 ppm Fe(II) only is also shown

slightly greater degree of removal than addition of both H_2O_2 and Fe(II). Model results indicate that this was to be expected as adequate H_2O_2 remained after the first dose of reagents and any subsequent addition resulted in partial scavenging of hydroxyl radicals.

Light-assisted ligand-mediated Fenton's processes

Fenton's reagent is only effective under acidic conditions (pH 2 - 4). The inactivity of Fenton's reagent at higher pH is usually attributed to the formation of ferric oxyhydroxide precipitates. The formation of ferric oxyhydroxides can be prevented by adding Fe(III) ligands such as oxalate and citrate (Aplin *et al.*, 2001). Photolysis of these Fe(III) complexes results in reduction of the metal centre to Fe(II) species (Eq. 13) which, in the presence of H_2O_2, will be reoxidised to Fe(III) with the concomitant production of hydroxyl radicals (or possibly similarly reactive ferryl species) (Sun and Pignatello, 1992; Safarzadeh-Amiri, 1993).

$$Fe(C_2O_4)_3^{3-} + hv \rightarrow Fe^{2+} + 2C_2O_4^{2-} + C_2O_4^{\cdot -} \tag{13}$$

This process, known as the modified photo-Fenton's process, is an effective means of degrading contaminants at higher pH (Sun and Pignatello, 1993; Safarzadeh-Amiri et al., 1996a, b, 1997). The process can be modelled but, as shown in the Appendix, the variety and number of possible reactions involved are large.

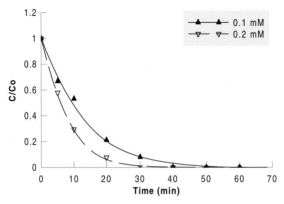

Figure 3. Effect of initial Fe(III) concentration on degradation of Reactive Red 235 by hv/Fe(III)/oxalate/H_2O_2. The initial molar ratio of oxalate to Fe(III) was 3:1. In each case $[dye]_o = 0.1$ g/L, $[H_2O_2]_o = 1$ mM and pH = 3

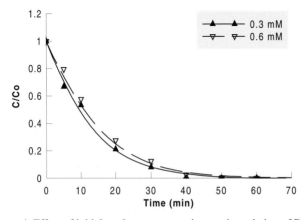

Figure 4. Effect of initial oxalate concentration on degradation of Reactive Red 235. In each case $[dye]_o = 0.1$ g/L, $[Fe(III)]_o = 0.1$ mM, $[H_2O_2]_o = 1$ mM and pH = 3

Models such as this have been found to adequately describe the degradation of reactive dyes as a function of solution conditions (see Figures 3 and 4), but many assumptions must be made in developing an effective model with considerable uncertainty surrounding the veracity of many of these assumptions (Aplin and Waite, 2002).

Given the complexity of such models and the uncertainty surrounding many of the key rate constants, one could question the uniqueness of the derived model and, ultimately, its value as a process optimisation tool. As demonstrated in the example below, however, development of models in a stepwise manner (from simple to complex) coupled with use of perturbation analysis provides considerable insight into both dominant pathways and, subsequently, to optimisation of operating conditions.

TiO$_2$-mediated processes

Recent experimental and computational investigations of the TiO$_2$/hv-mediated degradation of the blue green algal toxin microcystin-LR (MLR) within a matrix of ill-defined algal exudates (Feitz *et al.*, 1999; Feitz and Waite, 2002) provides an example of the use of modelling in probing dominant degradation pathways.

At low pH and low concentrations of toxin (such that all toxin resides on the TiO$_2$ surface), degradation is satisfactorily described (see Figure 5) using a surface-reaction mediated model (see part A model in Appendix, Table A 4).

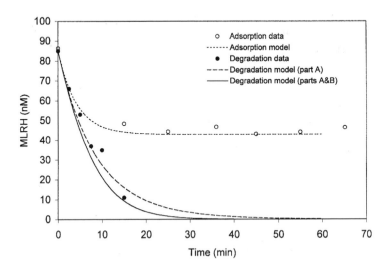

Figure 5. Loss of microcystin-LR from solution at pH 3.5 and low concentration. Ionic strength of 0.01 M and TiO$_2$ loading of 1 g•L^{-1}; "Degradation model (part A)" refers to the surface only degradation model; "Degradation model (parts A&B)" incorporates both surface and solution degradation components; "Adsorption model" excludes light based reactions

In order to examine the sensitivity of the model to variation in a particular rate constant, the square of residuals between the original (control) model output and the perturbed model output was calculated. The sum of squares of the residuals (SSR) was calculated as follows, where m_i and p_i represent the predicted MLRH⁻ concentrations in the control and perturbed models, respectively, at each data point:

$$SSR = \Sigma \, (m_i - p_i)^2/m_i$$

Sensitivity analysis was undertaken by varying the rate constant for a particular reaction whilst the other rate constants were unchanged. This approach assumes that the rate constants in the original model are close to the correct value, since a change in one of the rate constants may affect the sensitivity of the model to a change in another of the rate constants. The change in model output was considered by perturbing each rate constant by an order of magnitude (both increase and decrease). This type of analysis has been used to good effect recently by Rose and Waite (2002) in examination of the relative importance of the various reactions involved in Fe(II) oxidation in seawaters.

The results of applying perturbation analysis to the "surface reaction" degradation (part A) model are shown in Figure 6 and demonstrate that the rate of degradation of toxin is strongly influenced by the rates of adsorption/desorption of both MLRH⁻ and oxygen to the particle surface.

As can be seen from Figure 7, this surface-reaction based model is inadequate at higher initial toxin concentrations and solution-based degradation pathways must be introduced to adequately describe the data.

There are essentially two mechanisms by which solution phase degradation may be initiated: i) desorption of superoxide and inducement of a free radical degradation

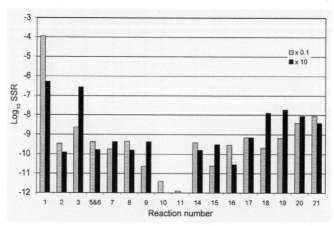

Figure 6. Log sum of squares of the residuals (log_{10}SSR) for one order of magnitude (increase and decrease) perturbation of each rate constant for degradation model (part A: surface degradation) at pH 3.5, $[MLRH^-]_{initial}$ = 85 nM, ionic strength = 0.01 M and TiO_2 = 1 g•L⁻¹

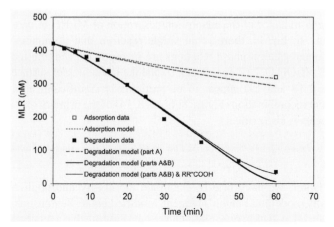

Figure 7. Loss of microcystin-LR from solution at pH 3.5 at high concentration. Ionic strength of 0.01 M and TiO$_2$ loading of 1 g•L^{-1}; "Degradation model (part A)" refers to the surface only degradation model; "Degradation model (parts A&B)" incorporates both surface and solution degradation components; "Adsorption model" excludes light based reactions; "Degradation model (parts A&B) & RR"COOH" incorporates both surface and solution degradation components but excludes oxygenated products from any further reactions

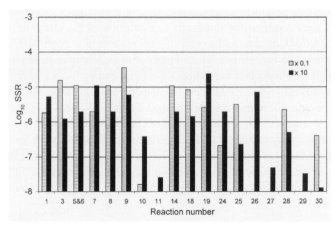

Figure 8. Log sum of squares of the residuals (log$_{10}$SSR) for one order of magnitude (increase and decrease) perturbation of each rate constant for degradation model (parts A&B) at pH 3.5, [MLRH$^-$]$_{initial}$ = 420 nM, ionic strength = 0.01 M and TiO$_2$ = 1 g.L^{-1}

pathway through reaction of superoxide with the bulk organic, and ii) assuming that the exudates radical is sufficiently long-lived to desorb from the surface into solution. While there is insufficient space here to present arguments in detail, the results of modelling suggest that the superoxide-mediated degradation pathway is of major importance (Feitz and Waite, 2002).

Unlike the dominance of the adsorption/desorption of MLRH in part A of the model (Figure 6; Eq. 1), there is no single reaction that dominates MLRH degradation once solution phase reactions are incorporated into the model. The desorption of $>TiOH:O_2^{\bullet-}$ (Figure 8; Eq. 9) and the quenching of surface radicals (Figure 8; Eq. 18 and 19) appear to be particularly significant, although the production and recombination (Figure 8; Eq. 5-8, 14) of the primary oxidising and reducing species is also critical.

Summary

In this chapter, we have briefly examined the ability and applicability of mathematically modelling Fenton's reagent, modified photo-Fenton's reagent and $TiO_2/h\nu$ advanced oxidation processes. Our results indicate that a relatively simple model with only one fitting parameter can be used to describe the Fenton's reagent degradation of trichloroethylene. The model can then be used to examine the impact of system conditions on contaminant degradation and has been used to demonstrate that the staged addition of reagents should result in considerable savings in both reagent costs and residence time required to achieve a desired level of degradation.

A substantially more complex model is required for description of the modified photo-Fenton degradation of a reactive dye. Despite the complexity, very few fitting parameters were required for satisfactory description of the effect of change in process conditions on rate and extent of degradation of a reactive textile dye. Nevertheless, it was necessary to make a number of assumptions in developing the model, which require further validation.

A model was also developed to describe the degradation of the blue green algal toxin microcystin-LR within a complex algal exudate. While it was necessary to include "best guesses" for a number of rate constants, the model was useful in assessing the likely importance of various possible reaction pathways. A "perturbation analysis" approach has been described which involves examination of the impact of change in rate constant assignment on model output and which provides particularly useful insights into dominant reaction pathways under particular system conditions.

References

Adams G. E., and Willson R. L. (1969) Pulse radiolysis studies on the oxidation of organic radicals in aqueous solution. *Trans. Faraday Soc.* **65**, 2981-2987

Aplin, R., Feitz, A. J., and Waite, T. D. (2001) Effect of Fe(III)-ligand properties on effectiveness of modified photo-Fenton processes. *Water Science and Technology* **44**, 23-30

Aplin, R., and Waite, T.D. (2002) Degradation of oxalate in a modified photo-Fenton (UV/ferrioxalate/H_2O_2) system: Effects of initial reactant concentrations and light intensity and wavelength. *Water Research* (in press)

Bielski, B. H. J., and Cabelli, D. E. (1995) Reactivity of HO_2/O_2^- radicals in aqueous solution. In *Active Oxygen in Chemistry* (eds C.S. Foote, J.S. Valentine, A. Greenberg and J.F. Liebman), p. 66, Blackie Academic & Professional, New York

Bielski B. H. J., Cabelli D. E., Arudi R. L., and Ross A. B. (1985) Reactivity of HO_2/O_2^- radicals in aqueous solution. *J. Phys. Chem. Ref. Data* **14**(4), 1041-1100

Bolton J. R., and Cater S. R. (1994) Homogeneous photodegradation of pollutants in contaminated water: an introduction. In *Aquatic and Surface Photochemistry* (eds G.R.H. Helz, R.G. Zepp and D.G. Crosby), pp. 467-490, CRC Press, Boca Raton

Braun, W., Herron, J. T., and Kahaner, D. K. (1988) ACUCHEM: Computer program for modelling complex reaction systems. *International Journal of Chemical Kinetics* **20**, 51-62

Buxton G. V., Greenstock C. L., Helman W. P., and Ross A. B. (1988) Critical review of rate constants for reactions of hydrated electrons, hydrogen atoms and hydroxyl radicals ($\cdot OH/\cdot O^-$) in aqueous solution. *J. Phys. Chem. Ref. Data* **17**(2), 513-886

Chen R. Z., and Pignatello J. J. (1997) Role of quinone intermediates as electron shuttles in Fenton and photoassisted Fenton oxidations of aromatic compounds. *Environ. Sci. Technol.* **31**(8), 2399-2406

Davis A. P., and Huang, C. P. (1993) A kinetic model describing photocatalytic oxidation using illuminated semiconductors. *Chemosphere* **26**, 1119 - 1135

Faust B. C., and Zepp R. G. (1993) Photochemistry of aqueous iron(III)-polycarboxylate complexes: roles in the chemistry of atmospheric and surface waters. *Environ. Sci. Technol.* **27**(12), 2517-2522

Feitz, A. J., Waite, T. D., Jones, G. J., Boyden, B. H., and Orr, P. T. (1999) Photocatalytic degradation of the blue-green algal toxin Microcystin-LR in a complex aqueous/organic matrix. *Environ. Sci. Technol.* **33**, 243-249

Feitz, A.J., and Waite, T.D. (2002). Kinetic modeling of TiO_2-catalysed photodegradation of trace levels of microcystin-LR. *Environ. Sci. Technol.* (in press)

Getoff N., Schwörer F., Markovic V. M., Sehested K., and Nielsen S. O. (1971) Pulse radiolysis of oxalic acid and oxalates. *J. Phys. Chem.* **75**(6), 749-755

Hoffmann, M. R., Martin, S. T., Choi, W., and Bahnemann, D. W. (1995) Environmental applications of semiconductor photocatalysis. *Chem. Rev.* **95**, 69 - 96

Huang C. P., Dong C., and Tang Z. (1993) Advanced chemical oxidation: its present role and potential future in hazardous waste treatment. *Waste Management* **13**, 361-377

Huang C. R., Lin Y. K., and Shu H. Y. (1994) Wastewater decolorization and TOC-reduction by sequential treatment. *Am. Dyestuff Rep.* **73**(10), 15-18

Morel F. F., and Hering J. G. (1993) *Principles and Applications of Aquatic Chemistry*. Wiley, New York

Mulazzani Q. G., D'Angelantonio M., Venturi M., Hoffman M. Z., and Rodgers M. A. J. (1986) Interaction of formate and oxalate ions with radiation-generated radicals in aqueous solution. Methylviologen as a mechanistic probe. *J. Phys. Chem.* **90**(21), 5347-5352

Neta, P., Huie, R. E., and Ross, A. B. (1990) Rate constants for reactions of peroxyl radicals in fluid solutions. *J. Phys. Chem. Ref. Data* **19**, 413 - 513.

Ollis D. F. (1993) Comparative aspects of advanced oxidation processes. *ACS Symposium Series*, **518** (Emerging Technologies in Hazardous Waste Management III), 18-34

Pelizzetti E. (1999) Solar water detoxification. Current status and perspectives. *Z. Phys. Chemie* **212**, 207-218

Pignatello J. J. (1992) Dark and photoassisted Fe^{3+}-catalyzed degradation of chlorophenoxy herbicides by hydrogen peroxide. *Environ. Sci. Technol.* **26**(5), 944-951

Rose, A.L., and Waite, T.D. (2002) A kinetic model for Fe(II) oxidation in seawater in the absence and presence of natural organic matter. *Environ. Sci. Technol.* **36**, 433-444

Ruppert G., Bauer R., and Heisler G. (1994) UV-O$_3$, UV-H$_2$O$_2$, UV-TiO$_2$ and the photo-Fenton reaction—comparison of advanced oxidation processes for wastewater treatment. *Chemosphere* **28**(8), 1447-1454

Safarzadeh-Amiri A. (1993) Photocatalytic method for treatment of contaminated water. U.S. Patent No. 5,266,214

Safarzadeh-Amiri A., Bolton J. R., and Cater S. R. (1996a) Ferrioxalate-mediated solar degradation of organic contaminants in water. *Solar Energy* **56**(5), 439-443

Safarzadeh-Amiri A., Bolton J. R., and Cater S. R. (1996b) The use of iron in advanced oxidation processes. *J. Adv. Oxid. Technol.* **1**(1), 18-26

Safarzadeh-Amiri A., Bolton J. R., and Cater S. R. (1997) Ferrioxalate-mediated photodegradation of organic pollutants in contaminated water. *Water Res.* **31**(4), 787-798

Scott J. P., and Ollis D. F. (1995) Integration of chemical and biological oxidation processes for water treatment: review and recommendations. *Environ. Prog.* **14**(2), 88-103

Sedlak D. L., and Hoigné J. (1993) The role of copper and oxalate in the redox cycling of iron in atmospheric waters. *Atmos. Environ.* **27A**(14), 2173-2185

Sun Y., and Pignatello J. J. (1992). Chemical treatment of pesticide wastes—evaluation of Fe(III) chelates for catalytic hydrogen peroxide oxidation of 2,4-D at circumneutral pH. *J. Agric. Food Chem.* **40**(2), 322-327

Sun Y., and Pignatello J. J. (1993). Activation of hydrogen peroxide by iron(III) chelates for abiotic degradation of herbicides and insecticides in water. *J. Agric. Food Chem.* **41**(2), 308-312

Tang W. Z., and Huang C. P. (1997) Stochiometry of Fentons reagent in the oxidation of chlorinated aliphatic organic pollutants. *Environmental Technology* **18**(1), 13-23

von Sonntag, C., Dowideit, P., Fang, X., Mertens, R., Pan, X., Schuchmann M. N., and Schuchmann, H. P. (1997) The fate of peroxyl radicals in aqueous solution. *Wat. Sci. Tech.* **35**, 9 - 16

Waite, T.D., Hug, S., and Feitz, A. J. (1999) Photocatalysed degradation of trace contaminants in complex aqueous media. In *Mineral-Water Interfacial Reactions: Kinetics and Mechanism* (eds D. L. Sparks and T. J. Grundl), p. 374 - 391, American Chemical Society: Washington D.C.

Walling C. (1975) Fenton's reagent revisited. *Acc. Chem. Res.* **8**, 125-131

Walling, C. (1995) Autoxidation. In *Active Oxygen in Chemistry* (eds C. S. Foote, J. S. Valentine, A. Greenberg, J. F. Liebman), p. 24 - 65, Blackie Academic & Professional, New York

Zuo Y., and Hoigné J. (1992) Formation of hydrogen peroxide and depletion of oxalic acid in atmospheric water by photolysis of iron(III)-oxalato complexes. *Environ. Sci. Technol.* **26**(5), 1014-1022

Appendix

Table A.1. ACUCHEM model of Fenton reagent-mediated degradation of TCE

Reaction	k or K	Ref.[a]
(1) ClCH=CCl$_2$ + OH$^\bullet$ → $^\bullet$CCl$_2$CHClOH	k = 4.2 × 10^9 M^{-1}s^{-1}	1
(2) Fe(II) + H$_2$O$_2$ → Fe(III) + OH$^\bullet$	k = 51 M^{-1}s^{-1}	2
(3) Fe(II) + OH$^\bullet$ → Fe(III)	k = 4.3 × 10^8 M^{-1}s^{-1}	3
(4) Fe(II) + HO$_2^\bullet$ (+H$^+$) → Fe(III) + H$_2$O$_2$	k = 1.2 × 10^6 M^{-1}s^{-1}	4
(5) Fe(II) + O$_2^{\bullet-}$ (+2H$^+$) → Fe(III) + H$_2$O$_2$	k = 1 × 10^7 M^{-1}s^{-1}	4
(6) Fe(III) + HO$_2^\bullet$ → Fe(II) + O$_2$ + H$^+$	k = 2.7 × 10^5 M^{-1}s^{-1}	b
(7) Fe(III) + H$_2$O$_2$ → Fe(II) + HO$_2^\bullet$	k = 0.02 M^{-1}s^{-1}	3
(8) Fe(III) + O$_2^{\bullet-}$ → Fe(II) + O$_2$	k = 1.5 × 10^8 M^{-1}s^{-1}	4
(9) H$_2$O$_2$ + OH$^\bullet$ → HO$_2^\bullet$ + H$_2$O	k = 2.7 × 10^7 M^{-1}s^{-1}	2
(10) HO$_2^\bullet$ + HO$_2^\bullet$ → H$_2$O$_2$ + O$_2$	k = 8.3 × 10^5 M^{-1}s^{-1}	4
(11) HO$_2^\bullet$ + O$_2^{\bullet-}$ (+H$^+$) → H$_2$O$_2$ + O$_2$	k = 9.7 × 10^7 M^{-1}s^{-1}	4
(12) OH$^\bullet$ + OH$^\bullet$ → H$_2$O$_2$	k = 5.3 × 10^9 M^{-1}s^{-1}	2
(13) HO$_2^\bullet$ ↔ O$_2^{\bullet-}$ + H$^+$	K = 1.6 × 10^{-5} M	4

Notes: [a]References: (1) Buxton *et al.* (1988); (2) Tang and Huang (1997); (3) Pignatello (1992); (4) Zuo and Hoigné (1992). [b] Fitted value.

Table A.2. Irreversible reactions and rate constants used in the kinetic model of modified photo-Fenton-mediated dye degradation.

Reaction	k	Ref.[a]
(1a) Fe^{2+} + H$_2$O$_2$ → FeOH^{2+} + OH$^\bullet$	76 M^{-1}.s^{-1}	1
(1b) Fe(C$_2$O$_4$)$_m^{2-2m}$ + H$_2$O$_2$ → Fe(C$_2$O$_4$)$_m^{3-2m}$ + OH$^-$ + OH$^\bullet$	3.1 × 10^4 M^{-1}.s^{-1}	2
(2a) H$_2$C$_2$O$_4$ + OH$^\bullet$ → C$_2$O$_4^{\bullet-}$ + H$_2$O + H$^+$	1.4 × 10^6 M^{-1}.s^{-1}	3
(2b) HC$_2$O$_4^-$ + OH$^\bullet$ → C$_2$O$_4^{\bullet-}$ + H$_2$O	4.7 × 10^7 M^{-1}.s^{-1}	3
(2c) C$_2$O$_4^{2-}$ + OH$^\bullet$ → C$_2$O$_4^{\bullet-}$ + OH$^-$	7.7 × 10^6 M^{-1}.s^{-1}	3
(3) H$_2$O$_2$ + OH$^\bullet$ → HO$_2^\bullet$ + H$_2$O	2.7 × 10^7 M^{-1}.s^{-1}	4
(4a) Fe^{2+} + OH$^\bullet$ → FeOH^{2+}	4.3 × 10^8 M^{-1}.s^{-1}	4
(4b) Fe(C$_2$O$_4$)$_m^{2-2m}$ + OH$^\bullet$ → Fe(C$_2$O$_4$)$_m^{3-2m}$ + OH$^-$	1 × 10^{10} M^{-1}.s^{-1}	
(5) Fe(C$_2$O$_4$)$_n^{3-2n}$ + hν → Fe(C$_2$O$_4$)$_{n-1}^{4-2n}$ + C$_2$O$_4^{\bullet-}$	b	
(6) C$_2$O$_4^{\bullet-}$ → CO$_2$ + CO$_2^{\bullet-}$	2 × 10^6 s^{-1}	5
(7a) Fe(OH)$_p^{3-p}$ + C$_2$O$_4^{\bullet-}$ → Fe(OH)$_p^{2-p}$ + 2CO$_2$	1 × 10^9 M^{-1}.s^{-1}	
(7b) Fe(OH)$_p^{3-p}$ + CO$_2^{\bullet-}$ → Fe(OH)$_p^{2-p}$ + CO$_2$	1 × 10^9 M^{-1}.s^{-1}	
(7c) Fe(C$_2$O$_4$)$_n^{3-2n}$ + C$_2$O$_4^{\bullet-}$ → Fe(C$_2$O$_4$)$_n^{2-2n}$ + 2CO$_2$	1 × 10^{10} M^{-1}.s^{-1}	
(7d) Fe(C$_2$O$_4$)$_n^{3-2n}$ + CO$_2^{\bullet-}$ → Fe(C$_2$O$_4$)$_n^{2-2n}$ + CO$_2$	1 × 10^{10} M^{-1}.s^{-1}	
(8a) C$_2$O$_4^{\bullet-}$ + O$_2$ → 2CO$_2$ + O$_2^{\bullet-}$	2.4 × 10^9 M^{-1}.s^{-1}	6
(8b) CO$_2^{\bullet-}$ + O$_2$ → CO$_2$ + O$_2^{\bullet-}$	2.4 × 10^9 M^{-1}.s^{-1}	6

Notes: Fe(C$_2$O$_4$)$_m^{2-2m}$ includes FeC$_2$O$_4$ and Fe(C$_2$O$_4$)$_2^{2-}$ (ie. m = 1 or 2); Fe(C$_2$O$_4$)$_n^{3-2n}$ includes FeC$_2$O$_4^+$, Fe(C$_2$O$_4$)$_2^-$ and Fe(C$_2$O$_4$)$_3^{3-}$ (ie. n = 1, 2 or 3); Fe(OH)$_p^{3-p}$ includes Fe^{3+}, FeOH^{2+} and Fe(OH)$_2^+$ (ie. p = 0, 1 or 2). [a]References: (1) Walling (1975); (2) Sedlak and Hoigné (1993); (3) Getoff *et al.* (1971); (4) Buxton *et al.* (1988); (5) Mulazzani *et al.* (1986); (6) Adams and Wilson (1969); [b]fitted value.

Table A.2. Irreversible reactions and rate constants used in the kinetic model of modified photo-Fenton-mediated dye degradation (continued)

Reaction	k	Ref.[a]
(9a) $Fe^{2+} + HO_2^{\bullet} (+H^+) \rightarrow Fe^{3+} + H_2O_2$	$1.2 \times 10^6 \ M^{-1}.s^{-1}$	7
(9b) $Fe(C_2O_4)_m^{2-2m} + HO_2^{\bullet} (+H^+) \rightarrow Fe(C_2O_4)_m^{3-2m} + H_2O_2$	$5 \times 10^7 \ M^{-1}.s^{-1}$	
(9c) $Fe^{2+} + O_2^{\bullet-} (+2H^+) \rightarrow Fe^{3+} + H_2O_2$	$1 \times 10^7 \ M^{-1}.s^{-1}$	8
(9d) $Fe(C_2O_4)_m^{2-2m} + O_2^{\bullet-} (+2H^+) \rightarrow Fe(C_2O_4)_m^{3-2m} + H_2O_2$	$5 \times 10^7 \ M^{-1}.s^{-1}$	
(10a) $Fe(OH)_p^{3-p} + HO_2^{\bullet} \rightarrow Fe(OH)_p^{2-p} + O_2 + H^+$	$3.1 \times 10^5 \ M^{-1}.s^{-1}$	7
(10b) $Fe(OH)_p^{3-p} + O_2^{\bullet-} \rightarrow Fe(OH)_p^{2-p} + O_2$	$1.5 \times 10^8 \ M^{-1}.s^{-1}$	8
(11) $Fe(OH)_p^{3-p} + H_2O_2 \rightarrow Fe(OH)_p^{2-p} + HO_2^{\bullet} + H^+$	$0.01 \ M^{-1}.s^{-1}$	8
(12a) $HO_2^{\bullet} + HO_2^{\bullet} \rightarrow H_2O_2 + O_2$	$8.5 \times 10^5 \ M^{-1}.s^{-1}$	7
(12b) $HO_2^{\bullet} + O_2^{\bullet-} (+H^+) \rightarrow H_2O_2 + O_2$	$9.7 \times 10^7 \ M^{-1}.s^{-1}$	7
(13) $Dye + OH^{\bullet} \rightarrow R + R^{\bullet}$	$9 \times 10^9 \ M^{-1}.s^{-1}$	
(14) $R + OH^{\bullet} \rightarrow R^{\bullet}$	$2 \times 10^9 \ M^{-1}.s^{-1}$	
(15) $R^{\bullet} + O_2 \rightarrow R + O_2^{\bullet-}$	$1 \times 10^9 \ M^{-1}.s^{-1}$	
(16a) $R^{\bullet} + Fe(OH)_p^{3-p} \rightarrow R + Fe(OH)_p^{2-p}$	$1 \times 10^9 \ M^{-1}.s^{-1}$	
(16b) $R^{\bullet} + Fe(C_2O_4)_n^{3-2n} \rightarrow R + Fe(C_2O_4)_n^{2-2n}$	$1 \times 10^9 \ M^{-1}.s^{-1}$	

Notes: $Fe(C_2O_4)_m^{2-2m}$ includes FeC_2O_4 and $Fe(C_2O_4)_2^{2-}$ (ie. m = 1 or 2); $Fe(C_2O_4)_n^{3-2n}$ includes $FeC_2O_4^+$, $Fe(C_2O_4)_2^-$ and $Fe(C_2O_4)_3^{3-}$ (ie. n = 1, 2 or 3); $Fe(OH)_p^{3-p}$ includes Fe^{3+}, $FeOH^{2+}$ and $Fe(OH)_2^+$ (ie. p = 0, 1 or 2). [a]References: (7) Bielski *et al.* (1985); (8) Chen and Pignatello (1997)

Table A.3. Reversible reactions and equilibrium constants used for kinetic modelling of modified photo-Fenton-mediated dye degradation.

Reaction	log K	Ref.[a]
(17) $H_2C_2O_4 \leftrightarrow HC_2O_4^- + H^+$	-1.25	1
(18) $HC_2O_4^- \leftrightarrow C_2O_4^{2-} + H^+$	-4.27	1
(19) $Fe^{3+} + OH^- \leftrightarrow FeOH^{2+}$	11.8	2
(20) $FeOH^{2+} + OH^- \leftrightarrow Fe(OH)_2^+$	10.5	2
(21) $Fe(OH)_2^+ + OH^- \leftrightarrow Fe(OH)_3$	16.5	2
(22) $Fe^{3+} + C_2O_4^{2-} \leftrightarrow Fe(C_2O_4)^+$	9.4	1
(23) $Fe(C_2O_4)^+ + C_2O_4^{2-} \leftrightarrow Fe(C_2O_4)_2^-$	6.8	1
(24) $Fe(C_2O_4)_2^- + C_2O_4^{2-} \leftrightarrow Fe(C_2O_4)_3^{3-}$	4.2	1
(25) $Fe^{2+} + C_2O_4^{2-} \leftrightarrow Fe(C_2O_4)$	4.3	1
(26) $Fe(C_2O_4) + C_2O_4^{2-} \leftrightarrow Fe(C_2O_4)_2^{2-}$	2.1	1
(27) $HO_2^{\bullet} \leftrightarrow O_2^{\bullet-} + H^+$	-4.8	3

[a]References: (1) Faust and Zepp (1993); (2) Morel and Hering (1993); (3) Bielski *et al.* (1985).

[a]Table A.4. (next page) References : (1) Feitz *et al.* (1999); (2) Waite *et al.*, (1999); (3) Hoffman *et al.*, (1995); (4) Davis and Huang (1993); (5) Bielski and Cabelli (1995); (6) Walling (1995); (7) Neta *et al.* (1990); (8) von Sonntag *et al.* (1997); [b]fitted value.

Table A.4. Reactions included in ACUCHEM model of microcystin-LR (MLRH⁻) degradation by TiO$_2$/hυ in the presence of excess algal exudate (RR"CH$_2$⁻).

Reaction		k or K	Ref.[a]
Part A: Surface-mediated degradation model			
(1)	$> TiOH + MLRH^- + H^+ \leftrightarrow > TiMLRH + H_2O$	Log K = 8	1
(2)	$> TiOH + RR"CH_2 \leftrightarrow > TiRR"CH_2 + OH^-$	$logK_c \sim 4$	2
(3)	$> TiOH + O_2 \leftrightarrow > TiOH : O_2$	LogK ~ 4	3
(4)	$> TiOH + h\nu \leftrightarrow > TiOH + h_{vb}^+ + e_{cb}^-$	See (5) & (6)	
(5)	$\Phi I \rightarrow h_{vb}^+$	k = 4x10⁻⁶ M⁻¹s⁻¹	
(6)	$\Phi I \rightarrow e_{cb}^-$	k = 4x10⁻⁶ M⁻¹s⁻¹	
(7)	$> TiOH + h_{vb}^+ + e_{cb}^- \leftrightarrow > TiOH + heat/light$	k= 4.6x10²³ M⁻¹s⁻¹	4
(8)	$> TiOH : O_2 + e_{cb}^- \leftrightarrow > TiOH : O_2^{\bullet-}$	k= 1.9x10¹⁰ M⁻¹s⁻¹	4
(9)	$> TiOH : O_2^{\bullet-} \leftrightarrow > TiOH + O_2^{\bullet-}$	b	
(10)	$HO_2^\bullet + O_2^{\bullet-} \rightarrow HO_2^- + O_2$	k = 9.7x10⁷ M⁻¹s⁻¹	5
(11)	$HO_2^\bullet + HO_2^\bullet \rightarrow H_2O_2 + O_2$	k = 8.3x10⁵ M⁻¹s⁻¹	5
(12)	$O_2^{\bullet-} + H^+ \leftrightarrow HO_2^\bullet$	LogK = 4.8	6
(13)	$HO_2^- + H^+ \rightarrow H_2O_2$	LogK = 11.7	6
(14)	$> TiOH + h_{vb}^+ \rightarrow > TiOH^{\bullet+}$	k= 2x10¹² M⁻¹s⁻¹	4
(15)	$> TiOH^{\bullet+} + > TiMLRH \rightarrow > TiOH_2^+ + > TiMLR^\bullet$	k = 10⁹ M⁻¹s⁻¹	
(16)	$> TiOH^{\bullet+} + > TiRR"CH_2 \rightarrow > TiOH_2^+ + > TiRR"CH^\bullet$	k = 10⁹ M⁻¹s⁻¹	
(17)	$> TiRR"CH^\bullet + > TiMLRH \rightarrow > TiRR"CH_2 + > TiMLR^\bullet$	k = 10⁸ M⁻¹s⁻¹	
(18)	$> TiRR"CH^\bullet + > TiRR"CH^\bullet \rightarrow > TiRR"CH^- + > TiRR"CH^+$	k = 10⁹ M⁻¹s⁻¹	
(19)	$> TiOH : O_2^{\bullet-} + > TiRR"CH^\bullet \rightarrow > TiOH : O_2 + > TiRR"CH^-$	k = 10⁷ M⁻¹s⁻¹	
(20)	$> TiMLR^\bullet + > TiRR"CH^\bullet \rightarrow > TiMLR^- + > TiRR"CH^+$	k = 10⁹ M⁻¹s⁻¹	
(21)	$> TiMLR^\bullet + > TiOH : O_2^{\bullet-} \rightarrow > TiMLR^- + > TiOH : O_2$	k = 10⁷ M⁻¹s⁻¹	
Part B: Solution-mediated degradation model			
(22)	$> TiRR"CH^\bullet + OH^- \rightarrow TiOH + RR"CH^-$	b	
(23)	$RR"CH^{\bullet-} + O_2 \rightarrow > RR"CHOO^{\bullet-}$	k = 2x10⁹ M⁻¹s⁻¹	7,8
(24)	$RR"CHOO^{\bullet-} \rightarrow O_2^{\bullet-} + RR"CH$	k = 10² s⁻¹	
(25)	$RR"CHOO^{\bullet-} + MLRH \rightarrow RR"CHOOH + MLR^{\bullet-}$	k = 10⁸ M⁻¹s⁻¹	
(26)	$RR"CHOO^{\bullet-} + RR"CH_2 \rightarrow RR"CHOOH + RR"CH^{\bullet 2-}$	k = 10⁸ M⁻¹s⁻¹	
(27)	$RR"CHOO^{\bullet-} + HO_2^\bullet \rightarrow RR"CHOOH^- + O_2$	k = 10⁸ M⁻¹s⁻¹	
(28)	$RR"CH_2^- + HO_2^\bullet \rightarrow RR"CH^{\bullet-} + H_2O_2$	k = 10⁵ M⁻¹s⁻¹	
(29)	$MLRH + HO_2^\bullet \rightarrow MLR^\bullet + H_2O_2$	k = 10⁵ M⁻¹s⁻¹	
(30)	$MLR^{\bullet-} + HO_2^\bullet \rightarrow MLR + HO_2^-$	k = 10⁷ M⁻¹s⁻¹	

see previous page for references

H.H. Hahn, E. Hoffmann, H. Ødegaard (Eds.)
Chemical Water and Wastewater Treatment VII, pp. 109-118
© IWA Publishing, London
ISBN: 1 84339 009 4

Taste and Odor Control for Drinking Water Supply: A Combined Solution of Chemical Oxidation and Powdered Activated Carbon Adsorption

S.S. Ferreira Filho, F.A. Lage Filho, R.L. Mendes and A.N. Fernandez*

*Escola Politecnica da Universidade de Sao Paulo,Departemento De Engharia Hidraulica e Sanitaria, Sao Paulo, Brazil
email: ssffilho@usp.br

Introduction

The City of Sao Paulo has eight drinking water systems. Among them the most complex is the ABV Water Treatment Plant (treating 16 m^3/s), with serious taste and odor problems since the eighties due to the presence of organic compounds (notably MIB (2-methylisoborneol) and Geosmin).

That situation led Sao Paulo's State Water Agency (SABESP) to build and operate a powdered activated carbon (PAC) dosing unit, which can dose up to 40 mg PAC/L into Guarapiranga reservoir's raw water at the intake and pumping station. It was expected that such a unit could handle MIB concentrations up to 100 ng/L. However, in recent months MIB concentrations in the range 200 to 600 ng/L were measured in the raw water.

The frequent scenario of high MIB concentrations makes it necessary to implement more advanced treatment technologies at ABV-WTP. Among them, an ozonation system alternative was considered feasible as a medium-long term treatment alternative. Consequently, there is a need to define design parameter numbers for the ozonation system and ozone dosages, and to evaluate the utilization of hydrogen peroxide alongside ozone in the oxidation of taste and odour causing compounds.

[109]

Objectives

To eliminate the current taste and odour problems at ABV-WTP, the main objectives of the work were to evaluate:

- Powdered activated carbon (PAC) adsorption kinetics to remove taste and odour (T & O) causing compounds from the raw water of Guarapiranga Water Supply System (GWSS). An estimate of the peak capacity of the PAC dosing unit and an evaluation of its operational capability as a function of raw water quality and finished water quality requirements were also included.
- The ozonation process (with and without hydrogen peroxide) in the removal of MIB from Guarapiranga System´s raw water.

Methods of analysis and experimental apparatus

Adsorption Experiments

Adsorption tests were conducted with raw water from Guarapiranga Reservoir, which supplies the ABV Water Treatment Plant. A flow rate of 16 m³/s is currently treated, which supplies about 30 % of the population in the Sao Paulo Metropolitan Region (SPMR).

Raw water from the Guarapiranga intake structure is pumped to the ABV- WTP. The hydraulic detention time between the intake and the rapid mix unit at the ABV-

Table 1. PACs utilised in the experimental investigation

PAC number	Commercial brand	Prime Material	Iodine number (mg/g)
1	NORIT (Hydrodarco B)	Mineral (Lignita)	530
2	CALGON (WPH)	Mineral (Bituminous)	870
3	CARBOMAFRA (CAG 118-90)	Vegetable (pine node)	841
4	BRASILAC (Delta V – CDV)	Vegetable (pine node)	742
5	CPL (WP9)	Mineral (Bituminous)	850
6	PAIOL (EB-10)	Vegetable (pine node)	860

Table 2. Summary of adsorption kinetics tests conducted in the experimental investigation

Test	Matrix	PAC dose (mg/l)	Adsorbate	Initial concentration of adsorbate in the liquid phase (ng/l)	PACs utilised
1	Raw water	10	MIB	100	1,2,3,4,5, 6
2	Raw water	20	MIB	100	1,2,3,4,5, 6
3	Raw water	40	MIB	100	1,2,3,4,5, 6
4	Raw water	10	Geosmin	100	1,2,3,4,5, 6
5	Raw water	20	Geosmin	100	1,2,3,4,5, 6
6	Raw water	40	Geosmin	100	1,2,3,4,5, 6

WTP is about 60 minutes. Since the detention time in the mains is long enough to assure maximum efficiency of PAC, a PAC dosing section was established at the reservoir intake. Therefore, PAC adsorption kinetics tests were conducted by means of static reactors with a contact time of 90 minutes.

In each test, raw water samples were collected after 15, 30, 45, 60 and 90 minutes of contact time. Membrane filtration (0.45 μm) was utilised immediately after sample collection to separate the PAC from the liquid phase. The filtrate was then transferred to chromatography flasks for analysis of MIB and Geosmin concentrations.

Since the adsorption process is highly specific regarding organic compound removal, six commercial brands of PAC were selected to evaluate their efficacy in MIB and Geosmin removal from raw water. Tables 1 and 2 summarise the main characteristics of the PACs utilised in the experimental investigation and the results of the adsorption kinetics tests, respectively.

Oxidation Experiments

The oxidation experiments were aimed at evaluating the efficacy of ozone, with or without hydrogen peroxide addition, to oxidize MIB in the settled water from the ABV Water Treatment Plant. The ozonation pilot plant was assembled at the ABV-WTP, owned by the São Paulo's State Water Agency, SABESP.

The pilot plant consisted of a conventional treatment train (coagulation-flocculation-settling-filtration) plus an ozonation unit. The ozone unit had a contact system with four columns in series. Each plexiglass-made column was cylindrical, with 150 mm inside diameter and 3.05 meters high, with four sampling points so that residual ozone concentrations could be determined.

All chemical oxidation tests were run with the ozone system operating in counter-current flow mode. Water flow rates in the columns were established so that the hydraulic detention times in the first through the last column were 3.6 minutes, 3.7 minutes, 3.8 minutes and 3.9 minutes, for a total of 15 minutes.

Table 3. Summary of chemical oxidation tests

| Test | Applied Ozone Dosage (mg/l) | | | | Applied H_2O_2 dosage (mg/l) |
| | Columns | | | | |
	1	2	3	4	
1	0.5	----	----	----	----------
2	0.5	0.5	----	----	----------
3	1.0	----	----	----	----------
4	1.0	1.0	----	----	----------
5	1.0	2.0	----	----	----------
6	2.0	----	----	----	----------
7	0.5	----	----	----	0.25
8	0.5	0.5	----	----	0.5
9	1.0	----	----	----	0.5
10	1.0	1.0	----	----	1.0
11	1.0	2.0	----	----	1.5
12	2.0	----	----	----	1.0

Each oxidation test was run twice, with and without hydrogen peroxide addition, in order to compare oxidation efficiencies. For all advanced oxidation tests, the mass ratio of peroxide (H_2O_2) to ozone (O_3) was set at 0.5 based upon the stoichiometry of the reaction for both oxidant agents. This ratio was used by others (Glaze *et al.,* 1988; Aieta *et al.,* 1988; Karimi *et al.,* 1997). For all tests, the pH conditions and the physical-chemical characteristics of the test water did not change significantly, making it possible to perform a statistical evaluation of the test results. Table 3 summarises the characteristics of the tests.

Assuming the MIB concentrations in the raw water are 600 ng/l and a removal efficiency of 90 % using a PAC dose of 40 mg/l is set, the resulting MIB concentration in the settled water should be about 60 ng/l. Thus, a MIB "dose" in the settled water of the same order of magnitude was selected for the chemical oxidation tests. MIB was then dosed into the water feed line of the first column of the pilot plant. For each test, the operation time was at least three times the total hydraulic detention time in the contact columns before samples were collected from each column in order to determine residual ozone concentration throughout them. The ozone concentration was then used to calculate the resulting parameter C•T (product of concentration and time) for the ozonation system. Ozone concentrations were measured by means of a HACH 2010 spectrophotometer at 600 nm.

A replicate sample was always taken from the feed water and the ozone system effluent. All MIB and Geosmin analyses were conducted by SABESP's Organic Chemistry Laboratory (APQ-SABESP) by GC-MS-PAT.

Results and discussion

Figures 1 and 2 depict the adsorption kinetics test results performed with different commercial brands and doses of PAC, with MIB as the adsorbate.

MIB was efficiently removed from the liquid phase with a contact time on the order of 30 minutes. Most of the removal occurred within 30 minutes, with only marginal MIB removal after that.

PAC dosage was extremely important in the removal of MIB and Geosmin from the liquid phase. PAC dosages of 10 mg/l removed about 60 % of MIB concentration in raw water, whereas dosages of 20 mg/l led to an 80 % removal. These results suggest that PAC dosage must be optimised as a function of MIB and Geosmin concentrations in the raw water to minimise taste and odour problems in the resulting potable water.

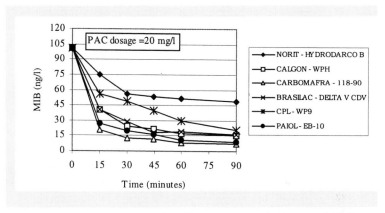

Figure 1. Adsorption kinetics for different commercial brands of PAC. PAC dosage: 20 mg/l

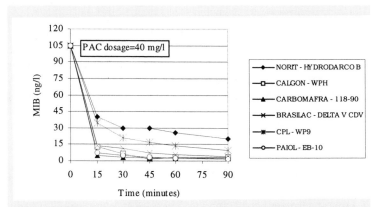

Figure 2. Adsorption kinetics tests with different commercials brands of PAC. PAC dosage: 40 mg/l

Since MIB and Geosmin concentrations in the raw water vary with time, the PAC dosage should be varied accordingly, in order to accomplish both technical and economic optimisation of the PAC adsorption process.

The experimental results corroborate the specificity of each PAC brand in the removal of specific organic compounds. Figures 1 and 2 show that the PAC brands PAIOL, CARBOMAFRA and CALGON had the highest removal efficiencies for both MIB and Geosmin.

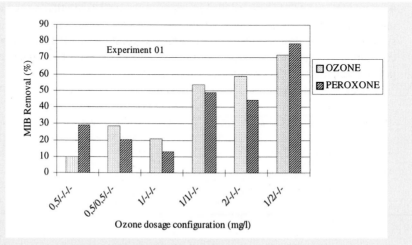

Figure 3. MIB removal efficiencies obtained with ozonation treatment with and without hydrogen peroxide addition (Group of experiments 01)

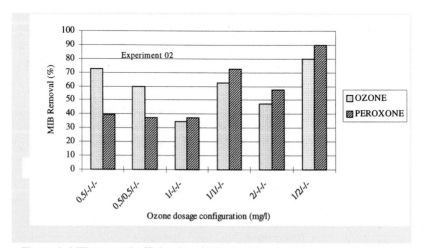

Figure 4. MIB removal efficiencies obtained with ozonation tests with and without hydrogen peroxide addition (Group of experiments 02)

The success of PAC adsorption for removing T & O causing organic compounds is thus related not only to an adequate contact time between PAC and the liquid phase but also to a correct choice of PAC brand. This implies that specific experimental adsorption tests should be carried out in the laboratory as often as possible.

Figures 3 and 4 present typical MIB removal results obtained with the chemical oxidation treatment of settled water from the ABV Water Treatment Plant.

Regarding ozone oxidation tests, MIB removal efficiencies were in the range 80 % to 90 % for the treatments in which ozone dosage configuration was set at 1.0 mg O_3/l in the first contactor/column and 2.0 mg O_3/l in the second column.

The results obtained with the same dosage configuration plus hydrogen peroxide addition were very similar. There was no significant difference between the results from treatments with and without peroxide addition.

Results presented in the reviewed literature are somewhat contradictory regarding ozone application with hydrogen peroxide addition for MIB and Geosmin oxidation. Studies by Ferguson *et al.* (1990) indicated that the application of ozone plus peroxide was much more efficient than ozone only. A 90 % MIB removal could be accomplished with an ozone dose of 4.0 mg/l, whereas only 2.0 mg/l of peroxone was needed to achieve the same removal efficiency.

On the other hand, Huck *et al.* (1995) showed that application of ozone and peroxide led to similar results in the oxidation of MIB and Geosmin, with no significant difference between them.

It is relevant to analyse the experimental results in parallel with the operational data from the ABV Water Treatment Plant for the same experimental period. During the testing period the settled water pH (effluent from the ozone system) was in the range 8.0 to 8.5, therefore alkaline. An increase in pH towards the alkaline side leads to a higher concentration of hydroxyl ions in the liquid phase, thus favouring the formation of hydroxyl free radicals ($\cdot$OH). Free radicals are much more powerful and much less selective oxidizing agents than ozone itself, thus enhancing MIB oxidation. On the other hand, the half-life of free radicals is much shorter than that of ozone.

Experimental MIB and Geosmin oxidation results from Lalezary *et al.* (1986) and Glaze *et al.* (1990) showed that the predominant oxidative mechanism is due to the action of the free radical $\cdot$OH, and not to the direct action of ozone itself. This would explain why the addition of hydrogen peroxide did not lead to a higher MIB oxidation efficiency when compared to oxidation efficiencies obtained with ozone alone.

For all chemical oxidation tests, it was observed that the residual ozone concentrations in the liquid phase were higher when ozone was the sole oxidant agent than when ozone and peroxide were applied together. Consequently, the ozone dosage configuration set for the ozone system is particularly important. The total ozone dose of 2.0 mg/l could be applied under two configurations: either as 2.0 mg/l applied solely in the first column/contactor or split 50 % - 50 % between the first and second columns. It was observed that the second configuration led to much higher residual ozone concentrations than the first configuration.

The highest residual ozone concentrations in the liquid phase were obtained with the total applied dose of 3.0 mg/l, configured as 1.0 mg/l and 2.0 mg/l in the first and second contactors/columns, respectively. Accordingly, with this ozone dosage configuration the highest C•T products were obtained in the range of 0.202 to 3.845 mg O_3 min/l. It should be stressed that the test conditions are critical, because the possible presence of naturally occurring organic compounds (which were not removed efficiently by coagulation flocculation-settling at the treatment plant) in the settled water would necessitate an additional ozone demand for their oxidation, thus further reducing the residual ozone concentrations in the liquid phase.

As reported by Urfer *et al.* (1999), coagulation optimisation with respect to naturally occurring organic compound removal is a fundamental condition for the ultimate goal of operational optimisation of the ozone system, particularly when one wants extra security regarding disinfection and to obtain higher C•T products. Thus the ozone dose applied in the first contactor/reactor of a series should satisfy an initial ozone demand, and so the dose in the second and subsequent contactors makes it possible to establish a higher, more stable residual ozone concentration.

Conclusions

The following conclusions can be drawn from the experimental data:

- The bench-scale PAC adsorption kinetics tests adequately simulated real-scale MIB and Geosmin adsorption kinetics observed for the Guarapiranga Water Supply System. Removal efficiency values in the bench tests were very similar to those measured in the real system with the same PAC dosages.

- Although the PAC Dosing Unit at Guarapiranga Reservoir had been originally designed for a raw water MIB concentration of 100 ng/L, MIB concentrations in the finished water could be reduced to values in the range of 25 ng/l to 50 ng/l for raw water MIB concentrations up to 400 ng/l, and PAC dosages up to 40 mg/l.

- Regarding MIB oxidation, removal efficiencies in the range of 80 to 90 % were obtained, with a total ozone dose of 3.0 mg/L (1.0 mg/L applied in the first contactor and 2.0 mg/L in the second contactor of the column series).

- No significant improvement in MIB oxidation was achieved by adding hydrogen peroxide to the ozone treatment. As a result, peroxone (ozone + peroxide application) was not considered as a treatment possibility for the Guarapiranga System.
- For all ozonation tests carried out under different contactor configurations, it was observed that the ozone residuals in the liquid phase were larger for ozone-only applications (no peroxide addition) than for the same ozone dose configuration with peroxide addition. Thus ozonation treatments without peroxide generated larger C x T product values than those with hydrogen peroxide addition.

Acknowledgements

The authors gratefully acknowledge the financial support of **SABESP** and the logistical support of the **Centro Tecnológico de Hidráulica (CTH-DAEE).**

References

Aieta, E.M., Regan, K.M., Lang, J.S., McReynolds, L., Kang, J.W., Glaze, W.H. (1988) Advance oxidation processes for treating groundwater contaminated with TCE and PCE: pilot scale evaluation. *Journal of American Water Works Association* (05), 64-72

Ferguson, D.W., Mcguire, M.J., Koch, B., Wolfe, R.L., Aieta, E.M. (1990) Comparing peroxone and ozone for controlling taste and odor compounds, disinfection by products and microorganisms. *Journal of American Water Works Association* (04), 181-191

Glaze, W.H., Kang, J.W. (1988) Advanced oxidation processes for treating groundwater contaminated with TCE and PCE: laboratory studies. *Journal of American Water Works Association* (05), 57-63

Glaze, W.H., Schep, R., Chaucey, W., Ruth, E.C., Zarnoch, J.J., Aieta, E.M., Tate, C.H., Mcguire, M.J. (1990) Evaluating oxidants for the removal of model taste and odour compounds from a municipal water supply. *Journal American Water Works Association* (05), 79-84

Huck, P.M., Anderson, W.B., Lang, C.L., Anderson, W.A., Fraser, J.C., Jasim, S.Y., Andrews, S.A., Pereira, G. (1995) Ozone and peroxone for geosmin and 2-methylisoborneol control: laboratory, pilot and modelling studies. In: *AWWA ANNUAL CONFERENCE,* California, 453-474

Karimi, A.A., Redman, J.A., Glaze, W.H., Stolarik, G.F. (1997) Evaluating an AOP for TCE and PCE removal. *Journal American Water Works Association* (08), 41-53

Lalezary, S., Pirbazary, M., Mcguire, M.J. (1986) Oxidation of five earthy-musty taste and odour compounds. *Journal American Water Works Association* (03), 62-69

Urfer, D., Huck, P.M., Gagnon, G.A., Mutti, D., Smith, F. (1999) Modelling enhanced coagulation to improve ozone disinfection. *Journal American Water Works Association* (03), 59-73

H.H. Hahn, E. Hoffmann, H. Ødegaard (Eds.)
Chemical Water and Wastewater Treatment VII, pp. 119-130
© IWA Publishing, London
ISBN: 1 84339 009 4

Removal of Arsenic from Contaminated Groundwaters using Combined Chemical and Biological Treatment Methods

A. Zouboulis and *I. Katsoyiannis*

*Aristotle University of Thessaloniki, Department of Chemistry, Division of Chemical Technology, Greece
email: zoubouli@chem.auth.gr

Abstract

Groundwater is an important source of drinking water, especially for small cities in rural areas. Although groundwater is usually sufficiently clean for drinking, its treatment is necessary when contamination from organic matter and metals occurs. In certain cases, groundwaters are often contaminated with elevated concentrations of soluble iron, manganese or arsenic ions. Especially arsenic contamination is of primary concern, due to its severe effects on human health, and the World Health Organization has listed arsenic as a first priority issue. This work focused on the removal of arsenic from groundwater using a combination of chemical and biological methods as an alternative to conventional physical-chemical treatment methods. In particular, arsenic was removed by adsorption onto biologically-produced adsorbents (iron oxides). As(V) was efficiently removed and residual concentrations were below the Maximum Contaminant Level of 10 μg/L, even though a wide range of initial arsenic concentrations was examined (50-200 μg/L). Contrary, the removal of As(III) was in the range of 70 - 75 %, when the redox potential was between 270-280 mV; however it was increased to over 90 %, when the redox potential was increased to values over 300 mV. This increase can be attributed to As(III) oxidation by microorganisms, which under these conditions might accelerate the conversion of As(III) to As(V), thus enhance the overall removal efficiency. The combined treatment approach can provide water free of iron and arsenic, resulting in an integrated treatment of contaminated groundwaters.

Introduction

Groundwater is a source of drinking water, especially in rural areas, in most countries in the world. Although groundwaters do not usually become contaminated by human activities, a degree of low-level contamination may occur due to dissolution of natural components of geogenic origin. Arsenic contamination of groundwaters often falls into the latter category and its effective removal before consumption is necessary to avoid severe effects on human health (Pontius *et al.,* 1994; Desesso *et al.,* 1998). Arsenic is known to cause several types of cancer, among other diseases, mainly due to chronic exposure. The latest EU directive has reduced the maximum concentration limit (MCL) of arsenic in drinking water to 10 µg/L (98/83/EU) and this limit has recently been adopted by the USEPA and will be in force by 2006.

Although several techniques have been developed to remove arsenic (Jekel, 1994; Kartinen and Martin, 1995), the best available technology would be efficient, of low cost and quite simple to use. The topic of this paper is the combination of an emerging technology for arsenic removal (adsorptive filtration using appropriate filtration media) and a biological treatment method, which relies on the use of indigenous microorganisms (such as *Gallionella* or *Leptothrix*) for the oxidation and removal of arsenic. The biological treatment method was previously applied in the oxidation-precipitation of iron and manganese from groundwaters, by using indigenous iron oxidizing bacteria (Mouchet, 1992; Dimitrakos *et al.,* 1992). The oxidized products are subject to precipitation and can be subsequently separated from water by (bed) filtration using appropriate media. If arsenic is also present, it can be simultaneously removed by adsorption on the produced/precipitated iron oxides (Lehimas *et al.,* 1999); meanwhile a possible oxidation of As(III) may also take place, leading to an increased overall arsenic removal (Seith and Jekel, 1992).

In general, the application of this combined method has been found to be very efficient for arsenic removal, as the residual arsenic concentrations are below the MCL, corresponding to percentage removals up to 95, over a long period of operation; it is very simple in use, as after set in operation, there is no need for close monitoring of a breakthrough point; it is quite economic as it does not require the addition of chemical reagents for the oxidation and removal of arsenic.

Materials and methods

Reagents and stock solutions

The physical-chemical experiments for the removal of arsenic were based on the coating of filter medium by iron oxides. The suspended iron hydroxides for the coating process were prepared by using $Fe(NO_3)_3 \cdot 9H_2O$ (0.025 M) (Merck), diluted in de-ionised water and pH was adjusted to 5.0 by adding HNO_3 (4N). As(V) stock solutions (100 mg/L) were prepared by dissolving sodium arsenate ($Na_2HAsO_4 \cdot 7H_2O$, Parneac) in de-ionized water. As(III) stock solutions (100 mg/L) were prepared by dissolving arsenite acid (As_2O_3, AnalaR) in de-ionized water. In order to dissolve As_2O_3 in water, it was necessary to heat the solution and add 10 ml/L HCl. The secondary standard solutions for spiking were freshly prepared for each experimental run from the stock solutions, by dilution with tap water. Regarding biological arsenic removal, the experiments were mainly performed at the Umweltbundesamt Research Institute in Berlin, Germany. The groundwater had an initial concentration of 2.8 mg/L iron and less than 1 μg/L arsenic. Arsenic was spiked in the groundwater, in order to investigate its removal. The main characteristics of the specific groundwater source are presented in Table 1.

Table 1. Physical-chemical characteristics of treated groundwater

Conductivity	1125 μS/cm
PH	7.2
Total hardness	5.6 mol/m³ CaCO₃
[Fe]	2.8 mg/L
[As]	<1 μg/L
Initial ORP (redox)	-150 mV

Methods

Adsorptive filtration

The treatment method is based on a two-stage operation. Initially, polystyrene beads were modified by coating their surface with iron hydroxides, and this modified material was then poured into a glass column (40 cm high, 2.86 cm diameter, total volume 200 ml). The polystyrene beads had a mean diameter of 4-5 mm, giving a mean bed porosity of 0.37. The preformed suspension passed through the column upwards under recirculation for 3 hours, to complete the coating process. Following this procedure, the bed was washed with 10 L of distilled water for the removal of the residual solids, which were not coated on the surface of the media but just filtered.

Following bed modification, aqueous solutions spiked with arsenic entered the treatment column in upflow mode, passed once through the modified media and samples were collected from the top of the column. A schematic diagram of the experimental set-up is presented in Figure 1.

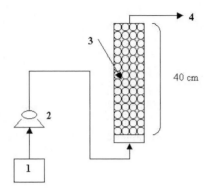

Column characteristics: area: 644.35 x 10⁻⁶ m², inner diameter: 28.65 mm, total bed volume: 200 mL, porosity: 0.37

1: Feed solution: Tap water spiked with soluble arsenic anions, 2: peristaltic pump, 3: filter containing polymeric beads coated with iron oxides, 4: effluent

Figure 1. Schematic diagram of adsorptive filtration unit

Whenever the breakthrough point was reached, the bed was regenerated by treatment with a strongly alkaline solution (NaOH, 1M). The breakthrough point was defined as the point when the effluent arsenic concentration was over the permissible maximum concentration limit of 10 μg/L.

Biological arsenic removal

The treatment method for As(III) removal (simultaneously with biological iron oxidation) was based on an upflow fixed-bed filtration unit. A schematic diagram of the experimental set-up is shown in Figure 2.

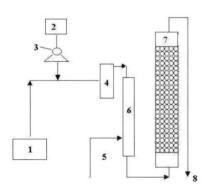

Column characteristics: active height: 1m, inner diameter: 68 mm, surface area: 0.0036 m², bed volume: 3.6 L, total bed porosity: 0.37

1: continuous flow of contaminated groundwater
2: arsenic stock solution
3: peristaltic (feeding) pump
4: influent sampling vessel
5: air injection
6: aeration column
7: filtration column
8: effluent

Figure 2. Schematic diagram of biological iron and experimental arsenic removal unit

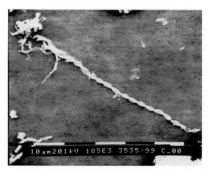 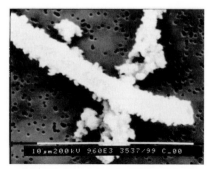

Figure 3. SEM micrographs of (a) *Gallionella ferruginea* and (b) *Leptothrix ochracea* entrapped in the iron sludge

The experimental device consisted of a Plexiglas column, which was filled with polystyrene beads used as filtration media. The indigenous microorganisms after 3 months of operation were deposited and accumulated in both filtration columns; this method does not require the maintenance of pure cultures, because the simultaneous oxidation of iron resulted in sludge formation in which both *Gallionella* and *Leptothrix* were entrapped (Figure 3).

Air injection made bacterial growth possible, which was essential for iron oxidation. The aeration was performed in a separate column, before filtration, in order to avoid the collision of bubbles with the deposited sludge, which may disturb the system and result in increased iron concentrations in the effluent stream. The column was backwashed every three days with a limited amount of treated water (3 liters).

Arsenic was analysed using hydride generation atomic absorption spectrophotometry (HG-AAS); iron was analysed by the photometric analysis of phenanthroline (Driehaus and Jekel, 1992).

Results and Discussion

Adsorptive filtration

During this experimental section (adsorptive filtration) solution pH and arsenic speciation were examined. The solution pH is an important factor for all water and wastewater treatment processes because it affects, among other things, the speciation of metals in water. Experiments were performed regarding both As(V) and As(III) removal. As(V) was found to be removed from the aqueous stream more efficiently than As(III) (Figure 4).

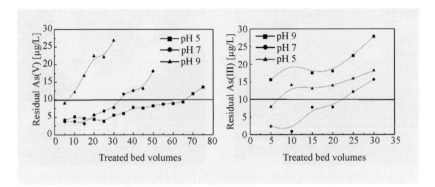

Figure 4. Arsenic removal as affected by pH: (a) As(V), (b) As(III) (experimental conditions: $[As]_o = 50$ µg/L, linear velocity= 0.7 m/h, $[Fe]_{coat} = 0.025$ M)

The optimal pH value for As(V) removal was pH 5.0, whereas that for As(III) was pH 7.0. Above pH 7.0 arsenic removal decreased for both As(V) and As(III). This observation could be well correlated with the point of zero charge (PZC) of iron oxides. Pure iron oxides, whether they can be identified as having a particular crystal structure or not, typically have PZCs in the pH range 7-9 (Benjamin *et al.*, 1996). At the PZC, iron oxides are present in the monomeric anionic form $[Fe(OH)_4^-]$, and thus cannot adsorb anionic components, as shown in Figure 5.

The difference in the removal of trivalent and pentavalent arsenic can be explained by the respective speciation differences. Above pH 3.0 As(V) is present in anionic form (Figure 6) and can therefore be effectively removed by iron hydroxides, which are present as cationic monomers $(Fe(OH)_2^+)$ over this pH range. On the other hand, As(III) is present in anionic form exclusively above pH 9.0 (Figure 6), whereas

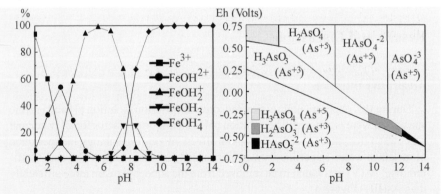

Figure 5. Speciation diagramme of Fe(III), obtained by Mineql⁺ software programme

Figure 6. Speciation diagramme of arsenic as affected by redox and pH

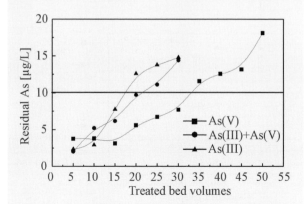

Figure 7. Arsenic removal as a function of speciation (experimental conditions: $[As]_o = 50$ µg/L, linear velocity= 0.7 m/h, pH= 7.0, $[Fe]_{coat} = 0.025$ M)

in the pH range 6-9 only a small percentage of H_3AsO_3 is dissociated. Thus for effective treatment of As(III), the pH of solution must be over 7.0.

In summary, these results indicate that when arsenic species are existing separately in aqueous solutions, the optimum pH for As(V) removal was 5.0, whereas for optimised As(III) removal, the respective pH value would be around 7.0. However, since these species usually occur simultaneously in most aqueous systems, the removal of a mixture, containing both species in 1:1 ratio, was subsequently examined (Figure 7).

As(III) comprised around 85 % of the total residual arsenic concentration, meaning that almost all As(V) was removed, whereas As(III) content was mostly remaining in the aqueous solution. This indicates the need for pre-oxidation of As(III) by the application of either physicochemical or biological methods, in order to increase the overall efficiency of arsenic removal. Following the oxidation step, As(V) may get in contact with iron oxides, forming surface complexes of ferric arsenates, which are highly insoluble in water, presenting a solubility product of 10^{-20} mol²/L² (Edwards, 1994).

$$M\text{-}FeOH + H_3AsO_4 \xrightarrow{\text{Sorption process}} M\text{-}Fe\text{-}H_2AsO_4 + H_2O$$

(Arsenates sorption)

Biological arsenic removal

The application of this technology involves the oxidation of trivalent arsenic by the indigenous microorganisms (biological means) and the adsorption of oxidized forms of arsenic on the biologically produced iron oxides (physical-chemical means). Therefore, this technology may be called "biological adsorptive filtration". The parameters examined were those affecting the performance of the treatment system,

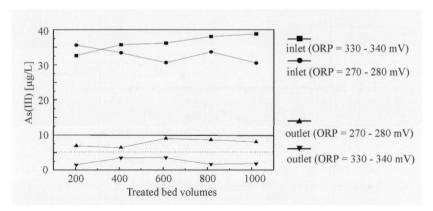

Figure 8. Arsenite removal as affected by redox potential variations

such as the redox potential, the dissolved oxygen, and the condition of the bacterial cultures (living, resting and dead cells).

The main parameter affecting the efficiency of specific treatment technology is the redox potential, which controls the performance of oxidation-reduction processes. During previous investigations, the redox potential was adjusted to 270-280 mV. Under these conditions the removal of iron was complete (> 95 %), whereas the removal of trivalent arsenic was found to be about 80 %, although it decreased to 65 % at higher initial concentrations (Katsoyiannis *et al.,* 2002). To improve the removal efficiency of As(III), the redox potential was increased by increasing the amount of dissolved oxygen in the influent stream.

When the redox potential was increased to values over 320 mV the removal of trivalent arsenic increased up to 95 % (Figure 8). At these redox values the residual arsenic was always below 5 μg/L, whereas at redox values below 300 mV the residual As(III) was close to the limit of arsenic in drinking water. During this operation the dissolved oxygen concentration was increased from 2.7 mg/L to approximately 3.7 mg/L. These results indicate that a possible oxidative action was involved in the removal of trivalent arsenic, as an increase in the redox potential produced a significant increase in arsenic removal. The Eh-pH speciation diagrammes for arsenic (Figure 6) showed that at a pH of 6.8 and a redox value of 320 mV, arsenic was present As(V), whereas for lower redox values As(III) species predominate. This transformation, which was probably responsible for the enhanced arsenic removal, might be due either to physical-chemical oxidation by the dissolved oxygen or to biological action from the microorganisms. Physical-chemical oxidation of arsenic by air is possible, and even thermodynamically favored, but it would take days, weeks or even months, depending on the specific conditions. With the residence time of just a few minutes (7 min) in these experiments, this conversion is highly unlikely. Thus it is more likely that the microorganisms play

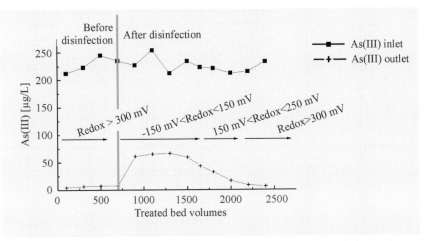

Figure 9. Comparison of arsenite removal by living and dead cells (before and after disinfection)

an important role in the removal of trivalent arsenic, probably by accelerating this conversion to the respective pentavalent form.

In order to clarify the influence of microorganisms in the oxidation process, control experiments were carried out with dead cells. The dead cells were obtained after filter disinfection with commercial chlorine bleach (NaOCl), which destroys bacteria. All other conditions were the same.

The results regarding arsenite removal, before and after disinfection, are illustrated in Figure 9, which shows the arsenite concentration as a function of the treated bed volumes. Changes in redox potential are given for every step of this sequential procedure.

Before disinfection, the removal of trivalent arsenic was efficient. For example, at high initial concentrations, up to 200 µg/L, the residual was below the maximum concentration limit of 10 µg/L. Directly after the addition of chlorine, the redox potential increased to over 600 mV, and there was no bacterial growth. Samples of the effluent were monitored for the presence of chlorine using potassium iodide. When all the chlorine was gone, arsenic and iron removal improved. The removal of arsenic initially decreased and followed closely the redox potential variations. The minimum removal was observed when the redox potential was −150 mV and continued to be low for redox values up to 200 mV. After several bed volumes had been washed to remove the chlorine, the bacteria started to grow again, as they were indigenous to the groundwater and the conditions (aeration) were favorable for their growth. This had a significant impact on the respective Eh values, which started to increase again. When the redox potential reached values over 300 mV, arsenite removal was again over 90 %, as before disinfection. These observations

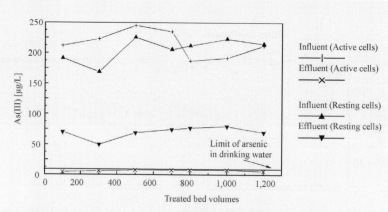

Figure 10. Arsenite removal by (a) living and (b) resting cells (a: redox potential >300 mV, b: redox potential = +150 mV, experiments performed without aeration)

indicate that the microorganisms were the major factor controlling the overall performance of the treatment system. While the bacteria are clearly involved in the oxidation of arsenite, it is still unclear what percentage of arsenite is actually oxidized by the bacteria.

In summary, removal of trivalent arsenic is primarily due to the action of iron-oxidizing bacteria, and not to adsorption onto preformed iron oxides, as under these pH and redox conditions arsenic is present in aqueous streams in its non-ionic form, arsenious acid. The bacteria accelerate the oxidation of trivalent arsenic to the pentavalent form, which is more easily removed from the stream.

When the bacteria were dead, the observed arsenic removals were attributed mainly to air injection. To test this hypothesis, the experiment was repeated without

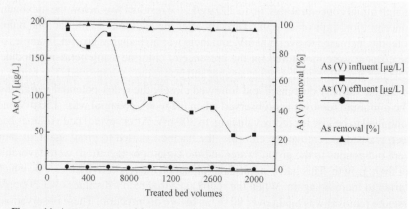

Figure 11. Arsenate removal from groundwater, as a result of biological oxidative filtration

aeration. Under anoxic conditions, the bacteria could not grow properly, as was shown by the formation of resting cells that could not act as oxidizing agents (Figure 10).

Without active bacterial cultures, the removal of both contaminants (Fe & As) was inefficient; iron was removed by 50%, whereas trivalent arsenic removal was around 75%. Arsenic removal was due to direct adsorption of trivalent arsenic on the preformed iron oxides. The redox potential was reduced from 330 mV to a constant 140-150 mV. These results indicate that controlled aeration is necessary to maintain an active bacterial culture in the column, so that conditions are optimal for iron and arsenic removal.

A possible mechanism for the removal of trivalent arsenic can be proposed based on these result

$$H_3AsO_3 + \frac{1}{2}O_2 \xrightarrow{\ Bacteria\ } H_3AsO_4$$

$$M\text{-}FeOH + H_3AsO_4 \xrightarrow{\ Sorption\ } M\text{-}Fe\text{-}H_2AsO_4 + H_2O$$

The results shown up to now were for the removal of trivalent arsenic. Under optimal conditions, which require controlled aeration, bacterial maintenance and redox potential values over 300 mV, the removal of arsenic was successful for a wide range of initial concentrations. While trivalent arsenic predominates in groundwaters due to anaerobic conditions, pentavalent arsenic may also be present, sometimes in equal ratios. Therefore, in order to obtain an integrated view of process effectiveness, experiments were performed by spiking pentavalent arsenic in the influent stream (Figure 11).

This treatment method is very effective for removing pentavalent arsenic. The treatment of a wide range of initial arsenic concentrations (50-200 µg/L) resulted in residual concentrations always below the MCL value of 10 µg/L. These results were expected as it is widely known that pentavalent arsenic can be removed more efficiently than trivalent arsenic by adsorption onto iron oxides, due to its speciation in water. At redox values over 300 mV and pH of 6.8, arsenate is present with in its anionic forms, which are more likely to be adsorbed on the precipitated iron oxides. Microorganisms and aeration are probably less important for the removal of arsenate, because it is removed by adsorption onto iron oxides, but microorganisms and aeration are needed to produce iron oxides, prerequisites for the adsorption process.

Conclusions

The combined chemical and biological treatment techniques used for iron removal were also effective for the simultaneous removal of As(III) and As(V) from contaminated groundwaters. As(III) removal was based mainly on arsenic sorption onto the pre-formed iron oxides. This sorption may take place either in the form of arsenite [As(III)], or after preliminary biological oxidation of As(III) to As(V) (as arsenate oxyanions) resulting in enhanced arsenic removal up to 95%. The combined process has the advantage that the adsorbents (iron oxides) were produced continuously by the oxidative action of microorganisms, which prevents breakthrough during the operation (a usual drawback of adsorption processes). Since iron is removed simultaneously, this method provides an integrated option for groundwater treatment.

References

Benjamin, M.M., Sletten, R.S., Bailey, R.P. and Bennet, T. (1996) Sorption and filtration of metals using iron-oxide coated sand. *Water Res.* **30**(11), 2609-2620

Desesso, J. M., Jacobson, C. F., Scialli, A. R., Farr, C. H. and Holson, J. F. (1998) An assessment of the developmental toxicity of inorganic arsenic. *Reprod. Tech.* **12**(4), 385-433

Dimitrakos, G., Martinez Nieva, J., Vayenas, D. and Lyberatos, G., (1992) Removal of iron from potable water using a trickling filter. *Water Res.* **31**(5), 991-996

Edwards, M. (1994) Chemistry of arsenic-removal during coagulation and Fe-Mn oxidation. *J. Am.Water Works Assoc.* **86**(9), 64-78

EC (1998) European Commission Directive, **98/83/EC**, related to drinking water quality intended for human consumption, Brussels, Belgium, 3/11/1998

Jekel, M. R. (1994) Removal of arsenic in drinking water treatment. In: Nriagu J.O.(ed). Arsenic in the environment, part I: Cycling and characterization. New York: Wiley-Interscience, pp.119-130

Kartinen, E. O. and Martin, C. J. (1995) An overview of arsenic removal processes, *Desalination* **103**, 79-88

Katsoyiannis, I., Zouboulis, A., Althoff, H. and Bartel, H. (2002) As(III) removal from groundwaters using fixed-bed upflow bioreactors. *Chemosphere*, In press

Lehimas, G. F., Chapman, J. I. and Bourgine, F. P. (1999) Use of biological processes for arsenic removal: a cost effective alternative to chemical treatment for As(III) in groundwater. Available at www.saur.co.uk/poster/html (Retrieved on 09/11/1999)

Mouchet, P. (1992) From conventional to biological removal of iron and manganese in France. *J. Am.Water Works Assoc.* **84**(4), 158-166

Pontius, F. W., Brown, G. K. and Chien, J. C. (1994) Health implications of arsenic in drinking water. *J. Am.Water Works Assoc.* **86**(9), 52-63

Seith, R. and Jekel, M. (1997) Biooxidation of As(III) in fixed bed reactors, *Vom Wasser*, **89**, 283-296

NOM-Removal in Drinking Water Treatment

H.H. Hahn, E. Hoffmann, H. Ødegaard (Eds.)
Chemical Water and Wastewater Treatment VII, pp. 133-142
© IWA Publishing, London
ISBN: 1 84339 009 4

Comparison of Iron and Aluminum Based Coagulants and Polymeric Flocculant Aids to Enhance NOM Removal

N. Lindqvist*, S. Korhonen, J. Jokela and T. Tuhkanen

*Tampere University of Technology, Institute of Environmental Eng. and Biotechnology,Tampere, Finland
Email: lindqvin@assari.cc.tut.fi

Abstract

This study examined the removal of natural organic matter (NOM) by coagulation from the water of Lake Roine, which is used as raw water in the Rusko Waterworks situated in Tampere, Finland. The ability of four coagulants and two coagulant/ flocculant aids to simultaneously reduce turbidity and remove dissolved organic matter (DOM) from the water was determined. The removal of NOM was examined using the following methods: turbidity, high-performance size exclusion chromatography (HPSEC) separation, dissolved organic carbon (DOC) and UV absorbance at $\lambda=254$ nm (UV-254). Ferric salts were superior to aluminum salts, and ferric chloride gave the greatest improvements in terms of coagulation at the Rusko Waterworks. The greatest difference was noticed in the removal of DOM having molecular weight (MW) of 1000-4000 Da. Turbidity removal was found to be highly pH dependent, and for all the coagulants the optimum pH for turbidity removals was higher than that for DOM removal. Poly-DADMAC as coagulant aid did not significantly improve the DOM removal whereas the use of polyacrylamide improved the removal of turbidity and DOM in the MW range of 500-1000 Da. The highest removal (30 %) of DOM in this MW range was achieved with ferric chloride and polyacrylamide.

Introduction

Surface waters in Finland usually contain a high amount of NOM compared to, for instance, surface waters in Central Europe. Up to 90 % of NOM can be composed of humic substances, like humic and fulvic acids (Woelki *et al.,* 1997). Humic material causes a brownish color and gives the water a muddy taste and odor. Besides these esthetic effects, humic substances also adsorb organic and inorganic micropollutants thus affecting their fate, transport and bioavailability in the environment (Dixon *et al.,* 1999). Since NOM is not entirely removed in the drinking water treatment process, these micropollutants can be carried through the process adsorbed onto NOM and eventually drift into the drinking water. Residual NOM also affects the formation of possible carcinogenic compounds in disinfection with chlorine (Kronberg, 1987) and contributes to microbial regrowth in distribution systems (Owen *et al.,* 1995). Therefore, enhancement of NOM removal in the drinking water treatment process is essential to produce a higher quality drinking water.

It has been hypothesized that polar micropollutants such as pyrene would be bound primarily by the largest MW NOM having a high number of aromatic rings in its structure (Chin *et al.,* 1997). Humic acids are usually larger in molecular size, more aromatic and have a lower charge density than fulvic acids, which makes them favorable molecules to bind to. Coagulation with coagulant doses commonly employed in water treatment is found to remove primarily high-molecular-weight and low charge density humic acids (Chow *et al.,* 1999). Thus, drifting of some micropollutants into the drinking water can be prevented by a proper optimization of the coagulation process. Coagulation as a pretreatment before granular activated carbon filtration also extends the operating time of the filters since it removes part of the organic matter load and especially poorly absorbing high-molecular-weight molecules introduced to the filters (Semmens *et al.,* 1986). By optimizing the coagulation process the performance of the subsequent processes and the quality of the treated water can also be improved.

The objective of this research was to study the ability of four coagulants and two coagulant/flocculant aids to simultaneously reduce turbidity and DOM content of the water. The efficiency with which various molecular sizes of DOM were removed was also examined.

Materials and methods

Raw water characteristics

The water used in the experiments was surface water from Lake Roine, which is used as raw water in the Rusko Waterworks, Tampere, Finland. The quality of the raw water was stable during the experimental part of this work. The raw water is typical Finnish lake water having low alkalinity (0.25 mmol l⁻¹) and very low turbidity (2.2 NTU). Organic carbon content is fairly low (5.2 mg l⁻¹) compared to typical Finnish surface waters. UV absorbance at l=254 nm (UV-254) is around 0.11 cm⁻¹. The full-scale process removes 55 %, 42 % and 60 % of NOM measured as UV-254, DOC and turbidity, respectively.

Experiments

Coagulation studies were performed in the laboratory as jar tests, a common procedure for studying coagulation (Gregor *et al.,* 1997, Bell-Ajy *et al.,* 2000). Tests were carried out at room temperature. The optimum coagulation pH and coagulant dose for aluminum sulfate ($Al_2(SO_4)_3 \times 18H_2O$) (purchased from J.T. Baker Chemicals B.V., Deventer, Holland), aluminum chlorohydrate, ferric chloride ($FeCl_3$) and ferric sulfate ($Fe_2(SO_4)_3$) (all received from Kemira Chemicals Oy, Finland) were determined. Polymers tested were cationic polydiallyldimethyl-ammonium chloride (poly-DADMAC) and anionic polyacrylamide (received from Kemira Chemicals Oy).

Jar test experiments on 1-L samples were made using a Kemira mini-flocculator. The procedure is shown in Figure 1. All the coagulants used were acidic and lowered the pH of the water. pH was adjusted before the coagulation step with 0.025 M sulfuric acid (H_2SO_4) (Reagecon, Shannon, Co. Clare, Ireland) or 0.05 M sodium hydroxide (NaOH) (Oy FF-Chemicals, Finland).

Optimized coagulation conditions (pH and coagulant dose optimized primarily for DOM removal) were compared to current plant conditions (referred to as baseline coagulation) at the Rusko Waterworks. To determine optimal pH the jar tests were conducted by keeping the coagulant dose constant and varying the pH

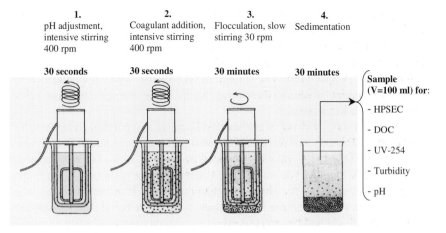

Figure 1. The four stages of the coagulation-flocculation-sedimentation process

in each jar. The pH at which the greatest amount of DOM was removed was considered to be the optimal pH. Optimal dose was determined by keeping the optimal pH constant and varying the coagulant dose in each jar.

Analytical methods

pH, turbidity, HPSEC separation, DOC and UV-254 were analyzed. These parameters were measured in the raw water and in the water after the coagulation/ flocculation/sedimentation procedure. HPSEC was carried out with a Hewlett-Packard 1100-series HPLC consisting of diode array UV detector operating at $\lambda = 254$ nm. A size exclusion column TSKgel G3000SW 7.5 mm (ID) x 30 cm and 0.01 M sodium acetate (Merck KGaA, Germany) eluent with flow rate of 1 ml min^{-1} were used (Vuorio *et al.* 1998). DOC concentration was determined according to the SFS-ISO 8245 standard (1989) and it was measured with Shimadzu TOC-5000 and ASI-5000 apparatus. UV-254 was determined with a Shimadzu UV-1601 UV-visible spectrophotometer using 1-cm length quartz cells. The turbidity was measured with a HACH Model 2100A turbidimeter.

Results and discussion

Optimum coagulation conditions

The optimum coagulation conditions, DOM, and turbidity removal efficiencies of the baseline coagulation and optimized coagulation are presented in the Table 1. Optimum coagulation pH values for DOM removal were 5.2 - 5.6 for aluminum sulfate, 6.6 - 6.8 for polyaluminum chlorohydrate, and 4.6 for ferric salts. For aluminum sulfate optimum pH values from 5 to 7 for NOM removal were reported, usually being closer to 5 than 7 (Huang and Shiu, 1996; Hundt and O'Melia, 1998). Optimum pH values for iron salts are usually below 5 (Bell-Ajy *et al.*, 2000).

Optimum doses for the coagulants were 1.2 mg Fe per mg DOC for ferric salts and 0.5 mg and 1.3 mg Al per mg DOC for aluminum sulfate and aluminum chlorohydrate, respectively. The amount of NOM in the water always affects the optimum coagulant dose, and there is much variation in the optimum dose values reported in the literature.

Aluminum salts did not significantly improve coagulation over baseline coagulation (Table 1), but ferric salts were superior to aluminum salts (Figure 2 and Table 1). DOM removal was enhanced by roughly 15 %. Turbidity removal was found to be highly pH dependent and its removal at the optimum DOM removal range was lower for all the chemicals than that in the baseline situation. In all cases, a higher pH was required for optimum turbidity removal than for optimum DOM removal. pH had the greatest effect on turbidity removal when aluminum chlorohydrate was used (Figure 2). At the pH range where DOM removal was

Table 1. Optimum conditions and NOM removal (%) for baseline coagulation and for different coagulants. Results from two parallel experiments were averaged for each coagulant. The results for the baseline coagulation represent the averaged values using 0.129, 0.140 and 0.152 mmol Al l^{-1}. PH was measured after coagulation. Sum of the peak heights refers to the total amount of DOM in the water (Matilainen *et al.*, 2001)

	Baseline	Aluminum sulfate	Aluminum chlorohydrate	Ferric chloride	Ferric sulfate
Optimum pH	5.9	5.6	6.8	4.6	4.6
Optimum dose (mmol Me^{3+} l^{-1})	-	0.15	0.23	0.12	0.11
Optimum dose (mg Me per mg DOC)		0.5	1.3	1.2	1.2
HPSEC (MW):					
>5 kDa	100%	100%	100%	100%	100%
4-5 kDa	100%	100%	98%	100%	100%
3-4 kDa	73%	75%	70%	91%	87%
1-3 kDa	20%	20%	18%	52%	40%
0.5-1 kDa	0%	9%	8%	16%	8%
sum of the peak heights	73%	73%	70%	84%	80%
DOC	48%	56%	51%	64%	60%
UV-254	67%	70%	69%	80%	76%
Turbidity	66%	24%	30%	39%	37%

optimal, only 15-20 % of the turbidity was removed. On the other hand, above pH 7, over 60 % of the turbidity was removed. This phenomenon was seen in previous studies (Dempsey *et al.*, 1984; Semmens and Field, 1980). PH affects both the composition of NOM in the water and the formation of different coagulant hydrolysis species. In the optimum pH range for NOM removal the formed NOM-metal hydroxide precipitates may still be too small to settle and therefore cause high turbidity after coagulation. The precipitates may also become less water-soluble as the pH increases, allowing the floc size to increase.

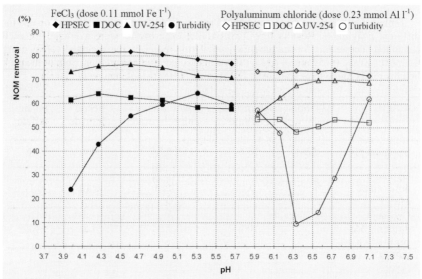

Figure 2. Comparison of NOM removals using ferric chloride and aluminum chlorohydrate

Effects of coagulation on different molecular weight fractions of DOM

The removal of various MW DOM was more dependent on the coagulant dose than the pH. Even low doses effectively removed DOM having MW of more than 4000 Da. This fraction is suggested to consist mainly of humic acids, which have low charge density and therefore require lower coagulant doses (Gregor et al., 1997). Different coagulants particularly affected the removal efficiency of DOM less than 4000 Da. Figure 3 shows the amount of DOM in various MW ranges in the raw water and the water after coagulation under optimal conditions.

Ferric salts were more effective than aluminum salts in removing DOM and especially the fraction with MW less than 4000 Da. This was also observed by other researchers (Bell-Ajy et al., 2000; Grozes et al., 1995). These results may suggest that ferric and aluminum salts have different removal mechanisms. Ferric chloride is reported to present roughly twice the number of active positive charges per dry weight unit of coagulant compared to aluminum sulfate (Grozes et al., 1995). NOM in the MW range of 1000 - 4000 Da is thought to consist of fulvic acids, which contain many carboxyl groups and thus have a high charge density. Because of its high charge density, ferric chloride may better be able to precipitate these molecules than aluminum sulfate. The character of aluminum and ferric hydroxide flocs (e.g., surface charge) is also different, which may affect the removal (Grozes et al., 1995).

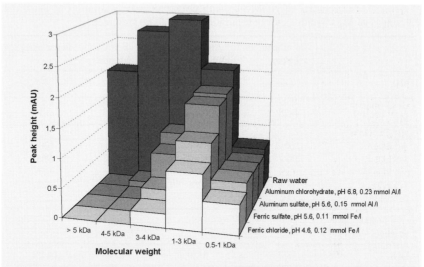

Figure 3. The amount of DOM in the various MW ranges in the raw water and water after coagulating with different coagulants at their optimum conditions

Effect of polymers

It was first expected that the use of cationic polymer (poly-DADMAC) as coagulant aid would enhance the removal of NOM since the surface charge of NOM in natural waters is usually negative. However, no significant improvement was observed. At pH values below 4.5 turbidity removal was slightly enhanced, whereas at pH values above 6.5 turbidity removal deteriorated considerably. The highest decreases in the DOM removal were detected at higher pH values. This may be because of the influence of pH on the degree of ionisation of the polymer, which again affects the degree of extension of the molecule and a polymer's ability to be adsorbed to NOM molecules (Gric and Lric, 1978).

The hypothesis that increasing the coagulant dose may cause restabilization of some fraction of the NOM and this fraction could be destabilized again with anionic polymer (polyacrylamide) was at least partly proved in this study. Table 2 shows the NOM removal values with polyacrylamide addition. The removal of DOM in MW range of 500-1000 Da was improved by 13-17 % compared to the coagulation without polymer addition in optimum conditions (see Table 1). Ferric chloride showed the best removal efficiency, ie., almost 30 % of the DOM having MW of 500-1000 Da could be removed. In full-scale process and baseline coagulation the removal of this fraction was less than 10 %. Use of polyacrylamide also resulted larger and better settling floc and improved removal of turbidity. This was most pronounced when coagulating with aluminum chlorohydrate.

Table 2. The removal of NOM (%) with polyacrylamide addition. For aluminum chlorohydrate the conditions for best turbidity removal are presented

	Aluminum sulfate	Aluminum chlorohydrate	Ferric chloride	Ferric sulfate
pH	5.6	7.6	4.5	4.5
Dose (mmol Me^{3+} l^{-1})	0.25	0.18	0.12	0.11
HPSEC (MW):				
>5 kDa	100%	100%	100%	100%
4-5 kDa	96%	93%	100%	96%
3-4 kDa	73%	57%	88%	83%
1-3 kDa	28%	20%	47%	42%
0.5-1 kDa	23%	23%	29%	25%
sum of the peak heights	69%	62%	79%	75%
DOC	58%	48%	62%	61%
UV-254	73%	75%	68%	77%
Turbidity	80%	55%	68%	77%

Conclusions

Ferric salts were found to be better in removing DOM than aluminum salts and the greatest improvements compared to baseline coagulation were achieved with ferric chloride. The removal of DOM in the MW range 500-1000 Da could be increased from lower than 10% up to 30% with simultaneous use of ferric chloride and polyacrylamide. Addition of this anionic polymer also significantly improved the removal of turbidity. Turbidity removal without polymer addition was found to be highly pH dependent and the highest removal was achieved at pH values higher than those for DOM removal. With the use of anionic polymer like polyacrylamide the turbidity removal can be maintained at high level and at the same time DOM removal can be enhanced.

References

Bell-Ajy, K., Abbaszadegan, M., Ibrahim, E., Verges, D., and LeChevallier, M. (2000) Conventional and optimized coagulation for NOM removal. *JAWWA* **92**(10), 44-58

Chin, Y-P., Aiken, G.R., and Danielsen, K.M. (1997) Binding of Pyrene to Aquatic and Commercial Humic Substances: The Role of Molecular Weight and Aromaticity. *Env. Sci. Tech.* **31**(6), 1630-1635

Chow, C.W.K., van Leeuwen J.A., Drikas, M., Fabris, R., Spark, K.M., and Page, D.W. (1999) The impact of the character of natural organic matter in conventional treatment with alum. *Wat. Sci. Tech.* **40**(9), 97-104

Dempsey, B.A., Ganho, R.M., and O'Melia, C.R. (1984) The coagulation of humic substances by means of aluminum salts. *JAWWA*, **76** (4) 141-150

Dixon, A.M., Myphuong, A.M., Larive, C.K. (1999) NMR Investigation of the Interactions between 4´-Fluoro-1´-acetonaphtone and Suwannee River Fulvic Acid. *Env. Sci. Tech.* **33**, 958-964

Gregor, J.E., Nokes, C.J., and Fenton, E. (1997) Optimising natural organic matter removal from low turbidity waters by controlled pH adjustment of aluminum coagulation. *Wat. Res.* **31**(12), 2949-2958

Gric, N.M., and Lric, B.D. (1978) Flocculation: theory and application. Mine and Quarry Journal, May, 1-8Grozes, G., White, P., and Marshall, M. (1995) Enhanced coagulation: its effect on NOM removal and chemical costs. *JAWWA*, 78-89

Huang, C. and Shiu, H. (1996) Interactions between alum and organics in coagulation. Colloids Surfaces A: *Physicochem. Eng. Aspects.* **113**, 155-163

Hundt, T.R., and O'Melia, C.R. (1988) Aluminum-Fulvic Acid Interactions: Mechanisms and Applications. *JAWWA*, **88**(4), 176-186

Kronberg, Leif. (1987) Mutagenic compounds in chlorinated humic and drinking water. Ph.D. thesis. Dept of Chemistry, Åbo Akademi. Turku

Matilainen, A., Lindqvist, N., Korhonen, S., and Tuhkanen, T. (2001) Characterization of the organic matter transformation in different stages of the water treatment process. Proc. IWA 2nd World Water Congress, October 15-19, Berlin, Germany

Owen, D.M., Amy, G.L., Chowdhury, Z.K., Paode, R., McCoy, G. and Viscosil, K. (1995) NOM characterization and treatability. *JAWWA* **87**, 46-63

Semmens, M.J., and Field, T.K. (1980) Coagulation: Experiences in Organics Removal. *JAWWA* **72**, 476-483

Semmens, M.J., Staples, A.B., Hohenstein, G., and Norgaard, G.E. (1986) Influence of Coagulation on Removal of Organics by Granular Activated Carbon. *JAWWA*, **78**(8), 80-84

SFS ISO 8245 standard. (1989) Guidelines for determination of total organic carbon (TOC). The Finnish Standard Association SFS. 6p

Vuorio, E., Vahala, R., Rintala, J., and Laukkanen, R. (1998) The evaluation of drinking water treatment performed with HPSEC. *Env. Int.* **24**(5/6), 617-623

Woelki, G., Friedrich, S., Hanschmann, G., and Salzer, R. (1997) HPLC fractionation and structural dynamics of humic acids. *Fresenius J Anal Chem.* **357**, 548-552

H.H. Hahn, E. Hoffmann, H. Ødegaard (Eds.)
Chemical Water and Wastewater Treatment VII, pp. 143-152
© IWA Publishing, London
ISBN: 1 84339 009 4

Metal Residuals in Contact Filtration of Humic Drinking Water

T. Saltnes[*], B. Eikebrokk and H. Ødegaard

*Norwegian University of Science and Technology, Dept. of Hydraulic and Environmental Engineering, Trondheim, Norway
email: torgeir.saltnes@bygg.ntnu.no

Abstract

The concentration of Al and Fe residuals in treated drinking water is regulated in the EU directive to be < 0.2 mg/l. This is also the limit in the new Norwegian standard, with a proposed guideline value of 0.15 mg Me/l in cases where metal coagulant is used in water treatment. Residual metal in treated water is controlled by the solubility of hydroxide precipitates, and the capability of the treatment process to remove precipitated metal. In general, the solubility of metals in water is dependent on pH, temperature and content of organic and inorganic substances.

In a pilot plant with two parallel filter columns receiving identical water the effects of different filtration parameters on residual metal were studied. One filter was a conventional dual media anthracite/sand, and the other was a more coarsely grained dual media clay aggregate (Filtralite) filter.

When the traditional metal salts aluminium sulphate (AS) or iron chloride sulphate were used to coagulate low turbidity humic water, residual metal problems were experienced for both filters. A high residual turbidity caused by insufficient filterability of the humic-Al-precipitates was observed with the coarser Filtralite filter, especially when AS was used for coagulation. For raw water with higher turbidity the particulate metal residuals were generally lower. The conflicting pH ranges for optimum turbidity and organic matter removal, and minimum solubility of precipitate were narrowed down when PAC was used for coagulation.

[143]

Introduction

Norwegian raw water sources are generally surface waters with relatively high levels of organic matter and low turbidity. Temperature varies with season and is normally between 2 and 15° C. When optimising for removal of humic substances a strict pH and coagulant dose control is needed, often with seasonal variations demanding elevations in coagulant dose. The content of organic matter and the water temperature are both important factors influencing the solubility of metal.

Metal added with the coagulant is removed in the filtration process as precipitated metal hydroxide, while the soluble metal fraction passes through the filter if no adsorption takes place.

Contact filtration of humic waters using metal salts for coagulation often results in high residual metal concentrations caused by suboptimal coagulation conditions. Coagulation at optimum pH conditions, i.e. at minimum solubility for the precipitates, may conflict with optimum removal of turbidity or organic matter (Eikebrokk, 1999).

Early breakthrough is experienced when metal salts are used for coagulation followed by contact filtration of humic waters. This is due to the high doses needed for coagulation, and the weak, voluminous hydroxide-humic flocs formed (Eikebrokk, 1999; Rebhun *et al.*, 1984).

Experimental

In a pilot scale coagulation/contact filtration treatment plant, experiments were carried out with two different filter beds operated in parallel. One filter bed was a conventional dual media anthracite/sand filter and the other a dual/mono media clay aggregate filter (Filtralite). The Filtralite filter consisted of clay aggregates only, in two different densities and size ranges. Both filters operated in down flow mode. In Figure 1 the two filters are shown together with a deeper Filtralite filter which was used in parts of the study. This set up allowed an analysis of the effect of grain size and bed depth on removal of particulate metal (precipitate creating turbidity).

Raw water with different levels of colour and turbidity was pH-adjusted with hydrochloric acid and coagulated in an in-line rapid mixer before it entered the filters. The raw water was made of tap water with the addition of a humic concentrate (regenerate from an ion-exchange plant) and/or bentonite clay suspension. Four different raw waters were tested, RW15 (about 15 mg Pt./l) and RW50 (about 50 mg Pt./l), both with turbidities < 0.2 NTU, and RW15/3 (RW15 + 3 NTU as bentonite) and RW50/3 (RW50 + 3 NTU as bentonite). RW15 and RW15/3 had a content of about 2.5 mg/l TOC, while RW50 and RW50/3 contained about 5.5 mg TOC/l.

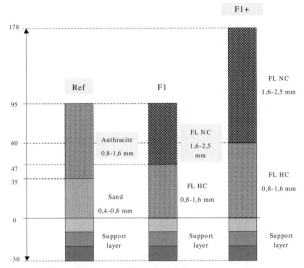

Figure 1. The three different filters used in this study

Coagulant type, dose and pH were varied in each run in order to optimise the filtration performance. Aluminium sulphate (AS), iron chloride sulphate (ICS) and poly-aluminium chloride (PAC) were used as coagulants.

Turbidity and pH were measured on-line in raw water and filtrate from the two filters. Water samples of raw water and filtrate were taken for each run and analysed in the laboratory for true colour, TOC, pH and turbidity. In addition, filtrate samples and some raw water samples were analysed for total metal content by ICP-MS (Inter Coupled Plasma-Mass Spectroscopy) after acidification.

To estimate the soluble Al and Fe fractions, samples of coagulated water were

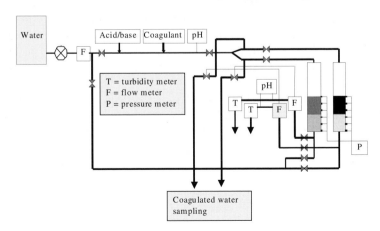

Figure 2. Contact filtration pilot plant

collected from the pilot plant (see Figure 2). These samples were withdrawn as the amount of HCl added was stepwise increased. Immediately after sampling, pH was measured in the lab, followed by membrane filtration (0.2 μm). Water temperature was about 10°C. The filtered samples were analysed for total Al or Fe in an ICP-MS. In water analysis, filtration through 0.45 μm pore size is usually used to separate soluble and particulate matter. However, colloidal precipitates from Al and Fe based coagulants can be smaller than this. Jekel and Heinzmann (1989) found considerable amounts of fine colloidal aluminium hydroxide in the size range of 0.05-0.45 μm. In the following discussion, the metal fraction passing through the 0.2 μm pore size membrane filter is referred to as the soluble fraction.

Results

Metal solubility

The sampling and filtration procedure described above was carried out on RW15 coagulated with AS and PAC, and on RW50 with AS, PAC and ICS. The coagulant doses that were used during the solubility experiments are shown in Table 1. The given doses were optimised with respect to turbidity, residual metal and NOM removal.

In Figure 3 the measured values of soluble Al and Fe are shown for the different raw waters. At low pH, about 5, most of the aluminium added as AS exists as

Table 1. Coagulant doses used during the solubility experiments

	RW15	RW50
Aluminium sulphate, AS (mg Al/l)	1.0	3.0
Poly-aluminium chloride, PAC (mg Al/l)	0.8	2.5
Iron chloride sulphate, ICS (mg Fe/l)	2.0	7.0

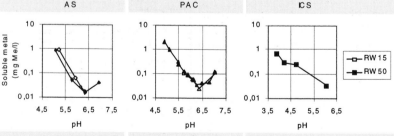

Figure 3. Solubility of Al and Fe when AS, PAC and ICS are used to coagulate RW15 and RW50

positively charged monomers (Al^{3+} and $Al(OH)^{2+}$). The dominating coagulation mechanism at this pH is adsorption and charge neutralisation. Normally, high NOM removal efficiency can be obtained at a pH of about 5.5. As can be seen from the Figure however, there is a problem with too high values of soluble Al at this pH. At a pH of about 6,4, aluminium added as AS or PAC is least soluble, i.e. maximum precipitation of hydroxide is obtained. Coagulation of organic matter is probably dominated by adsorption of humic substances onto colloidal metal precipitates. For pH values on the alkaline side of minimum solubility, the negatively charged monomer $Al(OH)_4^-$ dominates, which is less effective for coagulation of organic matter. When PAC is used for coagulation the effects of pH on Al-solubility follow the same trend as when AS is used, with minimum solubility at the same pH. When ICS is used, the solubility of Fe is about the same as that of Al. Theoretically, Fe is far less soluble than Al, and lower residuals were expected. This might be explained in part by the formation of small colloidal precipitates with a diameter less than 0.2 µm.

An Al fractionation study (Eikebrokk, 2000) of filtered water coagulated with aluminium sulphate revealed a monomer organic fraction of about 30-40 µg/l, which is aluminium associated with humic substances. This fraction was almost constant for different filtration parameters and time of filtration in a run. The monomer inorganic fraction, which is Al associated with inorganic substances, had almost the same value, and little variation was observed. The polymeric and colloidal fraction was generally the highest, and varied with time of filtration. This fraction is responsible for the ripening and breakthrough in turbidity that is observed in a filtration run.

Raw water organic matter

From Figure 3 it can be seen that the solubility of Al added as AS is higher in RW15 than in RW50, in the optimal pH range for contact filtration (5.5 - 6.5). The TOC values for the two raw waters are about 2.5 and 5.5 mg/l for RW15 and RW50, respectively. Experiments have shown that it is possible to remove almost 100 % of the colour causing organic matter, but only about 80 % of the TOC (Eikebrokk, 1999; Eikebrokk and Saltnes, 2001). Thus, the non-coloured TOC is more difficult to remove than the coloured TOC, presumably because this is low molecular weight and/or consists of neutral organic compounds.

The colour to TOC ratios for RW15 and RW50 are 5.2 and 9.1, respectively, indicating a higher fraction of less removable low molecular weight NOM in RW15. The higher solubility of Al in RW15 than in RW50 is probably related to this fraction of TOC. Humic substances in water have a tendency to form soluble complexes with cations, and surface complexes with precipitated metal hydroxides (Jekel and Heinzmann, 1989). The formation of complexes will therefore increase metal solubility, resulting in a fraction of the added metal that is not effective for coagulation purposes. More metal is then needed to coagulate the formed complexes, and obtain low metal residuals. Klute (1990) found that complex formation was very rapid for monomeric species, within microseconds, and within 1 second for polymeric species. Precipitation of aluminium hydroxide is slower and starts in the following 1-7 seconds. The species formed by adding Al as aluminium sulphate to water were studied by van Benschoten and Edzwald (1990), who found Al to be present almost entirely as the monomers Al^{3+}, $Al(OH)^{2+}$ and $Al(OH)_4^-$.

Looking at the results obtained with PAC, the difference between RW15 and RW50 is minimal. Poly-aluminium chloride forms other hydrolysis products in water than aluminium sulphate. Van Benschoten and Edzwald (1990) found that about 90 % of the aluminium species were polymeric when they studied hydrolytic reactions with poly-aluminium chloride. The PAC product used (PAX-16) had a basicity of 35 % (r = 1.05). For high basicity products a highly charged polymeric species is formed, $Al_{13}O_4(OH)_{24}^{7+}$ (Ødegaard et al., 1990). Both curves for PAC show a deviation (higher solubility) on the acid side of minimum solubility. This probably shows that a highly charged polymeric Al species is formed at this pH. Al solubility from PAC coagulation is less influenced by the content of organic matter in the water due to less formation of monomeric species, which is the species that first forms complexes.

Effect of coagulant type

In Figure 4 the experimental solubility curves for RW50 coagulated with AS, PAC and ICS are compared. These results show that the solubility of Al is higher when added as PAC than as AS. The reason for this is, as discussed previously, the formation of other hydrolysis products (polymeric) when PAC is used. This could also partly be caused by a larger fraction of colloidal precipitates smaller than 0.2 µm. The experimental solubility of iron when added as ICS is much higher than the theoretical solubility of iron. Even though small precipitates (less than 0.2 µm) are formed, these tiny particles are probably difficult to remove in a filter.

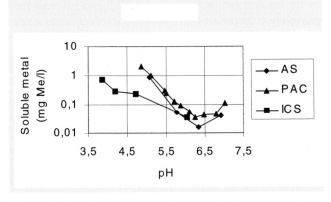

Figure 4. Solubility (< 0.2 µm) of Al and Fe in RW50 when added as AS, PAC or ICS

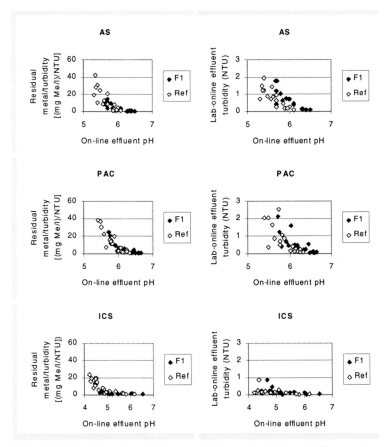

Figure 5. Effluent metal residuals per turbidity units (left) and difference between lab and on-line turbidity (right) for AS, PAC and ICS (RW15 and RW50

Filtration experiments

The coagulant type/-dose and pH optimisation experiments for the four raw waters provided important information on the influence of different doses and pH conditions. In Figure 5 the ratio of metal residuals per effluent turbidity unit measured on-line is presented for RW15 and RW50. Results for different doses are shown in the figure as a function of on-line effluent pH. The right column of Figure 5 shows the turbidity difference between lab and on-line measurements. The amount of metal per unit turbidity gives a picture of the content of metal associated with turbidity-forming particles. At low pH values the ratio is high for all three coagulants indicating a large soluble fraction of metal in the effluent, which is not removed in the filter. The difference between F1 and the reference filter is negligible at the same pH. However, pH in the effluent from F1 is generally higher due to a pH increase in the clay aggregate filter. When the more coarse grains in filter F1 produce the same ratio of Al/turbidity as the anthracite/sand filter, and the effluent turbidity from the reference filter is generally lower, the amount of aluminium in the effluent is dependent on pH (solubility) and the capability of the filter to remove turbidity (particles). The ratio is lower when ICS is used as coagulant than when AS or PAC is used. This is due to the lower solubility of Fe at low pH compared to Al.

The difference between turbidity measured in the lab and on-line is high at low pH. The reason for this is partly the pH increase in the filters due to alkaline filter material, and the increase in pH due to stripping of CO_2 upon sampling and analysis. These phenomena lead to post-precipitation of hydroxides and a corresponding increase in turbidity after filtration. For the aluminium based coagulants post precipitation is high at low pH, but high values are also obtained at pH about 6, especially in filter F1 using AS for coagulation. This implies that filter F1 removes less post-precipitation potential than the reference filter. This is probably a result of the coarser grains with less available adsorptive grain surface area in this filter bed. The pH increase when water passes through the somewhat alkaline filter material was higher in F1 than in the reference filter, and this may also affect the post-precipitation process. This shows that the difference in effluent turbidity is not only dependent on grain size, but also on the chemical characteristics of the filter material. This can be beneficial in a water treatment process, provided that the filter is able to remove the precipitates produced when there is a pH increase in the filter.

The results also show that post-precipitation is a problem when PAC is used. The difference in performance and filter effluent turbidity was, however, minimal with this coagulant

With ICS the post-precipitation was much less than for the Al-based coagulants, and similar results were obtained with both filters. This implies that iron is more filterable, and higher removals are obtained in the filters.

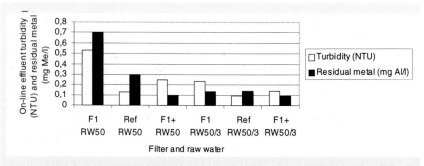

Figure 6. Performance of the three filters used (3.0 mgAl/l as AS, RW50 and RW50/3, filtration rate 7.5 m/h)

Effects of filter bed depth and raw water turbidity

In Figure 6 the performance of the three different filters used in this study is compared. RW50 and RW50/3 were coagulated with 3.0 mg Al/l as AS. The figure shows that the filterability of the aluminium-humic flocs depends on filter grain size. The coarse F1 filter is less able to produce a good water quality with AS and RW50. When there is turbidity in the raw water, residual turbidity and metal are lower for all filters. The clay particles in the raw water seem to make the flocs more filterable by increasing floc density and probably also strength. It can also be seen that the increased turbidity level in the raw water, and an increased bed depth, lowers the Al/turbidity ratio. The increased filter bed depth, significantly influences the residual metal in the effluent water. This might indicate that the precipitates formed in the filter, due to the pH increase, are removed further down in the filter bed. The deeper bed also provides more area for grain surface precipitation and adsorption.

Conclusions

The results demonstrate that NOM influence the dissolved fraction of residual metal, probably due to the formation of soluble complexes. However, with proper pH control and effective particle removal in the filter bed, low residuals were achieved. Compared to PAC, AS seemed to form less filterable flocs. There was lower effluent turbidity and metal residuals for PAC coagulation and contact filtration of humic waters. However, the soluble fraction of Al (< 0.2 μm) was higher when PAC was used as coagulant.

When ICS was used as coagulant, there was evidence of the formation of small iron precipitates with a diameter less then 0.2 μm. This was a minor problem compared to AS and PAC post-precipitation.

Increased raw water turbidity reduced the fractions of residual metal in the filtrate, probably due to enhanced precipitation and filterability.

An increased filter bed depth was able to compensate for coarser filter grains, and low levels of residual metal were achieved.

References

Eikebrokk, B. (1999) Coagulation-direct filtration of soft, low alkalinity humic waters. *Wat. Sci. Tech.* **40**(9), 55 - 62

Eikebrokk, B. (2000) Pilot scale testing of Filtralite as an alternative to anthracite in dual media filters treating coagulated humic waters. SINTEF report No. STF22 F00311, Trondheim, Norway

Eikebrokk, B., and Saltnes, T. (2001) Removal of natural organic matter (NOM) using different coagulants and lightweight expanded clay aggregate filters. *Wat. Sci. Tech.* **1**(2), 131 - 140

Jekel, M.R., and Heinzmann, B. (1989) Residual aluminium in drinking-water treatment. *J. Water SRT-Aqua*, **38**, 281 - 288

Klute, R. (1990) Destabilisation and aggregation in turbulent pipe flow. In: Chemical Water and Wastewater Treatment, Hahn. H.H, and Klute, R. (Eds.), Springer Verlag, Berlin and Heidelberg, 33 - 54

Ødegaard, H., Fettig, J., and Ratnaweera, H. (1990) Coagulation with Prepolymerized Metal Salts. In: Chemical Water and Wastewater Treatment, Hahn. H.H, and Klute, R. (Eds.) Springer Verlag, Berlin and Heidelberg, 189 - 220

Rebhun, M., Fuhrer, Z., and Adin, A. (1984) Contact flocculation-filtration of humic substances. *Wat. Res.* **18**(8), 963 - 970

Van Benschoten, J.E., and Edzwald, J.K. (1990) Chemical aspects of coagulation using aluminium salts-I. Hydrolytic reactions of alum and polyaluminum chloride. *Wat. Res.* **24**(12), 1519 - 1526

H.H. Hahn, E. Hoffmann, H. Ødegaard (Eds.)
Chemical Water and Wastewater Treatment VII, pp. 153-162
© IWA Publishing, London
ISBN: 1 84339 009 4

On the Importance of Aluminium Coagulant Basicity in Organic Matter Removal

M. Swiderska-Bróz and M. Rak*

*Institute of Environment Protection Engineering, Wroclaw University of Technology, Poland
email: broz@iios.pwr.wroc.pl

Introduction

The problem of how to increase the efficiency of coagulation for the removal of organic pollutants (which are precursors of oxidation/disinfection by-products and, also, contributors to the biological stability of the water) is still raising interest. So far, the mechanism that governs the removal of organic pollutants in the coagulation process has not been explained satisfactorily. Some investigators place great emphasis on the coagulants which destabilise colloids (Gray *et al.*, 1995) and neutralise the charge of the organic anions (Osterhus and Eikebrokk, 1994); others point to the presence of the polymeric forms of the hydrolysis products, which induce bridging coagulation (Eikebrokk and Fettig, 1990; Fettig *et al.*, 1990). Regardless of the mechanism, experience has shown that non-prehydrolysed coagulants, e.g. alum, provide the best organic matter removal in an acidic environment (pH < 6.0), which is generally obtained by acidifying the raw water or by applying increased doses of hydrolysing coagulants. Unfortunately, both of these methods have the disadvantage of increasing undesirably the corrosive tendency of the water. To minimise this adverse effect it is advisable to replace alum with prehydrolysed coagulants, e.g. polyaluminium chlorides (PACls), which contain polymerised products of the prehydrolysis of aluminium and have a high positive charge. But there is disagreement among investigators over the utility of PACls in removing organic pollutants. Eikebrokk (2000) and Edzwald *et al.*, (2000) believe that PACls are more effective than alum, but Diamadopoulos and Vlachos (1996) and Yao and O'Melia (1989) disagree. The goal of the present study was to try to clarify the role of aluminium coagulant basicity in the removal of organic substances from aquatic media.

Materials and Methods

The experiments were carried out with riverine water samples (collected from the Odra and the Olawa), which differed in organic pollution level (Table 1). The efficiencies of six aluminium coagulants were compared: that of alum (ALS) and those of five prehydrolysed PACls differing in basicity (Table 2). Coagulant doses (D_c), expressed in gAl/m^3, ranged between 2.12 and 10.58 gAl/m^3.

Volume coagulation was performed in 2 L riverine water samples, using 3 minute rapid mix (200 rpm; $G = 220$ s^{-1}) and 30-minute slow mix (30 rpm; $G = 20$ s^{-1}), followed by 1-hour sedimentation. The experiments involved a natural pH ($pH_n = 6.7 - 7.9$) or a pH adjusted to 6.0 to 8.5 (pH_a), using aqueous solutions of HCl, H_2SO_4 or NaOH. Water quality parameters were determined by Polish standard methods. TOC, DOC and colloidal organic carbon (COC) were measured using the thermal method and a 5050 TOC analyser. DOC was determined in water samples filtered through 0.45 μm pore diameter membranes. Water samples subjected to filtration on 1.2 μm pore diameter membranes were used to establish the sum of COC and DOC and thereafter to calculate COC. In the data analysis, the quotient of DOC to TOC concentrations was defined as coefficient A, whereas the quantity (g) of TOC, COC, DOC or DOC_p removed by 1g of aluminium was denoted as the coefficient of coagulant utilisation (CCU).

Table 1. Concentrations of organic pollutants in the investigated riverine water

Sampling site	COD_p	TOC	COC	DOC
	gO_2/m^3	gC/m^3	gC/m^3	gC/m^3
Odra	5.2-11.9	5.88-13.6	0.089-2.576	3.529-7.898
Olawa	4.1-6.1	6.05-8.71	0.808-1.664	4.862-5.067

Table 2. Basicity of the investigated aluminium coagulants (Zaklady Chemiczne Kemipol Ltd. 2000; Zaklady Chemiczne Zlotniki S.A. 1999)

Type of coagulant	Alum (ALS)	PACls				
		PAC	PAX-18	PAX-XL3	PAX-XL60	PAX-XL61
Basicity, %	0	35	41 ± 3.0	70 ± 5.0	40 ± 10	70 ± 10

Results and Discussion

Coagulation at pH$_n$

The results obtained evidence the significant contribution of the coagulant basicity to the removal efficiency for organic pollutants. Regardless of the type of water being treated and regardless of the value of coefficient A, all of the prehydrolysed coagulants tested were found to be more effective in removing COD$_p$ (Figure 1) and TOC (Figure 2) than was ALS.

Generally, as the basicity and dose of the coagulant increased, so did the removal efficiency (η) for COD$_p$ and the investigated organic carbon fractions. Relevant data for the Odra river water are plotted in Figure 3. Of the organic carbon fractions examined, COC was removed with the highest efficiency, according to expectations. But the most efficient coagulant was found to be the one with the highest basicity, i.e. PAX-XL61. This finding was substantiated by the comparison of the required coagulant doses (D$_{rc}$) providing removal efficiencies for the investigated organic carbon fractions comparable to that obtained with ALS (Table 3).

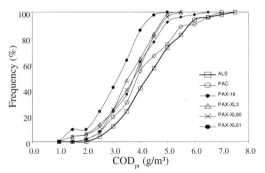

Figure 1. Cumulative frequency distribution of residual COD$_p$ values for samples treated at pH$_n$ (Number of water samples N = 305)

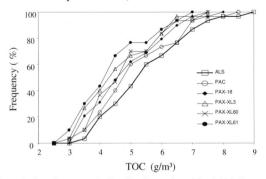

Figure 2. Cumulative frequency distribution of residual TOC concentration for the investigated aluminium coagulants (N = 179)

Table 3. Required coagulant doses (D_{rc}) providing comparable removal efficiencies for organic pollutants

Type of coagulant	D_{rc} (gAl/m³):		
	TOC	COC	DOC
ALS	5.31	5.31	5.31
PAC	$4.55 < D_{rc} < 5.31$	$4.55 < D_{rc} < 5.31$	$3.79 < D_{rc} < 4.55$
PAX-18	$3.79 < D_{rc} < 4.55$	$3.79 < D_{rc} < 4.55$	$2.27 < D_{rc} < 3.03$
PAX-XL3	$3.79 < D_{rc} < 4.55$	$3.03 < D_{rc} < 3.79$	$D_{rc} < 2.27$
PAX-XL60	$3.79 < D_{rc} < 4.55$	$3.03 < D_{rc} < 3.79$	$2.27 < D_{rc} < 3.03$
PAX-XL61	$3.03 < D_{rc} < 3.79$	$2.27 < D_{rc} < 3.03$	$D_{rc} < 2.27$

The polyaluminium products of prehydrolysis that were present in the PACl solutions removed more DOC (the most difficult fraction to remove) than did ALS (Figure 3c and Table 4).

As shown by these data, prehydrolysis was less effective in removing TOC and COC than DOC. This is of vital importance in terms of the biological stability of the water supplied to the users.

The favourable effect of increased coagulant basicity was also evident when the temperature of the water to be treated ranged from 5 to 12°C. The prehydrolysed coagulants yielded a higher removal of COD_p (Figure 4a) and TOC (Figure 4b) than did the ALS coagulant. These findings indicate that in the cold seasons, when aluminium hydrolysis efficiency is low, it is advisable to replace ALS with polyaluminium chlorides, especially with those having basicities equal to, or higher than, 70%.

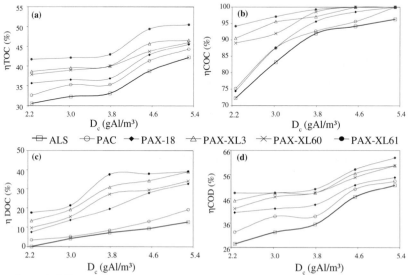

Figure 3. Effect of coagulant type and dose on the removal efficiency for (a) TOC, (b) COC, (c) DOC, and (d) CODp

Table 4. Coefficients of coagulant utilisation (CCU) in removing organic carbon fractions (D_c = 2.27 to 5.31 gAl/m³). Dc = coagulant dose added to water (gAl/m³)

Type of coagulant	ALS	PAC	PAX-18	PAX-XL3	PAX-XL60	PAX-XL61
CCU-DOC, gC/gAl	0-0.116	0.071-0.174	0.159-0.298	0.286-0.360	0.207-0.311	0.356-0.375
CCU-COC, gC/gAl	0.025-0.044	0.026-0.045	0.026-0.045	0.026-0.055	0.026-0.053	0.026-0.059
CCU-TOC, gC/gAl	0.59-1.01	0.629-1.07	0.636-1.17	0.650-1.263	0.643-1.239	0.706-1.366

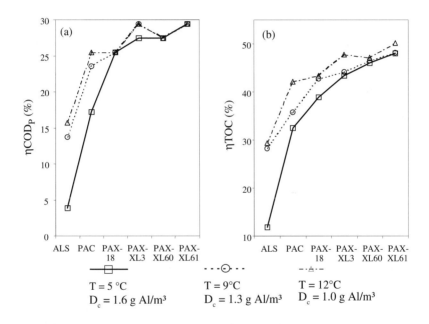

Figure 4. Effect of temperature, coagulant type and coagulant dose on the efficiency of (a) CODp and (b) TOC removal

The efficiency of TOC removal was affected not only by coagulant basicity, but also by the percentage of DOC in the total organic matter in the water to be treated. When the value of coefficient A increased, the efficiency of TOC removal decreased, although the doses of the investigated coagulants remained unchanged (Figure 5).

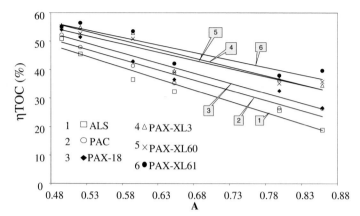

Figure 5. Effect of percentage of DOC in total organic matter on the efficiency of TOC removal (trend lines)

Contribution of pH adjustment

Acidic pH improved the ability of all the coagulants to remove organic pollutants. pH adjustment had the greatest impact on ALS efficiency and decreased with the degree of prehydrolysis in the case of polyaluminium chlorides (Figure 6). These

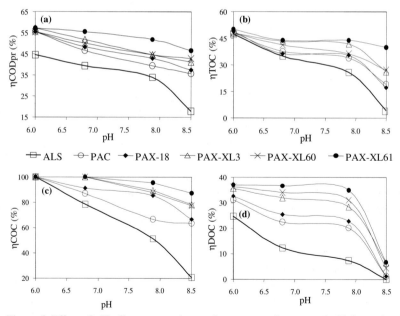

Figure 6. Effect of pH adjustment and coagulant type on the removal efficiency for (a) COD_p, (b) TOC, (c) COC, and (d) DOC in the Odra river water samples ($D_c = 3.1$ gAl/m^3)

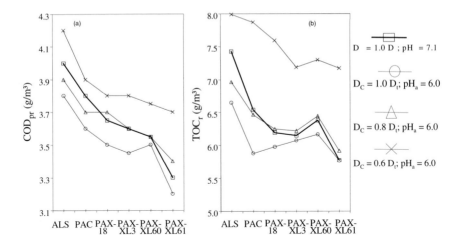

Figure 7. Effect of coagulant type, coagulant dose and pH on residual concentration of (a) COD_p and (b) TOC (Olawa river water); $D_t = 4.735$ gAl/m³; theoretical coagulant dose which was calculated from formula: $D_t = 6 \cdot B^{-0.5}$ (gAl/m³), where B = colour of raw water (gPt/m³)

results indicate that the polymeric forms of polyaluminium chlorides with high positive charges remain stable in a wider pH range than do the products of ALS hydrolysis. The highest removal efficiencies were achieved at $pH_a = 6.0$ to 6.8 and $pH_a = 6.0$ to 6.5 for the Odra river water (Figure 6) and Olawa river water (Figure 7), respectively, regardless of the coagulant type used. For the majority of the samples, the substitution of ALS with polyaluminium chlorides eliminated the need for pH adjustment to the acidic range prior to the coagulation process. The advantages of using PACls thus include reducing the corrosive tendency of the water under treatment and simplifying the treatment train itself.

The adjustment of pH to alkaline levels before coagulation reduced the efficiency of organic matter removal. This is possibly due to the following mechanism: The rise in pH is concomitant with the increase in the dissociation of organic pollutants on one hand; on the other hand, there is a decrease in the quantity of positively charged polymeric products of aluminium hydrolysis, which form in the course of coagulation. Since the amount of hydrolysing aluminium in the water under treatment (in the course of the coagulation process) decreased with the increasing basicity of the coagulants, the unfavourable effect of water alkalinity was the most distinct in the presence of non-prehydrolysed ALS, and pertained primarily to the efficiency of DOC removal. This problem can be visualised by comparing the quantity of DOC removed with 1 gram of Al at $pH_a = 8.5$ with that removed at $pH_a = pH_n$ (Table 5). However, irrespective of the pH of the water to be treated, when the basicity of the coagulants increased, so did the efficiency of DOC removal.

Table 5. Effect of water alkalinity on the utilisation of the coagulants in removing DOC ($D_c = 3.1$ gAl/m^3)

Type of coagulant	ALS	PAC	PAX-18	PAX-XL3	PAX-XL60	PAX-XL61
CCU-DOC gC/gAl $pH_a = pH_n$	0.107	0.293	0.301	0.411	0.450	0.505
CCU-DOC gC/gAl $pH_a = 8.5$	0	0.0184	0.019	0.069	0.063	0.096

Conclusions

The results may be summarised as follows:
- The removal efficiency for organic pollutants increased with the basicity and dose of the aluminium coagulants tested.
- Besides the non-dissolved fraction of organic carbon, a certain portion of the dissolved carbon fraction was removed in the course of the coagulation process. When the proportion of dissolved organic carbon in the total organic pollution load increased, the efficiency of total organic carbon removal decreased.
- Prehydrolysed coagulants were of greater utility than alum in removing not only non-dissolved, but also dissolved fractions of organic pollutants, and they were efficient over a wider range of pH and temperature.
- Adjustment of pH to 6.0 to 6.8 increased the efficiency of total organic carbon and dissolved organic carbon removal, but the effect of this adjustment on removal efficiency was more distinct in the presence of alum than in the presence of polyaluminium chlorides.
- Adjustment of pH to 8.5 had a negative effect on the removal of organic substances in the course of coagulation; at $pH_a = 8.5$, mainly non-dissolved fractions were removed, regardless of the coagulant type applied.

References

Diamadopoulos, E., Vlachos, C. (1996) Coagulation-filtration of a secondary effluent by means of pre-hydrolized coagulants. *Wat. Sci. Tech.* **33**, 193-208

Edzwald, D.K., Pernitsky, D.J., Parmenter, W.L. (2000) Polyaluminum coagulants for drinking water treatment: Chemistry and Selection, In Hahn,H.H. Hoffmann, E., Ødegaard, H. (Eds.) Chemical Water and Wastewater Treatment VI, Berlin, Springer, 3-14

Eikebrokk, B. (2000) Effects of coagulant type and coagulation conditions on NOM removal from drinking water, In Hahn,H.H. Hoffmann, E., Ødegaard, H. (Eds.) Chemical Water and Wastewater Treatment VI, Berlin, Springer, 211-220

Eikebrokk, B., Fettig, J. (1990) Treatment of Coloured Surface Water by Coagulation/Direct Filtration: Effect of Water Quality, Type of Coagulant & Filter Aids, In Hahn, H.H. and Klute, R. (Eds.) Chemical Water and Wastewater Treatment, Berlin, Springer, 361-376

Fettig, J., Ratnaweera, H.C., Ødegaard, H. (1990) Simultaneous Phosphate Precipitation and Particle Destabilization Using Aluminium Coagulants of Different Basicity, I In Hahn, H.H. and Klute, R. (Eds.) Chemical Water and Wastewater Treatment, Berlin, Springer, 221-242

Gray, K.A., Yao, C., O'Melia, Ch.R. (1995) Inorganic Metal Polymers: Preparation and Characterization, *JAWWA* **4**, 136-143

Østerhus S.W., Eikebrokk B. (1994) Coagulation and Corrosion Control for Soft and Coloured Drinking Water, In Hahn, H.H., and Klute, R. (Eds.) Chemical Water and Wastewater Treatment III, Berlin, Springer, 137-153

Yao, C., O'Melia, C.R. (1989) Required coagulant dosage, *JWSRT-aqua* **38**, 339-344

Zaklady Chemiczne Kemipol Ltd.(2000) PAX. Manufacturer's specification

Zaklady Chemiczne Zlotniki S.A.(1999) PAC and alum. Manufacturer's specification

Cryptosporidium Removal

Chapter Eleven. Research

H.H. Hahn, E. Hoffmann, H. Ødegaard (Eds.)
Chemical Water and Wastewater Treatment VII, pp. 165-182
© IWA Publishing, London
ISBN: 1 84339 009 4

Robust Drinking Water Treatment for Microbial Pathogens – Implications for *Cryptosporidium*

P. M. Huck* and B. M. Coffey

*University of Waterloo, Dept. of Civil Engineering, Waterloo, Canada
email: pm2huck@uwaterloo.ca

Abstract

It is important that drinking water systems be as robust as possible. That is, they should be capable of delivering excellent quality water under adverse conditions. Robustness is important for each of the five elements necessary for providing safe drinking water (a good source, adequate treatment, secure distribution, appropriate monitoring and appropriate response to adverse monitoring results). However, a given degree of overall system robustness can be achieved in varying ways.

The quantification of robustness facilitates its improvement in a rational way. The experimental investigations reported in this paper evaluated the application of a newly-developed robustness index for assessing granular media filtration. The focus of the assessment was on performance in relation to *Cryptosporidium* removal. The simplified index used showed promise, and, in the broader sense, provided an example of how robustness of one element of a drinking water supply system could be quantified.

Introduction

Water supply providers seek to provide their customers with high-quality drinking water at all times. However this can sometimes be challenging because of changing raw water quality or problems with treatment and distribution. It is therefore important that water supply systems be as robust as possible. A robust system is defined as one that provides excellent performance under normal conditions and deviates minimally from this during periods of upset or challenge.

[165]

Within the context of an appropriate regulatory framework, it is possible to identify five elements to providing safe drinking water (Huck, 2000):

- Start with the best possible source
- Design and operate appropriate treatment
- Provide secure distribution
- Conduct appropriate monitoring, and
- Respond appropriately to any adverse monitoring results.

While all of these elements are essential, a given final quality or degree of robustness can be achieved in different ways. For example a very secure source will require less intensive and robust treatment.

In a general sense the water industry has always been aware of the need for reliable treatment systems. However the advent of newer technologies for both treatment and monitoring as well as heightened public and professional awareness and expectations have created the need for addressing this matter in a more formal way.

This paper addresses the concept of robustness in general and reports on investigations to quantify the robustness of granular media filters with regard to *Cryptosporidium* removal.

Robustness of drinking water systems

General Considerations

In the short-term, improvements in drinking water system robustness must focus on treatment and monitoring. This is because most water treatment plants cannot easily change their raw water source, and steps to increase its security or quality (such as watershed protection or wellhead protection programs) are long-term measures that will not show immediate results. Although improvements in water quality at the consumer's tap can be achieved by distribution system investments, these cannot compensate for inadequate treatment.

The degree of robustness required in treatment is a function of both average raw water quality and the extent and rapidity with which quality varies. Key raw water quality parameters in this regard are microorganism concentrations (especially protozoan pathogens such as *Giardia* and *Cryptosporidium*), turbidity, total organic carbon (TOC), pH, temperature, and susceptibility to organic or inorganic contamination from upstream industry. For example a partial coagulation failure in a plant treating water from a pristine and protected watershed is probably of much less significance than in a plant on a heavily used river. It would be useful to develop metrics for raw water quality and its variation that would help define the appropriate degree of treatment robustness early in the design or plant upgrading process. Such robustness metrics would address individual process steps as well as the level of redundancy required in monitoring and chemical feed systems. The metrics would also help to define the level of operator skill required to ensure a robust system.

The concept of robustness is analogous to a vehicle suspension, which is designed to provide a smooth and safe ride by absorbing the impact from variable road conditions. Figure 1 (adapted from Coffey *et al.*, 1998) graphically depicts the concept of robustness. In this figure, the horizontal axis represents influent water quality or some other operating condition such as coagulant dose. The vertical axis represents a measure of treatment performance, in this case a filtration process. As shown, the performance of a robust system does not vary widely with changing operating conditions. Two examples of robust systems are shown: one (R1) in which the optimal performance is equal to that of the non-robust system, and the second (R2) in which the performance of the optimal filtration system is greater than that of the robust system, but limited to a narrow band of operating conditions. Depending on both the relative performance differences between the optimal and robust systems and the breadth of the operating conditions, the second robust system (R2) may be preferred to the optimal system. The ideal would of course be the first robust system (R1), where the optimal performance is maintained over a wider range. As part of an overall process robustness evaluation, a water utility would have to assess the relative cost-effectiveness (in comparison to increases in robustness elsewhere in the process) of going from system R2 to R1.

Robustness for *Cryptosporidium* Removal

Robust treatment processes are particularly important for *Cryptosporidium* removal. The possibility of *Cryptosporidium* appearing in any surface water supply can never be entirely excluded. Often, peak *Cryptosporidium* levels may be associated with challenging raw water quality conditions (e.g., Atherholt *et al.*, 1998), such as elevated turbidity and organic levels associated with heavy precipitation. Since it is impossible to monitor for *Cryptosporidium* in real-time, treatment processes such as filtration must use appropriate real-time surrogate parameters (Huck *et al.*, 2001a) to ensure adequate removal at all times.

Although inactivation using UV radiation has been rapidly showing promise in the last few years as a viable treatment technology for *Cryptosporidium* (e.g., Clancy *et al.*, 2000), physical-chemical removal by solids separation processes culminating in granular media filtration remains a crucial barrier. When the main disinfection step is located downstream of filtration, as is often the case, an important role of filtration is to provide a relatively constant input of microorganisms to the disinfection process. Maintaining filtration performance during peak raw water *Cryptosporidium* levels becomes critically important. Filtration also must protect disinfection from elevated turbidity, which may impair disinfection or inactivation. If filtration is downstream of the main disinfection step it must be able to handle anything "missed" by disinfection during a peak event. Despite the increasing popularity of membrane filtration, granular media filtration remains the filtration process in all but a small fraction of surface water treatment plants in the world.

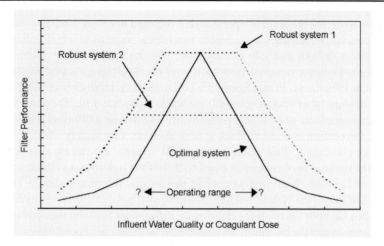

Figure 1. Representation of the filtration system robustness concept

By nature, granular media filtration is a process more vulnerable to upsets than technologies such as membrane filtration. Removals in granular media filtration, where virtually all of the particles being removed are smaller than the spaces between the media grains, are dependent on a complex combination of chemical and physical factors. The particles are "stored" in the filter during a run but they remain in contact with the flow, and the delicate force balance that keeps them on the media surface can be easily disrupted. Similarly, inadequate chemical pre-treatment can greatly reduce the efficiency of particle capture. The ripening and breakthrough periods at the beginning and end, respectively, of every filter cycle are known to be less efficient for particle removal.

Various design and operational approaches have been used to improve filter performance for particle removal. Although these strategies have not necessarily been viewed from the perspective of robustness, one of the results of implementing them has often been to improve robustness. Such strategies include the use of tri-media instead of dual-media filters, the use of a filter aid, the practice of filter-to-waste and the termination of a run prior to significant particle breakthrough.

Robustness of Filtration for *Cryptosporidium* Removal

Approach

Huck *et al.* (2001b) have recently reported on a major investigation of *Cryptosporidium* removal by granular media filtration. A significant emphasis of this study was to examine conditions under which filtration was not optimal. This included the ripening and breakthrough periods, as well as operational events such as suboptimal coagulation, a sudden increase in flow (hydraulic step) and a sudden change in influent water quality.

One of the objectives of this project was to develop and apply a filtration robustness concept for particle removal. The work focused on the development of a practical measure of a filter's ability to consistently maintain superior levels of particle removal. A number of filter performance indices, such as unit filter run volume (Trussell *et al.*, 1980) and filter performance index (Montgomery Consulting Engineers, 1985), have been used to compare filter performance during stable filter operating conditions. However, none of the indices currently available directly addresses the issue of performance robustness.

Various ways of quantifying robustness were explored. After some initial investigation, it was decided that approaches such as neural networks and advanced statistical techniques such as Monte-Carlo analysis were too sophisticated for easy implementation at most water treatment plants. Thus, a simpler approach was desired. The robustness index ultimately developed could also be applied to other treatment processes. It or similar quantitative measures of the robustness of a treatment step could be incorporated into an overall robustness factor for a complete treatment plant.

Methods

The experimental investigations in this study were carried out using a pilot plant located at the Metropolitan Water District (MWD) of Southern California's Weymouth treatment facility in LaVerne, California. The pilot plant received water that was low in turbidity and particles (average in the range of about 5000 particles/ mL (>2 μm)), and was operated to mimic as closely as possible the full-scale treatment plant at the same location. The pilot plant used a low coagulant dose (5 mg/L alum and 1.5 mg/L cationic polymer) for particulate removal only. Chlorine (~2 mg/L) was added at rapid mix as a pre-oxidant. The optimized coagulation conditions were selected to meet the 0.1 NTU turbidity goal of the Partnership for Safe Water, a voluntary treatment optimization program sponsored by the U.S. Environmental Protection Agency and the American Water Works Association.

Following rapid mix, the pilot plant had three-stage tapered flocculation. The overall flocculation hydraulic detention time was 20 minutes. The hydraulic detention time of the sedimentation step was 80 minutes. The filters contained 508 mm (20 in.) of anthracite over 203 mm (8 in.) of sand. The backwashing regime consisted of chlorinated water with surface wash. Further operating details for the pilot plant are provided elsewhere (Huck *et al.*, 2001b). Information concerning the demonstration-scale plant, from which some data were obtained, is provided by Stanley *et al.* (1998).

Initial Investigations

Three preliminary methods for evaluating the robustness of particle passage were applied to data collected at MWD's pilot plant as part of the larger (Huck *et al.*, 2001b) study. For the tests described here, both trains of the pilot plant were operated with dual-media filters, using chemical dosages of 5.0 mg/L alum, and 1.5 mg/L cationic polymer (not necessarily an optimized chemical dosage). One train of the pilot plant used ozone as a pre-oxidant and the other train did not use

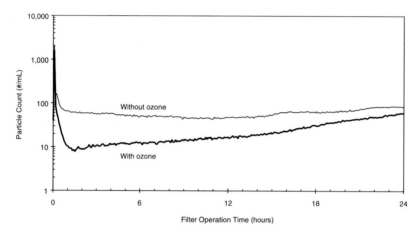

Figure 2. Time-series data from the MWD pilot plant comparing use of ozone as a pre-oxidant

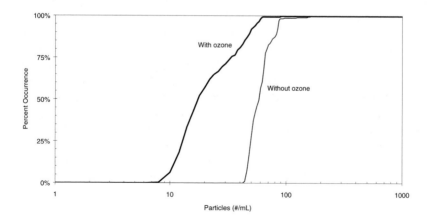

Figure 3. Percent occurrence for particle data from the MWD pilot plant

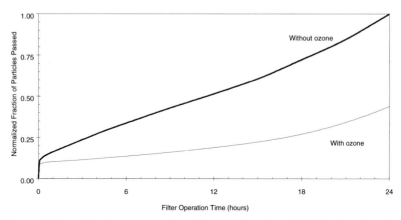

Figure 4. Normalized percentage of particles passing through MWD pilot plant filters (data normalized such that 100 percent equals the total particles from the filter without ozone)

any pre-oxidant. For these data, graphs depicting a time-series analysis, percent occurrence, and normalized particle passage all provided valuable insight into the treatment effectiveness.

Figure 2 shows time-series data from the pilot plant. Without ozone, more particles passed through the filter than for the train with ozone. However, with ozone, an increasing number of particles passed through the dual-media filter beginning after approximately one hour of operation. Though Figure 2 is certainly useful—at least qualitatively—in evaluating the relative filter performance, additional data analysis can provide both greater insight and quantitative means of expressing the differences between the two filters. Figure 3 (a percent occurrence graph) shows that the 90th-percentile particle occurrence—which may be a better predictor of the public health risk than the average particle concentration—was ~50 particles/mL with preoxidation and ~80 particles/mL without preoxidation.

A third method for evaluating particle passage was to use a normalized particle passage graph (Figure 4). This figure shows that the treatment with ozone passed less than half as many total particles as the treatment without ozone. The relative number of particles passed may provide an indication of a problem filter or treatment (assuming that the particle counters were carefully calibrated and maintained). Figure 4 also indicates some potential for treatment improvements. For example, using a filter-to-waste system for 1 hour would reduce the total number of particles passed by ~25 percent for the treatment with ozone, but only ~15 percent for the treatment without ozone. Thus, this type of analysis may help utilities determine where limited financial resources should be allocated for maximum treatment improvement.

Development of a Filtration Robustness Index

Figures 2 to 4 indicate that some combined measure of particle passage that incorporates both average quality and deviation from stable operation may be useful. One basic formula with several variations is evaluated herein for use as a robustness index. Using turbidity as the measurement, the robustness index was written as follows:

$$TRI_{95} = \frac{1}{2}\left[\frac{T_{95}}{T_{50}} + \frac{T_{50}}{T_{goal}}\right]$$ (Eq. 1)

where:
TRI_{95} = turbidity robustness index using the 95th percentile, dimensionless
T_{50}, T_{95} = 50th and 95th percentile turbidity, NTU
T_{goal} = filter turbidity goal, NTU

The "goal" term, T_{goal}, represents a utility- or plant-specific performance goal for turbidity. In addition to the turbidity robustness index, a similar equation may be used to characterize the particle performance of the filter (PRI). In this case, particle data are used instead of turbidity data and the particle robustness indices are labeled accordingly. For either turbidity or particles, other percentages such as the 90th or 99th percentile could also be used (Huck *et al.*, 2001b).

The physical significance of the robustness index may be seen by examining the two terms of the index. The first term, the ratio of the 95th percentile to the 50th percentile (median), represents the uniformity of the turbidity or particle performance over a specified duration (e.g., a single filter run or a 24-hour period). The second term—the ratio of the median turbidity/particle count to the goal—represents how well the filter is performing overall.

Using Equation 1, a lower value of the robustness index indicates a treatment process that is meeting the water quality goal with relatively low variation. For example, if the particle counts always met a treatment goal of 25 particles/mL and did not deviate from that goal, the value of the robustness index would be equal to 1.0. Thus, a higher robustness index value indicates that the treatment is either not meeting its goal (on average) or the variability is high. Note that depending on the selected treatment goal, the robustness index could be less than unity. The minimum possible value for the robustness index is 0.5, when the second term approaches zero because the median turbidity was much lower than the goal, i.e. when the turbidity value used for the goal was far too high.

A simplified robustness index consisting of only the ratio of a chosen percentile value (e.g. T_{90}) to the goal was also presented by Huck *et al.* (2001b).

Filtration Robustness Index Examples

Preoxidation

The pilot-plant data shown in Figure 2 were processed using an equation for the 95th percentile particle robustness index (PRI_{95}). The particle goal was 25 particles/mL (>2 μm). This particle goal was chosen because (1) it was measurable with reasonable precision and accuracy; (2) it was attainable during normal operation; and (3) it was occasionally exceeded. For both filters - with and without preoxidation - the PRI_{95} was 1.9 even though the filter with preoxidation always produced fewer particles than the filter without preoxidation. The PRI_{95} was equivalent in this case because the particle performance of the filter that received preoxidized water was less uniform. Thus, these two filters achieved equivalent robustness, as defined by the equations above.

Suboptimal coagulation

A second example was used to evaluate robustness using pilot-plant tests with optimal and suboptimal coagulation. For each experiment, the pilot-plant filters usually operated for 24 hours. Thus, for a 24-hour filter run, 90th, 95th, and 99th percentiles would capture the highest turbidity or particle data for 144, 72, and 14 minutes, respectively. Note that the data during these periods do not need to be consecutive. For the pilot-plant analysis, the turbidity and particle goals were 0.1 NTU and 25 particles/mL (≥ 2 μm), respectively.

The 90th or 95th percentile robustness equation may be more applicable when comparing major differences in treatment performance or possibly when evaluating

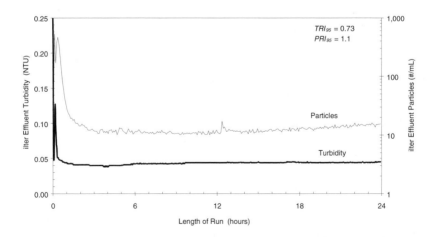

Figure 5. Time-series data for turbidity and particles from stable filter operation test at MWD pilot plant

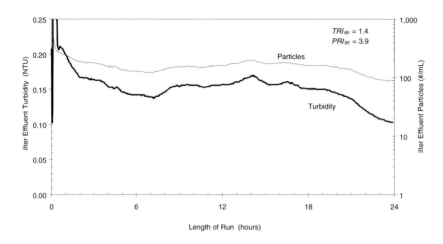

Figure 6. Time-series data for turbidity and particles from suboptimal coagulation test at MWD pilot plant

combined filter effluent samples. The 99th percentile robustness equation may be more desirable when the objective is to evaluate the effects of ripening or hydraulic surges because the duration of these events is short enough that their effect may be "diluted" when a period of 72 min/day is used. The disadvantage of using the 99th percentile is that it is heavily influenced by noise in the particle or turbidity data.

An example of a robust filter run is given in Figure 5 as a time series of turbidity and particles. This preoxidized (with ozone), optimal coagulation, dual-media filter achieved a TRI_{95} of 0.73 and a PRI_{95} of 1.1 which indicates that the filter was producing stable levels of turbidity and particles which consistently met the prescribed goals. Figure 6 shows an example of a much less robust filter run (using the same filter, pre-oxidant, and filter aid, but at decreased alum and cationic polymer doses). The TRI_{95} for this run was 1.4 and the PRI_{95} was 3.9. For this example, the turbidity and particle values were of poorer quality and less stable.

Application to Pilot Plant Results

The 95th percentile index values (TRI_{95} and PRI_{95}) were used to compare the robustness of various treatments investigated at MWD's pilot plant. Stable filter operation (baseline) and hydraulic step experiments using filters which received preoxidized water generally produced the lowest (best) robustness index values. The 25 percent filtration rate increase applied during hydraulic step experiments negligibly increased turbidity and particles. The highest (worst) robustness index values were produced by suboptimal coagulation and runoff (turbidity/TOC spike) experiments on all filters.

Although *Cryptosporidium* removals were not assessed during the robustness experiments (oocysts were not seeded in these tests), *Cryptosporidium* removals were evaluated in other experiments conducted at MWD as part of the overall study (Huck *et al.*, 2001b) where conditions were directly comparable to those tested in the robustness investigations. Suboptimal coagulation resulted in a major deterioration in *Cryptosporidium* removal capability, compared to stable or baseline operation. This was qualitatively correlated with a deterioration in particle removal (based on raw water and filter effluent particle counts). Although hydraulic step experiments with seeded oocysts were also conducted at MWD, their interpretation was complicated (Huck *et al.*, 2001b).

Tables 1 through 4 show the mean value and the statistical significance of the robustness indices for both turbidity and particles, for several comparisons of paired data. Statistical significance was determined using the paired Student's *t*-test. A significance level of five percent ($\alpha = 0.05$) and a hypothesized mean difference of zero were used to compare the data.

Figure 7 shows a box-and-whisker plot for the robustness indices for all of the experimental challenges for which robustness was evaluated at MWD. In this plot the small square represents the median value, the upper and lower edges of the 'box' represent the 75th and 25th percentile values respectively, and the upper and lower ends of the line represent, respectively, the maximum and minimum values observed.

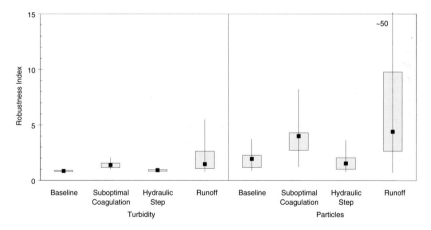

Figure 7. Box-and-whisker plot of turbidity and particle robustness indices comparing the different types of experiments at MWD pilot plant. Plot indicates minimum, 25th percentile, median, 75th percentile, and maximum values

Table 1. Statistical significance of robustness index for different types of experiments at MWD pilot plant

		Turbidity performance (TRI_{95})				Particle performance (PRI_{95})			
		Baseline	Suboptimal coagulation	Hydraulic step	Runoff	Baseline	Suboptimal coagulation	Hydraulic step	Runoff
All filters (with and without preoxidation)	Mean	0.85	1.4	0.91	2.2	1.8	3.7	1.7	8.5
	Different than baseline?	-	Yes	Yes	Yes	-	Yes	No	Yes
Dual- and tri-media filters with preoxidation	Mean	0.79	1.4	0.85	1.6	1.3	2.5	1.1	5.1
	Different than baseline?	-	Yes	Yes	Yes	-	Yes	No	Yes
Dual- and tri-media filters without preoxidation	Mean	0.91	1.4	0.96	2.8	2.3	5.0	2.3	12
	Different than baseline?	-	Yes	No	Yes	-	Yes	No	Yes

Table 1 compares the robustness index for the different types of experiments: baseline (stable filter operation), suboptimal coagulation, hydraulic step, and runoff (turbidity/TOC spike). Three groupings of data are presented for each experimental challenge: (1) all dual- and tri-media filters (with and without preoxidation); (2) only filters that received preoxidized water; and, (3) only filters that received water without preoxidation. The robustness indices for the suboptimal coagulation and runoff (turbidity/TOC spike) experiments were significantly greater than for the baseline experiments for all groupings. For the hydraulic step experiments, the robustness index was either not statistically different from that for the baseline tests, or the magnitude of the difference was quite small. Figure 7 shows these data graphically. Thus, decreasing coagulation effectiveness by reducing both the alum and cationic polymer dosages or by increasing turbidity and organic matter (and not changing the coagulant dosages) impaired treatment performance. Though preoxidation could reduce the effect of the treatment challenge (e.g., a PRI_{95} of 2.5 with preoxidation compared to a PRI_{95} of 5.0 without preoxidation), the treatment still deteriorated substantially from baseline operations.

Table 2 quantifies the effect of preoxidation. In this table, the robustness index for the train with preoxidation versus the train without preoxidation are grouped for all of the treatment challenges. For both turbidity and particle robustness indices, adding a pre-oxidant significantly improved treatment (reduced the particle and turbidity robustness index).

Table 2. Statistical significance of robustness index for effects of preoxidation at MWD pilot plant

	Turbidity (TRI_{95})		Particles (PRI_{95})	
	With preoxidation	Without preoxidation	With preoxidation	Without preoxidation
Mean	1.2	1.5	2.5	5.3
Statistical difference?	Yes		Yes	

Table 3. Statistical significance of robustness index for effects of filter aid at MWD pilot plant

		Turbidity (TRI_{95})		Particles (PRI_{95})	
		With filter aid	Without filter aid	With filter aid	Without filter aid
Dual- and tri-media (with and without preoxidation)	Mean	1.3	1.3	3.3	4.5
	Statistical difference?	No		No	
Dual- and tri-media filters (with preoxidation)	Mean	1.1	1.2	1.9	3.1
	Statistical difference?	No		No	
Dual- and tri-media filters (without preoxidation)	Mean	1.6	1.4	4.7	6.0
	Statistical difference?	Yes		Yes	

Table 3 compares results for the experiments with and without filter aid. The three groups of data were (1) all filters; (2) filters receiving preoxidized water; and, (3) filters receiving non-preoxidized water. For the first two groups, there was no effect of filter aid on turbidity and particle robustness indices. However, without preoxidation, there was a statistically significant (though small) treatment impairment as measured by turbidity when filter aid was used. However, the particle robustness index indicated that filter aid substantially improved treatment in the absence of a pre-oxidant.

A comparison of robustness based on media type (dual- versus tri-media) is shown in Table 4. Based on turbidity, the tri-media filters improved treatment. However, based on particle robustness, the improvement was not statistically significant. Though the mean difference between the dual- and tri-media filters were greater for the particle robustness than for the turbidity robustness, the variability of the particle data prevented the difference from being statistically significant.

Table 4. Statistical significance of robustness index comparing dual- and tri-media filters at MWD pilot plant

		Turbidity (TRI_{95})		Particles (PRI_{95})	
		Dual-media	Tri-media	Dual-media	Tri-media
All filters (with and without preoxidation)	Mean	1.4	1.3	4.4	3.5
	Statistical difference?	Yes		No	
All filters (with preoxidation)	Mean	1.3	1.1	2.8	2.2
	Statistical difference?	Yes		No	
All filters (without preoxidation)	Mean	1.6	1.5	5.9	4.7
	Statistical difference?	Yes		No	

Application to Demonstration Plant Results

The pilot-scale testing just described indicated that treatment robustness was heavily influenced by coagulation conditions. Demonstration-plant data—collected as part of a separate AWWA Research Foundation project (Stanley *et al.*, 1998)— also supported this result. In the pilot-scale tests, suboptimal coagulation was achieved by reducing both the alum and cationic polymer dosages by equal amounts (typically by ~65 percent) to achieve a turbidity goal of 0.2 - 0.3 NTU. For the demonstration-plant testing, combinations of six alum dosages (0, 1, 3, 5, 7, and 10 mg/L) and four cationic polymer dosages (0, 1, 2.5, and 5 mg/L) were applied to one dual- and one tri-media filter at a filtration rate of 8.6 m/h (3.5 gpm/ft^2). Limited tests also evaluated the use of chlorine and ozone as pre-oxidants.

Figures 8 and 9 show the 90th percentile turbidity and particle robustness indices, respectively, of a 37.2 m^2 (400 ft^2), dual-media filter operated at 8.6 m/h (3.5 gpm/ft^2) with chlorine used as a pre-oxidant. (For the demonstration-plant tests, the 90th and 95th percentile robustness indices were similar. The 90th percentile indices were chosen for the comparison with full-scale data discussed below.) The results in Figures 8 and 9 show that the use of multiple coagulants—in this case, alum and cationic polymer—improved treatment robustness. For example, for all of the tests that used 3 to 10 mg/L of alum and 1 to 5 mg/L of cationic polymer, the turbidity robustness index (TRI_{90}) varied from 0.7 to 1.1. When operating below these coagulant conditions (as was the case for the pilot plant suboptimal experiments), the robustness indices increased substantially.

Additional results presented in Huck *et al.* (2001b) show the effects of preoxidation on the turbidity and particle robustness indices for both dual- and tri-media filters. Each pre-oxidant was tested at three different alum doses (permanganate was not tested at the demonstration plant). In all of these tests, the cationic polymer dose

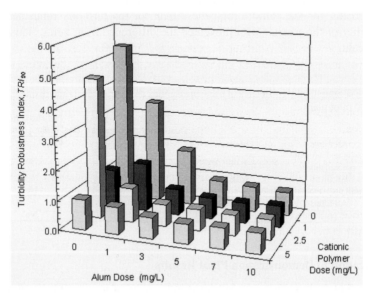

Figure 8. The effect of coagulation on turbidity robustness index at MWD's demonstration-scale treatment plant (dual-media filters operated at 3.5 gpm/ft^2; chlorine added as a pre-oxidant)

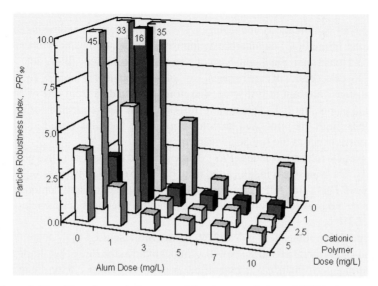

Figure 9. The effect of coagulation on particle robustness index at MWD's demonstration-scale treatment plant (dual-media filters operated at 3.5 gpm/ft^2; chlorine added as a preoxidant)

was fixed at 1.0 mg/L. The turbidity treatment robustness index in these tests was influenced by (in order of decreasing magnitude) alum dosage, pre-oxidant presence, and media configuration (dual-media vs. tri-media). The greatest improvement in the turbidity robustness index came from increasing the alum dose from 1 to 5 mg/L, with little further improvement at a dose of 10 mg/L. Except at the lowest alum dose, use of a tri-media filter brought essentially no measurable improvement in robustness. For the particle robustness index, the only factor having a measurable deleterious effect was the absence of a preoxidant.

It should be noted that the suboptimal coagulation conditions for the pilot plant in the experiments discussed above (which treated a water quality similar to that at the demonstration plant) were 1.8 mg/L alum and 0.53 mg/L of cationic polymer. Clearly, the pilot plant was being operated in a mode that was highly sensitive to the coagulant dosage.

Application to Full-Scale Results

The filtration robustness index was initially developed to compare the ability of various design or operating conditions during the pilot phase to achieve excellent particle removal with minimal variation. Necessarily, because of the large experimental matrix, only two coagulant dosages were tested at the pilot-scale (optimal and suboptimal). Data collected during the pilot testing lasted 24 hours (the length of one filter run). Testing at the demonstration plant allowed a more in-depth evaluation of coagulant combinations and dosages. Demonstration-plant testing established that the use of multiple coagulants may improve treatment robustness. Data shown for the demonstration plant consisted of two complete 24-hour filter runs.

At full-scale, control of experimental variables was obviously limited. To evaluate the robustness concept at full-scale, therefore, on-line data from one of MWD's full-scale plants, the F. E. Weymouth filtration plant, was collected for calendar year 1999. Both the pilot- and demonstration-scale facilities were fed with the same source water as the full-scale plant.

The particle robustness index (PRI_{90}) was calculated for two filters that performed similarly. The robustness index varied from 0.6 to ~6 (median value of 0.9) for the entire year. This value fell within the same range as the pilot- and demonstration-scale tests and proved useful in identifying potential treatment deficiencies (Huck *et al.*, 2001b).

The modified robustness index ($PRI_{90}/goal$) was also used to evaluate the full-scale data. This modified index showed a similar trend as the PRI_{90}, but the departure from normal performance was magnified. A possible goal of 2.0 for the modified robustness index was proposed (Huck *et al.*, 2001b). The physical significance of the index goal of 2.0 is that the 90th percentile particle concentration was double the goal. This simplified index may be more useful for plant operators who desire to modify plant performance to achieve the goal.

Summary

The robustness indices presented here represent a first attempt to develop an index that accounts both for average treatment performance and variability in filters. It was desired to develop an index, which could maintain physical significance, could be easily understood, and could be implemented at a water treatment plant with minimal sophistication.

The index was successfully applied to pilot-, demonstration-, and full-scale treatment. Within the parameters tested during these experiments, the following observations were made using the robustness index and standard statistical techniques:

- Chemical coagulation conditions controlled treatment performance, followed by the presence of a pre-oxidant, filter media configurations, and the presence of filter aid.

- The presence of dual coagulants improved the robustness of particle removal compared to single coagulants.

Because the robustness approach described herein deliberately valued simplicity, it suffered from some limitations. Some of these limitations are:

- Values of the robustness index were heavily influenced by the treatment goal term. Selecting a treatment goal that is too high minimizes the variability term in the index.

- The robustness index can be less than one. This effect results once again from setting a treatment goal substantially above normal operations.

- The relative weighting of the two components (average performance and variability) is equal.

The simplified robustness index which incorporated the ratio of the 90th percentile (or some other chosen percentile) into the goal represents an alternative approach that may be more easily implemented at full-scale.

Concluding remarks

The robustness of drinking water systems is very important. It is becoming increasingly so as performance requirements tighten and as the sometimes conflicting requirements of several standards (e.g. disinfection vs. disinfection by-products) increasingly constrain treatment operating ranges.

Quantification of robustness is a prerequisite to improving it. Improving robustness can contribute to reducing risk in a cost-effective and logical way. Improving robustness therefore provides a goal for water utilities and a quantifiable basis for continuous improvement.

Ultimately, quantification of robustness for each of the five components of providing safe drinking water (source, treatment, distribution, monitoring and response) will be required. For treatment plants, robustness will be a function of the robustness of individual treatment processes. Different ways of quantifying robustness may be appropriate for different types of treatment processes. For example, a disinfection robustness parameter may not be the most appropriate choice for filtration. Robustness of an entire system includes the robustness of non-physical components, such as response plans.

As an example of quantifying robustness, this paper has reported on the application of a robustness index for filtration. This index was developed by the second author within the context of a larger study (Huck *et al.*, 2001b). The index incorporated a performance goal, average performance, and some measure of deviation from average such as the 95th percentile turbidity or particle value achieved during a filter run. The index was applied specifically with a view to identifying ways to make filtration more robust with respect to *Cryptosporidium* removal.

The specific outcomes of the experimental investigation have been discussed above. In the broader context, the investigation showed that for the conditions tested, chemical challenges (such as suboptimal coagulation and a change in raw water quality) posed a greater challenge to filter robustness than did physical challenges to the process. By providing a common basis of comparison for various potential perturbers of filter effluent quality, the robustness index can assist process design and operational improvement. Further investigations should evaluate alternative ways of quantifying filtration robustness for both research-oriented investigations and for practice.

Acknowledgements

The experimental work and development of the filtration robustness index reported in this paper was conducted as part of a larger project for which funding was provided by the American Water Works Association Research Foundation (Project 490). The authors acknowledge the support and valuable input of the AWWARF Project Advisory Committee and of AWWARF Project Manager Kathryn Martin. In-kind support was provided by the Metropolitan Water District of Southern California. The authors are indebted to the valuable contribution of D. Maurizio and other members of the AWWARF project team.

References

Atherholt, T.B., LeChevallier, M.W., Norton, W.D., and Rosen, J.S. (1998) Effect of Rainfall on *Giardia* and *Crypto. Jour. AWWA* **90** (9) 66-80

Clancy, J.L., Bukhari, Z., Hargy, T.M., Bolton, J.R., Dussert, B.W., and Marshall, M.M. (2000) Using UV to Inactivate *Cryptosporidium. Jour. AWWA* **92** (9) 97-104

Coffey, B.M., Liang, S., Green, J.F., and Huck, P.M. (1998) Quantifying Performance and Robustness of Filters During Non-Steady State and Perturbed Conditions in Proc. AWWA Water Quality Technology Conference. Denver, Colo. AWWA

Huck, P.M. (2000) Testimony before the Walkerton Inquiry. Walkerton, Canada, October, 2000

Huck, P.M., Coffey, B.M., Anderson, W.B., Emelko, M.B., Maurizio, D.D., Slawson, R.M. Douglas, I.P., Jasim, S.Y., and O'Melia, C.R. (2001a) Using Turbidity and Particle Counts to Monitor *Cryptosporidium* Removals by Filters. Proceedings, IWA Berlin 2001 Conference, Berlin, Germany (in press)

Huck, P.M., Coffey, B.M., O'Melia, C.R., Emelko, M.B., and Maurizio D.D. (2001b) Filtration Operation Effects on Pathogen Passage. American Water Works Association Research Foundation and American Water Works Association, Denver, Colorado, 285 pp. ISBN 1-58321-170-5

Montgomery, J.M., Consulting Engineers, Inc. (1985) Water Treatment Principles and Design. New York: John Wiley & Sons

Stanley, S.J., Coffey, B.M., and Rector, D.W. (1998) Feasibility Study to Develop an Artificial Intelligence System for Optimization of Water Treatment Plant Operation. American Water Works Association Research Foundation, Denver, CO

Trussell, R., Trussell, A.R., Lang, J.S., and Tate, C.H. (1980) Recent Developments in Filtration System Design. *Jour. AWWA* **72**(12): 705-710

H.H. Hahn, E. Hoffmann, H. Ødegaard (Eds.)
Chemical Water and Wastewater Treatment VII, pp. 183-190
© IWA Publishing, London
ISBN: 1 84339 009 4

Effects of Coagulant Type and Conditions on *Cryptosporidium* and Surrogate Removal by Filtration

M. Emelko * and *T. Brown*

*University of Waterloo, Dept. of Civil Engineering; Waterloo, Canada
email: mbemelko@eawag.ch

Introduction

The considerable costs and practical limitations of adequate inactivation of pathogens such as *Cryptosporidium parvum* by traditional disinfection technologies have underscored the importance of multiple treatment strategies for inactivation and removal of *C. parvum* from drinking water. When operated properly, granular media filtration offers an excellent barrier against *C. parvum* passage into drinking water systems. Maintenance of appropriate chemical pretreatment is a critical component of ensuring proper filtration performance. Several investigations have demonstrated that sub-optimal chemical pretreatment can result in a deterioration of *C. parvum* removal by filtration (Huck *et al.*, 2001; Patania *et al.*, 1995).

While it is recognized that the interaction of *C. parvum* oocysts with chemical coagulants contributes to the removal of oocysts by filtration, only limited information regarding the charging mechanisms is available. Recent studies have suggested oocyst surfaces contain glycoproteins and trace amounts of fatty acids (Nanduri *et al.*, 1999) that can have ionizable groups such as carboxylates or phosphates (Karaman *et al.*, 1999). The interactions between the surfaces of oocysts and the chemical properties of coagulants used in drinking water treatment are likely therefore to be mechanistically different and may affect subsequent treatment processes differently.

Recent work by Bustamante *et al.* (2001) has suggested that enmeshment in precipitate may be the primary mechanism of oocyst removal when ferric chloride is used. In contrast, these authors also suggested that chemisorption of hydrolyzed aluminum species was an important mechanism when alum was used as a coagulant. These investigations suggest that the specific interactions between the alum and the oocyst surfaces might provide benefits in oocyst removal during water treatment. While some studies have demonstrated that improvements in *C. parvum* removal by filtration could be associated with specific coagulants (Yates *et al.*, 1997), other studies have underscored the general optimization of chemical pretreatment rather than specific coagulant selection (Patania *et al.*, 1995).

Huck at al. (2001) demonstrated comparable levels of C. *parvum* removal by filters treating alum-coagulated water at both warm (~20-26°C) and cold (~2-3°C) water temperatures. Similar studies conducted at the same pilot plant demonstrated that oocyst-sized polystyrene microspheres were reasonable surrogates for *C. parvum* removal by filtration (Emelko, 2001).

In the present study, pilot-scale filtration studies were performed to investigate the relative impact of coagulant type on the removal of *Cryptosporidium* and oocyst-sized polystyrene microsphere surrogates by granular media filtration. The impacts of in-line alum, ferric chloride, and chitosan coagulation on subsequent filtration were investigated. This paper reports on some of the alum and ferric chloride outcomes.

Materials and Methods

Pilot-scale treatment plant

Two glass filter columns (50 mm in diameter) were operated at a loading rate of ~10.4 m/h (~4.3 US gpm/ft^2) in a constant rate, rising head mode during the filter evaluations. Each of the filters contained 508 mm of anthracite (ES=0.98 mm, UC = 1.5) over 203 mm of sand (ES = 0.5 mm, UC = 1.5). The filters treated dechlorinated tap water with 3.5 NTU of kaolinite-induced turbidity. The raw water was coagulated in-line and then filtered. One filter treated alum-coagulated water at a dose of 5 mg/L alum (Al$_2$(SO$_4$)$_3$·18H$_2$O) at pH 6.9. A second filter treated ferric chloride-coagulated water at a dose of 3 mg/L at pH 6.9.

During the experiments described herein, formalin-inactivated oocysts and oocyst-sized polystyrene microspheres were added to the filter influent to yield concentrations of ~10^5 oocysts/L and microspheres/L respectively. The oocysts were added to the raw water and were subsequently coagulated. The addition of oocysts to the raw water did not substantially increase particle loading to the treatment system.

Operational conditions

The effects of alum and ferric chloride coagulation on *Cryptosporidium* and oocyst-size microsphere removal by filtration were investigated during three experimental conditions: stable operation, sub-optimal coagulation, and no coagulation. Stable operating conditions were periods of optimized treatment during which filter effluent turbidities did not exceed 0.1 NTU. Sub-optimal coagulation conditions represented a coagulant misfeed resulting in a 50 % reduction in applied coagulant dose. The no coagulation experiments represented a complete coagulation failure.

All of the experiments were conducted after 2 to 4 hours of stable operation after filter ripening. *Cryptosporidium* oocysts and oocyst-sized polystyrene microspheres were seeded into the raw water for one hour during each of the experiments. Filter influent and effluent samples were collected at four time points, each approximately 10 minutes apart, throughout the seeding period. Removal calculations were based on these influent and effluent concentration pairs.

Analytical Methods

Cryptosporidium parvum

Stock suspensions of formalin-inactivated *C. parvum* were vortexed for thirty seconds and then a small portion of the suspension (< 100 μL in total) was removed to enumerate the oocyst concentration. The stock concentration was determined by averaging triplicate counts using a hemocytometer and light microscopy. The entire grid (1 mm^2) was used for oocyst enumeration at 400× magnification (Nikon Labophot 2A, Nikon Canada Inc., Toronto, ON).

During the filtration investigations, *C. parvum* oocysts were measured in filter influent and effluent samples. Filter influents were analyzed in 2.5-mL volumes. Filter effluents were analyzed in volumes ranging from 5 mL to 1 L, depending on the operating condition studied. Sample volumes were chosen to yield between 10 and 2000 oocysts per membrane.

All of the samples were directly filtered through 25 mm, 0.40 μm polycarbonate membranes utilizing a previously described method and standard immunofluorescence assay (Emelko *et al.*, 2000). Presumptive microscopic analysis for *C. parvum* was performed at 400× magnification (Nikon Labophot 2A, Nikon Canada Inc., Toronto, ON). Recovery data from the water matrix indicated approximately 75% recovery of oocysts, comparable to results previously reported (Emelko *et al.*, 2000).

Polystyrene microspheres

Fluoresbrite™ carboxylated YG fluorescent-dyed, oocyst-sized polystyrene microspheres (Polysciences Inc., Warrington, PA) were used as non-biological surrogate indicators for *C. parvum* removal. The microspheres had an average diameter of 4.675 ± 0.208 μm and a density of 1.045 g/mL. The YG dye matches the fluorescence filter settings of fluorescein (*i.e.*, maximum excitation at 458 nm and maximum emission at 540 nm; similar to FITC for *C. parvum*). The microspheres were concentrated and enumerated concurrently with *C. parvum*, by the method generally described above. Recovery data from the water matrix indicated approximately 75 % recovery of microspheres, comparable to results previously reported (Emelko *et al.*, 2000).

On-line parameters: turbidity and particle counts

Turbidity was monitored at the filter influent and effluent locations using on-line turbidimeters (Model 1720C, Hach Co., Loveland, CO). The turbidimeters were calibrated using dilute formazin solutions as specified by the manufacturer. An IBR particle counter (IBR, Grass Lake MI) measured total particles from 1-150 μm at the filter effluent location.

Results and Discussion

C. parvum removals by filtration preceded by in-line alum and ferric chloride coagulation during stable operation, sub-optimal coagulation, and coagulation failure are presented in Figure 1. This figure is a box-and-whisker plot in which the dash in the center represents the median removal (50th percentile). The lower and upper

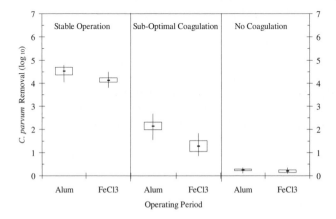

Figure 1. Box-and-whisker plot of *C. parvum* removal by in-line alum coagulation and filtration

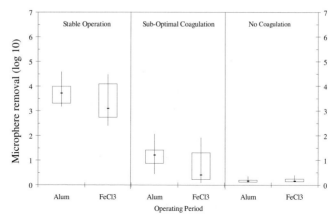

Figure 2. Box-and-whisker plot of *polystyrene microsphere* removal by in-line ferric chloride coagulation and filtration

portions of the box respectively indicate the 25th and 75th percentile removals. The lower and upper portions of the line (whisker) respectively indicate the minimum and maximum removals observed.

The importance of maintaining proper coagulation to ensure good filtration of *Cryptosporidium* is clearly demonstrated in Figure 1. These data also suggest that under the conditions investigated, oocyst removals by filtration preceded by alum coagulation are similar to, perhaps slightly lower than, those obtained by when ferric chloride is used (median removals were 4.5-log and 4.1-log respectively). Compared to alum, ferric chloride may also result in slightly lower *C. parvum* removals by filtration during sub-optimal coagulation conditions (median removals were 2.1-log and 1.3-log respectively). Further analysis is necessary to determine whether these differences are statistically significant. As proposed by Bustamante *et al.* (2001), a possible explanation for such differences between oocyst removal by filtration may be the different mechanisms of interaction between the coagulants and the oocysts. In agreement with previously reported alum data (Emelko, 2001; Huck *et al.*, 2001), this study demonstrates almost no oocyst removal by filtration during complete coagulation failure, regardless of coagulant type (median removals were 0.3-log and 0.2-log by filtration preceded by alum and $FeCl_3$ respectively).

Oocyst-sized polystyrene microsphere removals by filtration preceded by in-line alum and ferric chloride coagulation during stable operation, sub-optimal coagulation, and coagulation failure are presented in Figure 2. The microsphere data are similar to the oocyst data presented in Figure 1, however, the microsphere removal data are considerably more variable. The overall trends regarding the impact of coagulant type and coagulation conditions are similar between Figure 1 and Figure 2. Microsphere removals by filtration preceded by alum filtration are

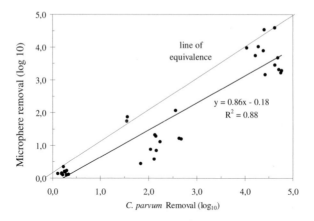

Figure 3. Relationship between C. *parvum* oocyst and oocyst-sized polystyrene microsphere removal during in-line alum coagulation and filtration

similar to, perhaps slightly higher than, those obtained by filtration preceded by ferric chloride filtration (median removals were 1.2-log and 0.4-log respectively). Again it appears that, compared to alum coagulation, ferric chloride coagulation may also result in slightly lower removals of microspheres by filtration during sub-optimal coagulation conditions (median removals were 3.7-log and 3.1-log respectively). When no coagulant was present, there was little removal of microspheres by filtration (median removals were 0.2-log by filtration preceded by each alum and $FeCl_3$).

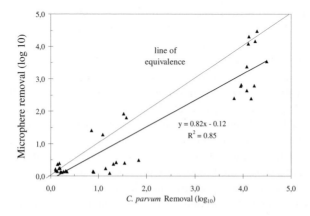

Figure 4. Relationship between C. *parvum* oocyst and oocyst-sized polystyrene microsphere removal during in-line ferric chloride coagulation and filtration

Oocyst and microsphere surface charge and compressibility may contribute to the observed differences in oocyst and microsphere removals by filtration. In the present study, the overall relationship between *C. parvum* oocyst and polystyrene microsphere removals during the range of operational conditions investigated was fairly linear. As indicated in Figure 3, least squares linear regression yields a coefficient of determination (R^2) of 88 % for the relationship between C. *parvum* and oocyst-sized polystyrene microsphere removals by filtration preceded by in-line alum coagulation. Microsphere removals were similar to oocyst removals, though often slightly lower, as evidenced by most of the data points falling below the line of equivalence.

Similar results are demonstrated in Figure 4, which presents the relationship between *C. parvum* oocyst and polystyrene microsphere removals observed by filtration preceded by ferric chloride coagulation. In this case, least squares linear regression yields a coefficient of determination (R^2) of 85 %. As observed in Figure 3 (alum coagulation), microsphere removals in Figure 4 (ferric chloride coagulation) were also similar to oocyst removals, though often slightly lower suggesting that microsphere removals were good, somewhat conservative indicators of *C. parvum* removals by filtration. Overall, Figure 3 and Figure 4 are very similar and suggest that, under the conditions investigated, coagulant type does not affect the somewhat linear overall relationship between C. *parvum* and oocyst-sized polystyrene microsphere removals by filtration.

Conclusions

Coagulation effects on *Cryptosporidium* oocyst and oocyst-sized polystyrene microsphere removals by filtration were investigated. Specifically, the impacts of alum and ferric chloride coagulation during stable (optimized) operating conditions, sub-optimal coagulation (50 % reduction in coagulant dose), and coagulation failure were studied. The pilot-scale results from this work indicated that:

- Alum and ferric chloride coagulation generally resulted in similar removals of *Cryptosporidium* oocysts and oocyst-sized microspheres during optimized operating conditions when filter effluent turbidities were consistently below 0.1 NTU. The median oocyst removals were 4.5-log and 4.1-log during filtration with alum and ferric chloride respectively; median microsphere removals were 3.7-log and 3.1-log respectively.
- Sub-optimal coagulation conditions with both alum and ferric chloride coagulation resulted in deteriorated *Cryptosporidium* and microsphere removal by filtration (relative to stable operation). Median *C. parvum* removals were 2.1-log and 1.3-log by filtration preceded by alum and ferric chloride respectively; median microsphere removals were 1.2-log and 0.4-log respectively.

- The observed differences in *Cryptosporidium* and microsphere removal during sub-optimal coagulation conditions (and possibly during stable operation) may be associated with the different mechanisms of alum and FeCl$_3$ interaction with oocysts during filtration. Further analysis is necessary to determine if these differences are statistically significant.
- Oocyst-sized polystyrene microspheres appeared to be reasonable indicators of *Cryptosporidium* removal by filtration, regardless of ferric chloride or alum coagulation.

References

Bustamante, H.A., Shanker, S.R., Pashley, R.M., and Karaman, M.E. (2001) Interaction between *Cryptosporidium* oocysts and water treatment coagulants. *Wat. Res.* **35**(13), 3179-3189

Emelko, M.B. (2001) Removal of *Cryptosporidium parvum* by granular media filtration. Ph.D. thesis, Dept Civil. Eng., Univ. of Waterloo, Waterloo, Ontario, Canada

Emelko, M.B., Huck, P.M., Douglas, I.P., and Van Den Oever, J. (2000) *Cryptosporidium* and microsphere removal during low turbidity end-of-run and early breakthrough filtration. In *Proc. AWWA Water Quality Technology Conference*. Salt Lake City, 5-9 November

Huck, P.M., Emelko, M.B., Coffey, B.M., Maurizio, D.D., and O'Melia, C.R. (2001) Filter Operation Effects on Pathogen Passage. American Water Works Association Research Foundation and American Water Works Association, Denver, CO

Karaman, M.E., Pashley, R.M., Bustamante, H.A., and Shanker, S.R. (1999) Microelectrophoresis of *Cryptosporidium parvum* oocysts in aqueous solutions of various inorganic and surfactant cations. *Colloids Surf.* **146**(1-3), 212-221

Nanduri, J., Williams, S., Aji, T., and Flanigan, T.P. (1999) Characterization of immunogenic glycolax on the surfaces of *Cryptosporidium parvum* oocysts and sporozoites. *Infect. Immunity.* 2022-2024

Patania, N.L., Jacangelo, J.G., Cummings, L., Wilczak, A., Riley, K., and Oppenheimer, J. (1995) Optimization of Filtration for Cyst Removal. American Water Works Association Research Foundation and American Water Works Association, Denver, CO

Yates, R.S., Green, J.F., Liang, S., Merlo, R.P., and DeLeon, R. (1997) Optimizing Direct Filtration Processes for *Cryptosporidium* Removal. In *Proc. AWWA Water Quality Technology Conference*. Denver, 9-13 November

H.H. Hahn, E. Hoffmann, H. Ødegaard (Eds.)
Chemical Water and Wastewater Treatment VII, pp. 191-200
© IWA Publishing, London
ISBN: 1 84339 009 4

Naturally Occurring Autofluorescent Particles as Surrogate Indicator of Sub-Optimal Pathogen Removal in Drinking Water

O. Bergstedt and *H. Rydberg*

*Chalmers University of Technology, Water Environment Transport, Göteborg, Sweden
email: olof.bergstedt@vaverket.goteborg.se

Abstract

Sub-optimal pathogen reduction in conventional water treatment plants (WTPs) could pose a danger to public health, but monitoring of drinking water for pathogens such as *Cryptosporidium* with as low infection doses is not feasible. Different surrogates have been studied previously and several of those that have been seeded to the water during well-controlled conditions have shown removals similar to those of *Cryptosporidium*. As full-scale operations can have a large variability in treatment performance, seeding is normally not advisable in full scale, and turbidity and particle counts are influenced by chemical precipitation itself. Naturally-occurring surrogates such as spores and algae have showed promising results, but require large sampling volumes and/or time consuming laboratory methods. Our aim was to explore the potential of naturally occurring surrogates using the autofluorescence (FL) of parasite-sized algae. A bench scale study showed that these FL particles were destabilized by chemical precipitation in a way similar to *Cryptosporidium* oocysts. They were also quite resistant to the chemicals added in full-scale treatment. Historical operational data contain many variables making evaluation of optimal conditions difficult, but factorial full-scale experiments with small changes in a couple of key variables might be useful for optimization.

Introduction

There have been several waterborne, disease-causing outbreaks reported in Sweden, with hundreds or thousands of people infected (Hult, 1998), in spite of the abundance of water from relatively pristine watersheds. Protozoan parasites, *Giardia* and *Cryptosporidium*, were among the causing agents that were identified, but in most cases the agent remains unknown. A microbiological risk assessment (Westrell *et al.*, 2001) showed that the multi-barrier approach of chemical precipitation and chlorination might be insufficient for pathogens that are insensitive to chlorination, e.g. *Cryptosporidium*.

Dependence on particle removal as the only functional barrier makes the treatment even more sensitive to sub-optimal conditions. Studies have pointed out the variability of particle removal performance as a possibility of optimization (Hijnen, 2000).

A study of 19 Swedish chemical precipitation WTPs (Hernebring 1980) showed that the wrong precipitation pH and too low an initial energy input in flocculation had a major impact on treatment performance. The sub-optimal conditions led to poor settling and, in spite of low surface loads in sedimentation, the filters had to take most of the load. Obvious problems were high alum contents, increased cost of filter operation, and possible interference with disinfection. Low temperature and increased surface loads worsened the problems and over-dosages of alum were not enough to compensate. The worst examples had turbidity removals from raw water to drinking water ranging from 0.2 to 0.4 log.

Seeding studies in both pilot and full scale (Nieminski and Ongerth, 1995) showed that turbidity removal can work as a rough estimate of oocyst removal and that 4 - 7 μm particles could be quite a good surrogate, while heterotrophic plate count removal was an ineffective one.

The impact of sub-optimal coagulation for the removal of *Cryptosporidium* and surrogates by filtration was studied in pilot scale (Huck *et al.*, 2000). Two different pilot plants with pre-chlorination and coagulation with alum and coagulation aids were investigated. Filters were seeded with jar-coagulated suspensions of formalin-inactivated oocysts and non-inactivated *Bacillus subtilis* spores. Experiments were performed without coagulation, with optimized coagulation and with sub-optimal coagulation. Removal of spores and 2 μm particles was less than oocyst removal, but under sub-optimal conditions the decreases in removal were similar.

Removal of sulphite-reducing clostridia spores has been suggested as a simple and cheap potential surrogate for protozoan (oo)cyst-removal (Hijnen *et al.*, 2000). Removal was studied in 8 full scale WTPs during two summer and two winter weeks. Samples of up to 500 L each were collected and analysed by standard microbiological methods. Removals were found to be similar to actual (oo)cyst-removals in other studies.

Algae have also been studied as surrogates for removal of *Cryptosporidium* oocysts (Akiba *et al.,* 2001). Algae found in raw water sources such as lakes and rivers are known to have physical properties, such as zeta potential, that are similar to those of oocysts. In laboratory experiments comparing coagulation and filtration characteristics of algae and oocysts, the authors found algal removal to be similar to oocyst removal.

Our aim was to explore the potential of even simpler naturally occurring surrogates and their usefulness as indicators of sub-optimal treatment of low turbidity raw waters.

Materials and Methods

Bench-scale flocculation

Bench-scale flocculation trials were made on three different surface waters that supply the Göteborg area (Table 1). Raw waters were seeded with labeled *C. parvum* oocysts and flocculated under optimal and sub-optimal conditions with regard to flocculant dose. Flow cytometry and turbidity were used to monitor removal of parasites and surrogates.

General setup. A Flocculator 2000 system (Kemira AB), consisting of 2 x 6 cylindrical 1 L glass beakers with microprocessor controlled stirrers, was used to conduct jar tests. Raw water pH was adjusted with 0.1M NaOH prior to addition of aluminium sulphate $(Al_2(SO_4)_3 \bullet H_2O)$. Both solutions were prefiltered. Flocculator settings (time, speed and G-value) were chosen to give precipitation results similar to those in the full-scale process: alum mixing phase (0.3 min, 400 rpm, 292 s^{-1}) followed by two flocculation steps (20 min, 46 rpm, 11 s^{-1} and 20 min, 28 rpm, 5 s^{-1}) and sedimentation (45 min). Experiments were made at approximately 10°C and all jars were kept in flow-through water baths, to prevent a rise in temperature. Optimal doses of aluminium sulphate (mg/L) were calculated for each raw water as (colour P_t mg/L • 0.7) +15. Adjustments of flocculation pH were made individually to give a final pH of 6.5 independent of raw water and alum dose.

Table 1. Raw water quality data for Feb. 2001 to Jan. 2002

	Göta river				Lake Delsjön				Lake Rådasjön			
	n	min	med	max	n	min	med	max	n	min	med	max
Turbidity FNU	51	2.5	4.5	10.5	161	0.59	0.84	1.7	10	0.50	1.2	2.1
Colour mg P_t/l	51	15	25	35	155	15	20	30	10	40	50	70
E. coli/100ml	163	1	39	460	105	< 1	< 1	700	47	1	< 2	30
Crypto /10l	5	< 1	2	< 5	5	< 1	< 1	2	4	< 1	< 1	3
Giardia /10l	5	< 1	< 1	< 5	5	<1	< 1	2	4	< 1	< 1	< 1

Cryptosporidium. Preparations ($1.25 \cdot 10^8$/ml) of formalinized (5 % formalin with 0.01 % Tween 20) *C. parvum* (Waterborne Inc) were used to seed raw waters. Oocysts were labeled with Cy-5 (Nycomed Amersham, Ltd) conjugated antibodies (Crypt-a-glo, Waterborne Inc), in solution and incubated for 40 minutes at 35 °C. Miniaturized control experiments (30 ml raw water) were made to check fading/dissociation, due to light exposure and precipitation chemicals used. The potential of non-specific binding of free antibodies to existing raw water particles was evaluated. This also included the risk of microfloc formation, incorporating enough free antibodies to generate new fluorescent particles. Control experiment samples were analysed by flow cytometry and visually verified by fluorescence microscopy (Carl Zeiss Axiovert S100; filter set GLF Visual Cy-5, Chroma Tech).

Experimental. The three raw waters were examined individually in duplicate trials. Each trial consisted of three jars seeded with *C. parvum* oocysts and three control jars. The alum doses used were optimal dose, 40 % less than optimal and 60 % less than optimal. A control experiment was also made to evaluate the effect of formalin/Tween 20 on precipitation results. Parasite addition (~$3 \cdot 10^7$ oocysts/ l raw water) and pH adjustment was made prior to the mixing phase. Alum (10 g/l) was then added during rapid mixing.

Sampling/analyses. Temperature, pH and turbidity were analysed for raw water samples, before each trial, and after 45 min sedimentation. Flow cytometry samples (1ml in particle free microcentrifuge tubes) were taken throughout the experiments: from the raw water, after parasite addition, during alum mixing, and during and after sedimentation (15, 30 and 45 minutes). At the end of each trial the floc was resuspended by rapid mixing and analysed by flow cytometry.

C. parvum oocysts were analysed directly by flow cytometry (Microcyte, Optoflow AS). The instrument, using a 635 nm diode laser, measured forward scatter (particle size ~0.4-20 µm) and particle-associated fluorescence (650-900 nm). Water samples were also analysed for two surrogate parameters: total particles 1 - 15 µm/ml, defined by 1 and 15 µm polystyrene standards, and autofluorescent particles 1-15 µm/ml (FL). Particles showing autofluorescence were mainly algae, probably due to 635 nm excitation of photosynthetic pigments.

Calculations. Log-removals of *Cryptosporidium* and surrogates were calculated as -log (C/C_0), where C = concentration at the end of sedimentation (45 min) and C_0 = raw water concentration. For *Cryptosporidium*, initial concentration was calculated from the alum mixing phase, and final concentration was not compensated for with the controls. Thus, removal was conservatively expressed as minimum reduction. Relationship between variables was tested by rank correlation (Spearmans r_s).

Full scale

Only 12 of the more than 2000 WTPs in Sweden regularly monitor raw water for *Cryptosporidium* and *Giardia* (Bergstedt and Andersson, 2001). One of those, the Lackarebäck WTP was chosen for a full-scale study. This WTP treats low turbidity raw water (Lake Delsjön, Table 1) using the most common process in Sweden, coagulation with alum, sedimentation and rapid filters.

Process monitoring. Throughout 2001 flow cytometry samples were analysed for surrogate parameters (total particles/autofluorescent particles 0.4-15 μm). Detailed sampling profiles, through the entire process, were also checked at periods with and without pre-chlorination, 0.3 g/m.[3]. Pre-rinsed centrifuge tubes (FALCON 50 ml) were used for all samples. A control experiment was set up to test the influence of treatment chemicals (chlorine, $CaCO_3$ and alum) on autofluorescent particles.

Factors that are supposed to influence the particle removal efficiency of chemical precipitation and the response in on-line particle counts of particles >1μm were checked for the full 12-month period and for 3-month periods. The mean values before and after changes in the process were also compared using nearby periods as external reference distributions.

Experimental. A simple 2^2-factorial design for the variables coagulation dosage and pH was used in a preliminary experiment. Changes in random order were made at a set time on four consecutive days with sampling on two occasions per day. Particle removals from raw water to filtered water were used as response factors. Other factors such as surface loads were kept as constant as possible by avoiding changes in raw water flow and the regular cleaning of sedimentation tanks. Samples were taken in the same part of the filter cycle, after ripening well before expected breakthrough, since backwash occurred on a set time each day. The changes were limited to +/- 1g/m[3] of aluminium sulphate and +/- 0.1 pH-units compared to the current target values.

Results

Bench-scale flocculation

The general flocculation setup gave consistent results. The temperature change throughout the experiments was less than 1°C and final pH varied between 6.3 - 6.7. Minimum reductions of *Cryptosporidium* were very similar for each raw water, at optimal and 40 % less than optimal dose. Removals typically ranged between 1-3 log. Reducing the dose further (60 % less than optimal) could generate flocculation, but on most occasions it did not. In both cases *Cryptosporidium*

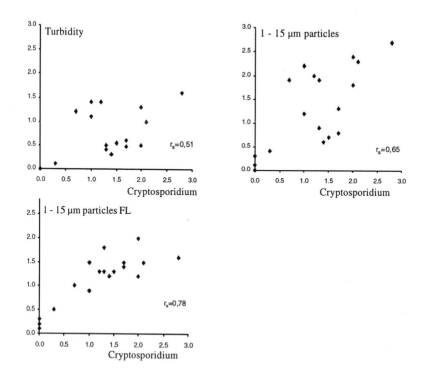

Figure 1. Log-removal of surrogates compared to removal of *Cryptosporidium*

removal was close to zero. The strongest association was seen between removal of *Cryptosporidium* and autofluorescent particles 1-15 µm (Figure 1).

Both experimental and control jars were in good agreement concerning turbidity after sedimentation, and formalin/Tween 20 had no general effect on precipitation results in the control experiment. Miniaturized control experiments showed no fading/dissociation of labeled *Cryptosporidium* during 5 h exposure to low light and the chemicals used. Too high a concentration of free antibodies initially generated new fluorescent particles, mainly in the < 1µm region, as soon as alum was added. Thus, the amount of free antibodies was decreased in the jar tests.

Fullscale

Process monitoring. Removal of surrogates differed significantly over the year (Figure 2). Removal of particles 0.4-1µm and 1-15 µm was 1.5 and 1.6 log, respectively. In the summer, when pre-chlorination was initiated, removal of autofluorescent particles 1-15 µm ranged between 2.2 and 3.8 log compared to 1.2 and 3.0 log in periods with no pre-chlorination.

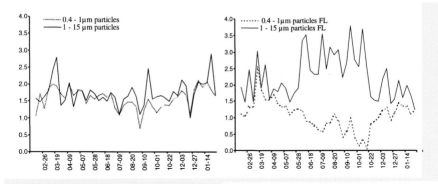

Figure 2. Log-removal of surrogate parameters Feb. 2001 to Jan. 2002

Full-scale variations. Even though there were substantial variations in the key parameters (Table 2), no strong relationship between any of the parameters and the total number of particles >1μm according to the on-line counter was found for either the 12-month or the 3-month period. The mean values before and after changes in the process were significantly different only between pre-chlorination and no pre-chlorination. Pre-chlorination typically reduced turbidity and on-line particle counts after filtration.

Table 2. Variations in key parameters at Lackarebäck WTP, Feb.2001 to Jan.2002

	Min. day	Max. day	Max changes day to day
Alum dosage g/m^3	28	40	5
Coagulation pH	6.2	6.7	0.2
Raw water flow m^3/s	2943	4313	799
Water temperature °C	2.5	19	1.2
Pre-chlorination g/m^3	0	0.3	0.3

Factorial experiment. Even though the changes were very small, the responses were quite clear. The worst particle removal was achieved when coagulation pH was increased and the alum dosage was decreased (Table 3).

The turbidity of filtered water varied between 0.02 and 0.07 FNU (Formazin Nephelometric Units). Log-removals were generally lowest for turbidity and highest for fluorescent particles 1-15 μm. The difference between the two daily samples was smaller for 0.4 - 15 FL than for the other parameters. A high particle reduction after sedimentation was not connected to a high total reduction, especially not for turbidity.

Table 3. Log-removals in raw water and filtered water at small changes of coagulation pH (+/- 0.1 pH-unit) and dose (+/- 1 g/m³)

Day	pH	Alum	Turbidity	0.4-15 µm FL	1-15 µm FL
0	0	0	1.10	1.32	1.52
1	+	-	1.09	1.13	1.45
2	-	+	1.36	1.38	1.88
3	-	-	1.53	1.22	1.64
4	+	+	1.18	1.35	1.52

Particle profiles. Detailed sampling profiles in the course of the treatment process revealed a significant increase in turbidity from 1 FNU in raw water to 5 FNU in the alum mixer. This was due to a large increase in particle content mainly in the 1 µm region (Figure 3).

Autofluorescent particles 0.4 - 15 µm were not sensitive to the addition of treatment chemicals. In a 9 h control experiment, when autofluorescent particles 0.4-15 µm were exposed to chlorine, chlorine + $CaCO_3$ and chlorine + $CaCO_3$ + alum, respectively, the mode of fluorescence intensity decreased ~13.5 % from approximately 14800 to 12800 molecular equivalents of allophycocyanine. This did not cause a significant loss of autofluorescent particle content due to detector sensitivity.

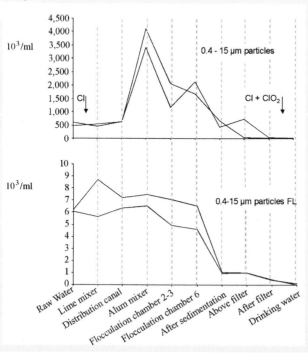

Figure 3. Profiles through Lackarebäck WTP during pre-chlorination

Discussion

The bench-scale results confirmed the work of previous investigators that turbidity under estimates oocyst removal and that particles of similar size may be a better surrogate. The use of *Cryptosporidium*-sized autofluorescent particles looks promising as a tool for optimization. The particles can be easily sampled/analysed and are naturally occurring tracers of raw water origin.

Any association between removal of *Cryptosporidium* and surrogates should be made carefully. In the jar tests oocyst concentration did not reflect occurrence in natural surface waters. However, sufficient flocculation was required for oocysts to settle. Under suboptimal conditions (60% less than the optimal dose) *Cryptosporidium* did not settle although flocs were formed. Jar tests were made with inactivated oocysts, for practical reasons, which may have influenced flocculation properties. The use of formalinized *Cryptosporidium* was investigated by Emelko (2001). An apparent effect of scale was the loss of seeded oocysts, probably due to binding on walls and stirrers, resulting in a lower initial concentration than expected. Despite this, reduction of *Cryptosporidium* seemed to model removal, since resuspension of floc recovered the initial concentration. Criteria for making a rank correlation may not be fulfilled completely. Removal of autofluorescent particles should be used as an operational optimization tool rather than for estimating true *Cryptosporidium* removals. However, further work on filtering is needed.

More than a year of analyses on raw water and filtered water from full scale operation showed that there were enough fluorescent particles to avoid repeated zero counts that would make evaluation of removals less reliable. Throughout the year at least 3 log removal could be monitored and excitation at 635 nm gave very low levels of background fluorescence due to non-algal material. Profiles with particle counts throughout the treatment process showed that numbers of autofluorescent (FL) particles were quite constant during pH changes, low dosage pre-chlorination, lime addition, alum dosage, coagulation and flocculation. Another advantage was the insensitivity to air bubble formation, which can be a problem with conventional particle counters. This is promising since studies can be made independent of the changes in turbidity and normal particle counts created by the process itself. It can be especially useful when comparisons are made between different parts of the process. Even though autofluorescent particles seem to withstand disinfectants at moderate doses, annual variation in algal composition may well affect sensitivity. At the moment, the most stable count includes both FL particles smaller than 1 μm and larger FL particles. This means that even if FL particles are a good surrogate for pathogen removal, they are far from specific to *Cryptosporidium*-sized particles, since smaller particles may be removed in part by different mechanisms. Further work will include pre-treatment of samples to distinguish between the algae that were of oocyst size from the beginning and smaller algae that have grown in the course of treatment.

Conclusion

Naturally occurring autofluorescent particles may be surrogate indicators of sub-optimal pathogen removal and an optimization tool in drinking water treatment.

References

Akiba, M., Kunikane, S., Kim, H.-S., and Kitazawa, H. (2001) Algae as surrogate indices for the removal of Cryptosporidium oocysts by direct filtration. In *Proc. of the IWA 2nd World Water Congress*, Berlin, 15-19 Oct. 2001

Bergstedt, O., and Andersson, Y. (2001) Health related water microbiology: waterborne pathogens, National Report Sweden. In *Proc. of the IWA 2nd World Water Congress*, Berlin, 15-19 Oct. 2001

Emelko, M.B. (2001) Removal of *Cryptosporidium parvum* by granular media filtration. *PhD Thesis, University of Waterloo, Ontario, Canada*

Hernebring, C. (1980) Operational studies of water treatment plants using chemical precipitation. Report in Swedish, Dept WET, Chalmers University of Technology, Göteborg, Sweden

Hijnen, W.A.M., Wilhelmsen-Zwaagstra, P., Hiemstra, P., Medema, G.J., and van der Kooij (2000) Removal of sulphite-reducing clostridia spores by full-scale water treatment processes as a surrogate for protozoan (oo)cyst removal. *Wat.Sci.Tech.* **41** (07), 165-172

Huck, P.M., Coffey, B.M., Emelko, M.B., and O'Melia, C.R. (2000) The importance of coagulation for the removal of Cryptosporidium and surrogates by filtration. In Hahn, H.H., Hoffmann, E., Ødegaard, H. (Eds.) Chemical water and wastewater treatment VI, Springer, Berlin, 191-200

Hult, A. (1998) The drinking water in Sweden. Report in Swedish, Swedish Water, Stockholm, Sweden

Nieminski, E.C., and Ongerth, J.E. (1995) Removing Giardia and Cryptosporidium by conventional treatment and direct filtration. *J. AWWA* Volume 87, (09), 96-106

Westrell T., Bergstedt O., Heinecke G., and Kärrman E. (2001) A system analysis comparing two drinking water systems—central physical-chemical treatment and local membrane filtration. In *Proc. of the IWA 2nd World Water Congress*, Berlin, 15-19 Oct. 2001

Wastewater Treatment

H.H. Hahn, E. Hoffmann, H. Ødegaard (Eds.)
Chemical Water and Wastewater Treatment VII, pp. 203-212
© IWA Publishing, London
ISBN: 1 84339 009 4

Characterisation of Particulate Matter in Municipal Wastewater

A. F. van Nieuwenhuijzen and A. R. Mels*

*Delft University of Technology ,Dept. of Sanitary Engineering, Delft, The Netherlands
email: A.F.vanNieuwenhuijzen@CiTG.TUDelft.NL

Introduction

Since a major part of the organic material and the nutrients in municipal wastewater are related to particulate material, advanced particle removal by physical-chemical pre-treatment results in a lower pollutants load to subsequent treatment steps. Thus the total treatment system can be designed smaller and operated in a more energy efficient way (Ødegaard, 1998; Nieuwenhuijzen *et al.*, 1998; Mels *et al.*, 1999). Traditionally, the choice of treatment methods is mainly based on the effluent standards as well as on practical experience with the various techniques by engineering companies and waterboards. Particularly the choice of adding pre-treatment steps, mainly a primary sedimentation tank, ahead of activated sludge systems was not, in general, based on technical knowledge and wastewater characteristics, but was made according to traditional practice and on plant size (Ødegaard, 1999).

Earlier studies (Levine *et al.*, 1991; STOWA, 2001) showed that a more detailed characterisation of the particulate matter in wastewater is necessary for a better understanding and prediction of removal efficiencies of physical-chemical treatment techniques and the application of optimal chemical dosages. Such a characterisation should include the distribution of contaminants over various particle sizes. Intensive wastewater characterisations have been conducted world-wide to determine biological fractions for activated sludge models (Henze, 1992), but few data are available on the size distribution of wastewater pollutants.

This article describes experimental research that aimed to determine how contaminants in wastewater are distributed over different particles sizes. From these data, the effect of various particle removal techniques on the wastewater composition can be predicted and the requirements and efficiencies of the necessary post-treatment can be determined. The results will also indicate whether a certain wastewater can be treated efficiently with physical-chemical pre-treatment methods and to what extent.

[203]

Calculations were made to identify the effect of advanced particle removal in the first treatment step on the biological post-treatment and total wastewater treatment system. The environmental criteria evaluation model DEMAS⁺ was used to calculate energy and cost potentials of increasing particle removal (Nieuwenhuijzen, 2002).

Experimental set-up

To determine how contaminants in municipal wastewater are distributed over particle size, a size fractionation was carried out on eight wastewaters from different sources. The different wastewater fractions were analysed for the components COD, BOD, nitrogen, phosphorus, suspended solids, turbidity and conductivity.

Sampling and fractionation

Wastewater grab samples were taken at eight WWTP (Apeldoorn, Arnhem, Bennekom, Berkel, Boxtel, Haarlem, Hoek van Holland and Vlaardingen).

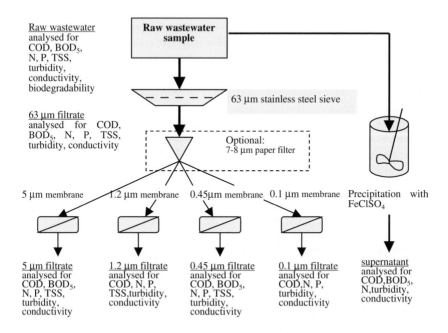

Figure 1. Schematic presentation of the fractionation procedure

The samples were collected after screening, just in front of the pre-treatment facility, since these are the desirable data to be collected. Duplicate samples were taken on three different days at the same location at the same time and preferably during the same weather conditions. In total, 48 samples were taken.

Grab sampling was preferred over flow proportional longterm sampling since processes like adsorption, degradation or destruction could cause changes in the wastewater composition regarding particle size distributions. On the other hand, it was recognised beforehand that grab samples would only give an instantaneous picture of the total wastewater composition on any given day.

The samples were immediately analysed for temperature and pH and stored in polyethylene containers of 3 to 5 litres. The samples were sent directly to the laboratory for fractionation and further analysis. In case of delay during transport and after arrival at the laboratory, the samples were kept at a constant temperature (4 °C) to prevent changes in the wastewater composition.

The samples were fractionated over clean, pre-flushed sieves (stainless steel) and membrane filters (cellulose nitrate composite, fabricated by Sartorius) with five different pore sizes (63, 5, 1.2, 0.45 and 0.1 μm) as shown in Figure 1.

The filtrates were analysed for COD, BOD_5, nitrogen, phosphorous, suspended solids, turbidity, conductivity and biodegradability via extensive BOD analyses on day 1, 2, 3, 5 and 10 ($BOD_{1,2,3,5,10}$). The particle size range and analysed parameters for each fraction are summarised in Table 1.

Table 1. Particle size range and parameters determined for each fraction

Fraction	Particle size range	Analysed parameters
raw wastewater	whole range	COD, BOD_5, $BOD_{1,2,3,5,10}$, N, P, TSS, turbidity, T, pH, conductivity
Settleable-suspended	> 63 μm	COD, BOD_5, N, P, TSS, turbidity, T, pH, conductivity
suspended	5 – 63 μm	COD, BOD_5, N, P, TSS, turbidity, T, pH, conductivity
supra colloidal	1.2 – 5 μm	COD, N, P, TSS, turbidity, T, pH,
colloidal	0.45 – 1.2 μm	COD, N, P, turbidity, T, pH, conductivity
semi-dissolved	< 0.45 μm	COD, BOD_5, N, P, turbidity, T, pH conductivity
dissolved	< 0.1 μm	COD, N, P, T, pH, conductivity
dissolved*	supernatant*	COD, BOD_5, N, turbidity, T, pH

*Fraction produced by sampling supernatant after precipitating raw wastewater with 20 mg Fe^{3+}/l iron-sulphate. The addition of Fe" could lead to changes in the ratio of dissolved/particulate matter

Results and discussion

Average wastewater fractions

From the fractionation data from the eight sampled wastewater treatment plants an 'average' fractionated wastewater composition was determined, as shown in Table 2. In total, 44 data sets per fraction were used for this calculation. Of the original 48 samples, four samples from the WWTP of Hoek van Holland were eliminated because of specific calamities that influenced the composition. The average composition provides a general picture of the size distribution of the various contaminants in the different wastewater sources. However, as can be seen from the standard deviation (expressed in %), results should not be used as generally valid, since local conditions may influence the wastewater composition and particle distribution. To determine the actual particle distribution for a specific wastewater, the influent should be investigated in detail on site.

In the average wastewater fractionation, the percentage of oxygen consuming components related to the settleable particle fraction was low, with a maximum of 21% for COD. A major part of the total BOD (44%), COD (38%) and phosphorous (35%) was present in suspended and supra-colloidal particle fractions with particle diameters between 1.2 and 63 μm. For nitrogen, only 4% can be related to settleable particles and 13% to colloidal and suspended fractions; so 83% of the nitrogen is present in the soluble form. The conductivity of the wastewater (not shown) decreased insignificantly with increasing particle removal. Figure 2 shows the

Table 2. Average fractionated wastewater composition (with standard deviation (%) in brackets), calculated from data-analysis of experimentally derived influent fractionations

Fraction	Dissolved	Supra dissolved	Colloidal	Supra colloidal	Suspended	Settleable	Raw influent
Parameter	(<0.1 μm)	(0.1–0.45μm)	(0.45-1.2μm)	(1.2 – 5.0 μm)	(5.0 – 63 μm)	(> 63 μm)	
TSS (mg/l)	n.d.	n.d.	n.d.	n.d.	68	62	130
Turbidity (NTU)	n.d.	n.d.	8	12	72	24	116
BOD_5 (mg O_2/l)	82	n.t.	n.t.	24	51	14	170
COD (mg O_2/l)	166	14	9	51	124	97	461
N_{total} (mg N/l)	36.3	0.4	1.3	1.7	2.2	1.7	43.7
P_{total} (mg P/l)	4.08	0.23	0.23	0.39	2.31	0.46	7.7
TSS	-	-	-	-	52 (±18)	48 (±18)	
Turbidity	-	-	7 (±4)	10 (±7)	62 (±17)	21 (±12)	
BOD_5	48 (±12)	-	-	14 (±6)	30 (±8)	8 (±4)	
COD	36 (±10)	3 (±4)	2 (±2)	11 (± 6)	27 (±11)	21 (± 9)	
N_{total}	83 (±25)	1 (±1)	3 (±3)	4 (±2)	5 (±3)	4 (±4)	
P_{total}	53 (±18)	3 (±3)	3 (±1)	5 (±2)	30 (±12)	6 (±3)	

n.d. = not detectable; n.t. = not tested

water quality components related to particle size (average distribution = solid line; 90th percentile = dotted line).

When the particle removal increased, the BOD/N ratio declined since more BOD than nitrogen is related to particles. The average BOD/N ratio decreases from 4.1 in the raw wastewater to 3.6 due to the removal of settleable particles down to 63 μm. For total particle removal, the BOD/N ratio decreased to 2.4. In diluted wastewater (by rain events) the BOD/N-ratio was already low (2.4) in the influent and decreased to below 1 after the complete removal of particles. This will negatively influence the denitrification potential of the remaining wastewater. However, experimental investigations as well as model calculations indicate that the effluent N_{total} concentration of an activated sludge plant can comply with the effluent standard (N_{total} < 10 mg N/l) even when a very intensive pre-treatment is applied (Mels, 2001).

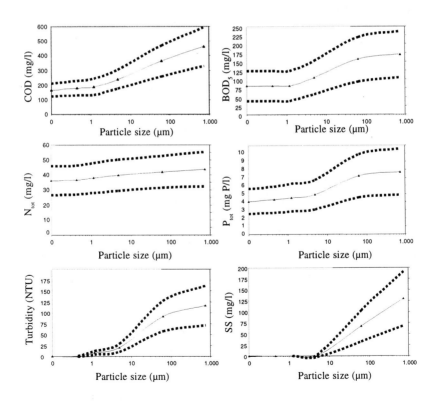

Figure 2. Wastewater quality parameters related to particle size

Energy calculations based on wastewater fractionations

Calculating the potential energy consumption of the remaining wastewater after removal of certain particle fractions can show the effect of advanced particle removal on the energy efficiency of a WWTP.

First, the energy consumption in a theoretical activated sludge system (load from 100000 p.e.) with COD and N removal was calculated based on the wastewater fractionation

With the evaluation model DEMAS+ (Nieuwenhuijzen, 2002), the total energy balance of the water treatment and sludge handling system for 100000 p.e. and a flow of 20000 m³/day was calculated, resulting in Figure 3d. Since nitrogen will hardly be removed by particle removal and COD is strongly related to particles, Figure 3c looks like Figure 3a. The total energy balance (Figure 3d) shows a steeper decline in energy consumption with more advanced particle removal, due to the fact that sludge production and handling depend strongly on particles and can possible lead to a positive energy input. Based on Figure 3, the highest energy saving is possible by the removal of the settleable fraction down to 63 μm (energy savings of 364 MWh/y, 34% of the total energy savings) and the suspended particle fraction between 63 μm and 5 μm (458 MWh/y, 46% of the total energy savings).

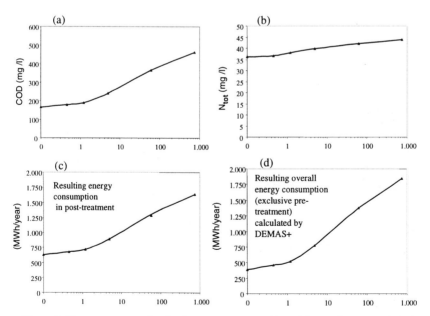

Figure 3. Energy consumption in the post-treatment (c) and the total water and sludge treatment (d), based on a 100000 p.e. WWTP with a daily average flow of 20000 m³/day (particle size in μm on x-axis)

Due to the removal of the remaining colloidal particles between 5 and 0.1 μm again 20 % (equal to 284 MWh/y) of the energy can be saved.

The energy consumption of each particle removal technique has to be compared with the overall energy saved by removing specific particle sizes. Thus the application of a certain pre-treatment technique can be counterbalanced with the resulting overall energy savings in the WWTP.

Cost calculations based on wastewater fractionation

In addition to energy calculations, space requirements, chemical use, final sludge treatment, and even effluent quality can be calculated based on wastewater characterisation by particle size. Together this can be summed up in the total treatment costs (see Figure 4) calculated as net present values (Nieuwenhuijzen *et al.*, 1998).

Figure 4 shows how the financial effective investment costs per pre-treatment technique can be derived. For break-even, a technique that removes particles down to 63 μm (like primary sedimentation) may cost about € 5 million. For a pre-treatment technique that removes particles down to 5 μm, € 13 million may be invested to compete with the primary sedimentation. An additional removal to 0.1 μm allows an additional investment of less than € 4 million.

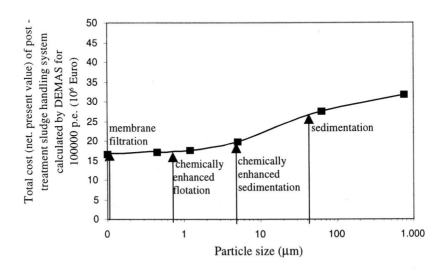

Figure 4. Relationship between treatment costs and size of removed particles (with pre-treatment techniques and particle size removal range)

Pre-treatment potential and optimal chemical dosage

As described above, wastewater fractionation based on particle size makes it possible to predict the applicability and effect of a certain pre-treatment technique on a specific wastewater, the so-called pre-treatment potential. From the fractionation test not only the primary effluent composition but also the quantity and quality of the primary sludge can be derived. This provides an opportunity to create specific, for example, polluted and non-polluted, sludge streams.

Another optimisation step using wastewater fractionation based on particle size could be the determination and control of chemical dosages. By using online fractionation data, an optimal coagulant or flocculant dosage could be applied to remove the desired number of particles or specific components from the wastewater. Mels and Nieuwenhuijzen (2002), for example, recommend using online turbidity-controlled polymer dosing to set particle removal to the level that allows biological treatment to operate optimally. This process control technique would even be optimised by applying (online) wastewater fractionation data and particle size counting.

Conclusions and recommendations

In the 'average' wastewater fractionation, the percentage of oxygen consuming components related to the theoretically determined *settleable* particle fraction is low, with a maximum of 21% for COD. A major part of BOD (44%), COD (38%) and phosphorous (35%) is present in *suspended* and *supra-colloidal* particle fractions with particle diameters between 1.2 and 63 μm. For nitrogen, only 4% can be related to settleable particles and 13% to colloidal and suspended fractions; that leaves 83% of the nitrogen present in soluble form.

By means of wastewater fractionation, the effect and applicability of a certain physical-chemical particle removal technique, the so-called 'pre-treatment potential' of a specific wastewater, can be derived. In addition, energy and cost calculations show the importance of particle removal in the pre-treatment down to approximately 5 to 1 μm, depending on the wastewater, to save energy and costs in the total wastewater treatment system.

From wastewater fractionation experiments, the primary effluent composition and the amount and composition of the produced primary sludge for a specific wastewater could be derived. The specific coagulant or flocculant dosage to gain the most efficient particle removal in the pre-treatment can be determined by wastewater fractionation based on particle size and could be controlled by turbidity-related dosing of organic polymers and online particle size counting.

The applied testing procedure was promising and should be carried out on a broader scale. More advanced (online) sampling, separation, fractionation and analysing techniques should be implemented to optimise the fractionation and characterisation tests.

References

Henze, M. (1992) Characterization of wastewater for modelling of activated sludge processes. *Wat. Sci. Tech.* **25**(6), 1 15

Levine, A.D., Tchobanolous, G., and Asano, T. (1991) Size distributions of particulate contaminants in wastewater and their impact on treatability. *Wat.Res* **25**(8), 911 922

Mels, A.R., Nieuwenhuijzen, A.F. van, Graaf, J.H.J.M. van der, Koning, J. de, Klapwijk, A., and Rulkens, W.H. (1999) Sustainability criteria as a tool in the development of new sewage treatment methods. *Wat. Sci. Tech.* **39**(5), 243-250

Mels, A.R. (2001) Physical-chemical pretreatment as an option for more sustainable municipal wastewater treatment plants. *Ph.D. Thesis*. Department of Environmental Engineering, Wageningen University, Wageningen, The Netherlands

Mels, A.R., and Nieuwenhuijzen, A.F. van (2002) Turbidity-related dosing of organic polymers to control the denitrification potential of flocculated municipal wastewater. In Hahn, H.H., Hoffmann, E., Ødegaard, H. (Eds.) *Chemical Water and Wastewater Treatment VII.* . IWA Publishing

Nieuwenhuijzen, A.F. van, Mels, A.R., Graaf, J.H.J.M. van der, Klapwijk, B., Koning, J. de, and Rulkens, W.H. (1998) Identification and evaluation of wastewater treatment scenarios based on physical-chemical pretreatment. In Hahn, H.H., Hoffmann, E., Ødegaard, H. (Eds.) *Chemical Water and Wastewater Treatment V.* Springer Verlag Berlin, pp. 351-362

Nieuwenhuijzen, A.F. van (2002) Scenario Studies into Advanced Particle Removal in the Physical-chemical Pre-treatment of Wastewater. *PhD Thesis*. Department of Sanitary Engineering, Delft University of Technology, Delft, The Netherlands. 2001

Ødegaard, H. (1998) Optimised particle separation in the primary step of wastewater treatment. *Wat. Sci. Tech.* **37**(10), 43 53

Ødegaard, H. (1999) The influence of wastewater characteristics on choice of wastewater treatment method. In *Pre-print Proceedings of the Nordic Conference on Nitrogen Removal and Biological Phosphate Removal*. Oslo, Norway, 1999

STOWA (2001) Physical-chemical pre-treatment of wastewater—explorative research into wastewater fractionation and biological post-treatment (in Dutch). *STOWA-report 2001-20*. Utrecht, The Netherlands

H.H. Hahn, E. Hoffmann, H. Ødegaard (Eds.)
Chemical Water and Wastewater Treatment VII, pp. 213-222
© IWA Publishing, London
ISBN: 1 84339 009 4

Particle and Microorganism Removal in Treated Sewage Using Several Types of Coagulants

A. Chávez*, B. Jiménez and L. Gilberg

*Instituto de Ingeniería Grupo: Tratamiento and Reúso, UNAM. Coyoacan,Mexico
email: acm@pumas.iingen.unam.mx

Abstract

This study evaluated the efficiency of four coagulants [$Al_2(SO_4)_3$ and three PACs] with different degrees of basicity each, in the separation of particles and microorganisms, such as fecal coliforms and helminth ova, present in sewage to be used for irrigation. A physical-chemical process was simulated in a jar test with applied dosages of 100, 200 and 300 μmolAl/L.

The average volume of particles found in the effluent was 2,624 mL particles/ m^3 of wastewater with a size range between 0.7 and 100 μm, a fecal coliform content of $3.3 \cdot 10^8$ MPN/100 mL, a *Salmonella* content of $2.0 \cdot 10^7$ MPN/100 mL and a helminth ova content of 7.6 HO/L. PAX-X 13 (medium-high basicity) was the most efficient coagulant tested and a dosage of 100 μmolAl/L was enough to remove 79 % of the particles ranging from 0.7 to < 5 μm in size, 97 % of the particles with a size of > 5 to 20 μm, 91.5 % of the particles with a size of > 20 to < 80 μm, and 100 % of the particles with a size > 80 μm. This coagulant also effectively removed 94 % of the fecal coliform and 90 % of *Salmonella*. The success of PAX-X may be due to its degree of basicity (medium-high), which allows for a greater exchange of charges between microorganisms and the system. There was good correlation between particle number and volume for helminth ova; a particle volume of 150 mL/m^3 with particle size mostly < 5 μm corresponds to a concentration of less than 1 HO/L, a value which complies with Mexican standards for reuse in agriculture. The use of a higher coagulant dosage increased particle removal up to 98 %, mainly in the case of the smallest particles (< 5 μm), and also increased removal of FC and *Salmonella*.

[213]

Introduction

The design and interpretation of the performance of a wastewater treatment system is typically defined in terms of BOD, COD, turbidity and TSS, which provide quantitative, but non-specific information on the substances present in sewage. These parameters express the presence of a heterogeneous mix of compounds between 0.001 μm and 100 μm with specific molecular weights (Levine, 1991 and Adin, 1999). These compounds have very different properties among them, which influence their behaviour and, thus, their removal efficiency in the sewage treatment processes (O´Melia *et al.*, 1997).

In Mexico, the main objective of any sewage treatment system that provides water for irrigation purposes is to reduce the concentration of helminth ova (ranging in size between 20 and 80 μm) and fecal coliforms (ranging in size from 0.7 to 1.5 μm) to levels that do not affect the health of the field workers or the consumers. Advanced Primary Treatment (APT), a physical-chemical process, has proven itself to be a good option (Jiménez *et al.*, 1999). In order to enhance the effectiveness of this process, several different coagulants $Al_2(SO_4)_3$, $FeCl_3$, $Ca(OH)$ and more recently the polyaluminum chlorides (PACs) have been used successfully. The effect of these coagulants on the removal of microorganisms and different-sized particles, however, is unknown. The goal of this study was therefore to evaluate the efficiency of $Al_2(SO_4)_3$ and three PACs with different degrees of basicity to remove microorganisms, and to correlate each coagulant with the size of the particles removed in treated sewage.

Background

In Mexico, due to the lack of water and high content of organic matter and natural fertilizers (N and P) in domestic wastewater, raw wastewater was reused for irrigation since 1896, but it was not until 1920 that people started to see the economic advantages. The reuse of sewage was a spontaneous development that was neither considered nor planned in the wastewater drainage works of the Valley of Mexico, which irrigates the Mezquital Valley. This custom significantly increased productivity of the main agricultural products (corn, alfalfa, wheat and barley) from 50 to 150 % over the national average, in an area of approximately 900000 ha, through a distribution network of 575 km (CNA *et al.*, 1998; Jiménez *et al.*, 1999).

Before sewage can be used in this area, it has to travel approximately 70 km at a speed of less than 1 m/s, which changes the water quality by promoting anaerobic conditions and sedimentation along the way. These characteristics make its treatment more difficult due to its high microbial content (CNA *et al.*, 1998 and Jiménez *et al.*, 1999).

The density and diversity of the helminth ova found in wastewater varies according to the intensity and prevalence of parasitic diseases in the communities.

At a national level, this situation is critical, as Tay *et al.* (1991) report that 33 % of the Mexican population has helminthiasis and in the Mezquital Valley, with a population of 500000, 10 children die each day from gastrointestinal diseases and the lack of resources to treat them. In the case of helminth infections, death results from complications with anemia, cysts in vital organs (kidney, brain or lungs), or bursting of the intestine.

Studies carried out by Jiménez *et al.* (1999) and Chávez and Jiménez (2000), determined that wastewater used in agriculture in Mexico is characterized by having high levels of a large variety of pathogens (Table 1). This wastewater must be treated in such a way that the effluent poses no health risks. The most common genera of helminths are *Ascaris spp.* (in 90 % of the cases), *Hymenolepis nana* (6 %), *Toxocara sp.* and *Trichuris sp.* (2 % each).

Advanced Primary Treatment (APT) is one of the *ad hoc* alternatives for treating this type of wastewater, since the installation of oxidation ponds (recommended by WHO, 1989) is difficult where there isn't enough space, where land is scarce and costly or where water evaporation is high.

Table 1. Comparison of the microbiological content in wastewater from Mexico and other countries. Adapted from: Chávez and Jiménez (2000)

Parameters	Size μm	Concentration	Country	Reference
Helminth ova (HO), HO/L	20-80	116-202	Brazil	Blumenthal, *et al.*, 1996
		840	Morocco	Schwarzbrod, *et al.*, 1989
		2-375	Mexico	Jiménez *et al.*, 1997, 1999
		28.3	Ismalia, Abu Attwa	Stott *et al.*, 1996
		9	France	Schwarzbrod, *et al.*, 1989
		1-8	USA	USEPA, 1992
		6	Ukraine	Ellis *et al.*, 1993
Fecal coliforms (CF), MPN/100 mL	0.7-1.5	10^7-10^9	Mexico	Jiménez *et al.*, 1997, 1999
		10^4-10^5	United Kingdom	Sttot *et al.*, 1996
		10^5	Egypt	Sttot *et al.*, 1996
		10^3-10^5	USA	USEPA, 1992
Salmonella spp., MPN/100 mL	0.7-1.5 width	10^6-10^9	Mexico	Jiménez *et al.*, 1997, 1999
		10^2-10^4	USA	USEPA, 1992
	2.5-5.0 length	10^3	United Kingdom	Sttot *et al.*, 1996
Protoozoa cysts, organisms/L	4-8 or	979-1817	Mexico	Jiménez *et al.*, 1997, 1999
	10-12	1500	United Kingdom	Hall T. And Croll B., 1997
	or	28	East of the USA	USEPA, 1992
	4-6	*0.9 (13 max.) **1.6 (10.8 max)	Germany	Karanis P., 1998

Entamoeba histolytica, E. coli, Giardia lamblia*, Balantidum coli and Cyptosporidium**

Furthermore a close correlation has been determined between the helminth ova (HO) content and the total suspended solids (TSS) in wastewater. Jiménez *et al.* (1999) suggest that such a correlation should be performed for every specific case. In Mexico it is very important to evaluate the APT efficiency based on the change and distribution of particles and thus optimize and correlate it quickly and directly with the content of HO and fecal coliforms (FC).

HO determination is too time-consuming and too expensive (a delay of 5 days at a cost of €80), so in this case it is necessary to correlate particle size distribution (a delay of 10 minutes at a cost of €15) present in water treated with an APT system, with HO content as an indirect way to determine HO and fecal coliform content.

Materials and methods

Sewage sampling. The sewage examined came from one of the most important drainages of Mexico City, the Central Emitter, which carries approximately 20 m^3/s, which corresponds to 45 % of the total wastewater from Mexico City. The samples were taken at constant intervals over the course of three months on the same day, at the same hour, and in the same place.

Coagulants. The following coagulants were compared:

- (A) $Al_2[SO_4]_3$

and three polyaluminum chlorides with different degrees of basicity, a property that results in varying degrees of efficiency

- (B) PAX XL 60, 40% basicity (low basicity)
- (C) PAX XL 13, 68% basicity (medium-high basicity)
- (D) PAX XL 19, 84% basicity (high basicity)

The dosages applied in each case were 100, 200 and 300 μmol Al/L.

Coagulation method. To simulate the physical-chemical process, one liter of wastewater and an automatic flocculator were used. Table 2 shows the speed used in each stage. These conditions were optimized in preliminary tests for treating this type of wastewater. After the sedimentation phase, 100 mL of supernatant was taken to determine physical-chemical and microbiological properties.

Table 2. Speed stages used in the jar tests

Stages	Mix time	G s^{-1}
Coagulation	30 s	380
Flocculation	10 min	15
Sedimentation	15 min	0

Evaluation Parameters. Table 3 shows the parameters determined for untreated and treated sewage, as well as the techniques used in each case. All parameters were determined between 1 and 8 h after the treatment.

To determine particle size and distribution (PSD), a Multisizer Coulter Counter II equipped with 20 and 100 µm capillary tubes was used. Particles with diameters from 0.7 to 80 µm were analyzed in four size intervals, with limits of up to 5 µm (FC and *Salmonella*), 6.4 µm, 20 µm (protozoan cysts), and >20-80 µm (helminth ova). Prior to this analysis, samples were filtered through a membrane filter with 20 and 100 µm pores, respectively, and diluted 5/200 with an electrolytic solution. After the sedimentation process, a small dosage of formaldehyde (0.33 mL/100 mL sample) was immediately added to prevent the possible regrowth of microorganisms in each sample.

Table 3. Parameters and techniques used to evaluate untreated and treated sewage

Parameters	Technique
Turbidity, NTU	HACH 2100, without filtration
PO_4-P, mg/L	Colorimetric method using the HACH technique
pH	Potentiometer, technique 4500-HB*
TSS, mg/L	Gravimetric, technique 2540-B*
Temperature, °C	Thermometer, technique 2550 B*
Conductivity, µS/cm	Conductimeter, technique 2510 B*
NH_4,-N,mg/L	Colorimetric, technique of silicylate HACH
Fecal coliforms MPN/100 mL	Membrane filtration, NMX-AA-042-87
Salmonella, MPN/100 mL	Membrane filtration, technique 9260D*
Helminth ova, HO/L	Sedimentation or density, NMX-AA-113-99

* Standard Methods of Examination of Water and Wastewater

Results and discussion

Due to the particular conditions of the Mexico City drainage system, sewage quality varied with the sampling date. The average particle volume determined was 2,624 ml of particles/m^3 of wastewater, that is, 0.26 % of the total pollutants. Of these, 9.1% fell in the size range of < 0.7 µm; 60.4 % between 5 and < 20 µm; 24.3 % between 20 and 80 µm and only 6.2 % > 80 µm.

In the jar test, the coagulant used determined the particle removal efficiency (Figure 1). Coagulant C (medium-high basicity) removed the highest percentage of particles in all size classes, while coagulants B (low basicity) or D (high basicity) were far superior to Coagulant A.

In sewage, the fecal coliform content was 3.3 • 10^8 MPN/100 mL and *Salmonella* was 2.0 • 10^7 MPN/100 mL. Coagulant *C* (PAX-XL13) removed the highest number of microorganisms (94% FC and 90% *Salmonella* or 1.6 and 1.2 log units,

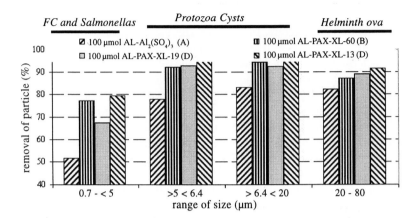

Figure 1. Percentage of particles removed by the four coagulants at a dosage of 100 μmolAl/L, based on particle size

respectively) presumably due to the fact that this is a medium-high basicity coagulant and thus performs better than the other coagulants during the removal process. Even though Coagulant *C* (medium-high basicity) removed up to 1.6 logarithmic units (of FC $8.35 \cdot 10^6$ MPN/100 mL), these removals were not enough to comply with Mexican legislation which is <1000 MPN/100 mL. In order to meet these limits, it is necessary to include UV or chlorine disinfection processes. The large amount of remaining particles in the effluent treated through APT will have a low size of 5 μm, value which should not affect the performance of the disinfection system, especially if used with a UV system.

There was a good correlation between the volume of the particles and HO content. The highest particle number and volume determined in each one of the effluents was found in the size class < 5 μm, which means there is a high probability of not finding HO (Figure 2). In fact, when the effluent had a particle volume of < 150 mL/m³ (coagulant *C*) and sizes < 5 μm, the HO content was always < 1 HO/L. This value guarantees a water quality with no restrictions on its use according to the NOM- 001-ECOL-96. Correlating particle volume with HO content would allow us to determine a system's expected microbial quality in less than 5 minutes instead of 5 days, making it possible to adjust the treatment regime quickly and appropriately.

Figure 3 shows that as the dosage of coagulant *C* (medium-high basicity) was increased, particle removal was enhanced, especially in the smallest range (< 5 μm, corresponding to *FC* and *Salmonella*). Higher dosage did not lead to enhanced particle removal in the 20-80 μm size class, possibly due to the resuspension of the particles.

TSS and turbidity were other parameters used to evaluate the performance of the APT. TSS in the effluent was 245 mg/L. TSS could be correlated to PSD

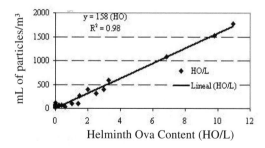

Figure 2. Correlation between the particle volume and the HO content in treated and untreated sewage

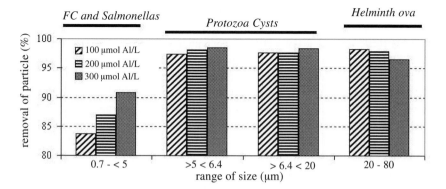

Figure 3. Effect of the coagulant dosage (coagulant C) on the removal of different size classes of particles

with an $r^2 = 0.91$. TSS and PSD were determined for dosages of 100, 200 and 300 μmol Al/L of coagulant C, whereby higher dosages resulted in lower particle concentrations and lower TSS.

Turbidity was also very closely related to the number of particles in treated water; as turbidity increased, the number of particles increased. The best turbidity removal was achieved with Coagulant *C* (medium-high basicity), with removal rates of 79 %, 87 % and 90 % when a dosage of 100, 200 and 300 μmol Al/L, respectively, was applied. During the treatment, TSS and turbidity removal behaved similarly, so it was possible to do a correlated series. The correlation between particle number and turbidity [9E+10 (turbidity) +1E+11] with an $r^2 = 0.93$ was very important especially in on-line systems where turbidity always serves as a reference.

In terms of nutrient content, the APT process removed at most 25 % of the nitrogen, regardless of the applied dosage, and aluminum sulfate removed less than the PACs. With regard to PO_4-P, the removal efficiency achieved depended on the type and dosage of coagulant, with the following preferences: Coagulant $A > B > C > D$, and with a dosage of 100 μmolAl/L the removals were 21%, 25 %, 13 % and 12 %, respectively. The greatest removal was found when PAC was dosed into the sewage at a lower pH (7.16) since low pH promotes $AlPO_4$ precipitation.

Conclusions and recomendations

In an APT, particle and microorganism removal depend on the type and dosage of the applied coagulant. The PAC with medium-high basicity (PAX-XL13, 100 μmol Al/L) removed the highest number of particles, up to 79 % of particles smaller than 5 μm. $Al_2(SO_4)_3$ removed only 51% of them. Following this comparison, 97 % and 83 % of particles between 5 and 20 μm were removed, while 91.5 % and 82 % of particles between 20 and 80 μm were removed, under the same selectivity order. Close to 100 % of particles of > 80 μm were withdrawn from the system in all cases. On the other hand, a high basicity coagulant produced particle resuspension which led to lower removal efficiencies for larger particles.

Higher dosages of coagulant enhanced the removal of particles, mostly those smaller than 5 μm, and also enhanced removal of fecal coliforms and *Salmonella* (1.6 and 1.2 log units for coagulant C). In spite of the significant improvement in water quality, it was not possible to achieve the 99.99 % removal needed to comply with the legislation. Disinfection with UV or chlorine is necessary to destroy the remaining microorganisms.

There was good correlation between particle number and volume for helminth ova; a particle volume of 150 mL/m^3 with particle size mostly < 5 μm corresponds to a concentration of less than 1 HO/L, a value which complies with Mexican standards for reuse in agriculture. Furthermore, a larger number of data is recommended, especially to evaluate what the particle volume is when the influent has from 4 to 10 HO/L of helminth ova.

These studies were performed in a batch system, where every single condition was controlled in each unit operation. In future work, similar studies should be performed in continuous systems, because changing conditions will change the results.

References

Adin A. (1999) Particle Characteristics: A Key Factor in Effluent Treatment and Reuse, *Water Science and Technology* **40** (4-5), 67-74

American Public Health Association, Water Works Association and Water Environment Federation (1995) Standard Methods for the Examination of Water and Wastewater. 19th ed. Washington D.C

Chávez A., and Jiménez B. (2000) Particle Size Distribution (PSD) Obtained in Effluent from and Advanced Primary Treatment Process Using Different Coagulant. Chemical

Water And Wastewater Treatment VI: Proceedings of the 9th International Gothenburg Symposium, (Hahn, H.H., Hoffmann, E. and Ødegaard, H.(eds)) Springer-Verlag, Berlin, pp 257-268

National Water Commission, British Geological Survey, London School of Hygiene and Tropical Medicine and University of Birmingham. (1998) Effects of Wastewater Reuse on Groundwater in the Mezquital Valley, Hidalgo State, Mexico. Final Report, November 1998. BGS Technical Report WC/98/42

Jiménez B., Chávez A., and Hernández C. (1999) Alternative Treatment for Wastewater Destined for Agricultural Use in Mexico. *Water Science and Technology* **40**(4-5), 355-362

Levine A., Tchobanoglous G., and Asano T. (1991) Particle Contaminants in Wastewater: A Comparison of Measurement Techniques and Reported Particle Size Distributions, *American Filtration Society* **4** (2), 89-105

Official Mexican Standard. NOM-001-ECOL/1996. The Official Federal Gazette. January 6, 1997, 67-81

O'Melia C., Hahn M., and Chen C. (1997) Some Effects of Particle Size in Separation Processes Involving Colloids, *Water Science and Technology* **36** (4), 119-126

Tay, Z, Aguilera L. Quiróz G and Castrejón V (1991) Parasitología Médica de Francisco Méndez Cervbefore. pp 6-327. In Spanish

World Health Organization (1989) *Health Guidelines for the Use of Wastewater in Agriculture and Aquaculture*, Technical Report Series No. 778. WHO, Geneva

H.H. Hahn, E. Hoffmann, H. Ødegaard (Eds.)
Chemical Water and Wastewater Treatment VII, pp. 223-232
© IWA Publishing, London
ISBN: 1 84339 009 4

Optimisation of Dual Coarse Media Filtration for Enhanced Primary Treatment of Municipal Wastewater

Z. Liao* and H. Ødegaard

*Norwegian University of Science and Technology, Department of Hydraulic and Environmental Engineering, Trondheim, Norway
email: Liao.Zuliang@bygg.ntnu.no

Introduction

Various authors (Levine, *et al.,* 1985 and 1991; Ødegaard, 1999 and 2000; Nieuwenhuijzen, 2002; Liao, 2002) have demonstrated that treatment of municipal wastewater depends to a very large extent on successful separation of particulate substances coming from the raw wastewater. Settling as an "old, good" process for primary treatment has inherent disadvantages such as low overflow rate and low removal efficiency. Coarse media filtration as an alternative process to primary settling for enhanced primary treatment has emerged over the last years. An earlier investigation (Ødegaard *et al.,* 1998) demonstrated the feasibility of the process. The most important challenge is to determine suitable media and filter bed configuration together with proper chemicals (polymers) in order to achieve good SS removal (> 75 %) together with acceptably long filter run times (> 6 h) at high filtration rates (10 - 30 m/h). Two different Kaldnes biofilm carrier media, the smaller K1 and the bigger K2, were compared as filter media in single media coarse filters (Liao and Ødegaard, 2001; Liao, 2002). It was concluded that both the K1 and K2 media filters were feasible to be used in enhanced primary treatment. The K1 media filter resulted, however, in higher SS removal but rather high head loss rates, while the K2 media filter resulted in lower SS removal but rather low head loss rates. In order to optimise the process, dual media filters combining two Kaldnes media could be expected to give an optimal configuration in order to optimise the process. This was evaluated in the research that is presented in this paper.

[223]

Experimental set-up and arrangement

The two Kaldnes media K1 and K2 (see Figure 1) were made of materials with two different densities, polyethylene with a density of 950 kg/m$^3_{material}$ (floating-denoted as L for "light"), and PVC with a density of 1450 kg/m$^3_{material}$ (sinking-denoted as H for "heavy"). The K1 media is formed as a short cylinder with inner cross fins and 18 short (1 mm) outer fins, while the K2 media is formed as two concentric cylinders with inner partitions and 12 long (5 mm) outer fins. So the four types of media of various sizes and densities are K1L, K2L, K1H, and K2H, respectively. The three filter bed configurations for dual media filters proposed for the experiments are shown in Figure 1. The experimental set-up is shown in Figure 2 together with the dimensions of two sizes of media.

The experimental set-up shown in Figure 2 is for the upflow filters. For down-flow K1H/K2H filter, the set-up (not shown here) was modified by introducing the influent to the top of the filter column and recovering the effluent from the bottom. The polymer dosing point was changed accordingly. Raw wastewater from the grit chambers of Ladehammeren Wastewater Treatment Plant (LARA) in Trondheim, Norway was pumped to the feed tank and distributed to each filter column through a measurement column of diameter 100 mm. Polymer was dosed and mixed through an in-line mixer.

Experiments were first carried out in order to compare the floating K1L/K2L filter and the partly floating, partly sinking K1L+K2H filter. Filtration rates were kept at 10-30 m/h and various dosages of the polymer FO4440SH (high MW, medium charge density, cationic polymer supplied by Floerger via Norsk Hydro AS) were applied. Later the sinking K1H/K2H filter was compared with the single K1H filter as reference (results not shown) at a filtration rate of 10 m/h. Finally, the K1H/K2H filter was compared with the K1L/K2L filter at a filtration rate of

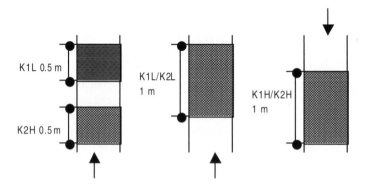

Figure 1. Bed configurations of dual media filters: (left) Upflow K1L+K2H filter, (middle) Upflow K1L/K2L filter, and (right) Downflow K1H/K2H filter

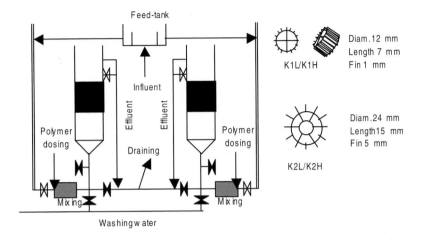

Figure 2. Experimental set-up (left) for dual media filters based on Kaldnes media K1 and K2 (right) (redrawn from Ødegaard *et al.*, 2002)

20 m/h. The three dual media filters were also backwashed properly based on previous work on floating filters (Liao, 2002). Backwashing consisted of a two-step procedure which involved batch air-scouring for 3-5 min with draining from the bottom of the filter, followed by a continuous air and water washing with draining from the top of filters. Optimal backwash parameters for the different dual media filters were determined and the optimal procedure was then performed.

Comparison of filtration performances in dual media filters

Comparison of the K1L+K2H filter and the K1L/K2L filter

Two parameters, SS removal efficiency (SS%) and specific head loss (dH/SS$_a$), were used for comparison of the filter performance. Specific head loss is defined (in a filter run to 1 m head loss) as the head loss (mm) divided by SS accumulation per unit cross section per 1 m filter bed depth (mm/(kg/m^3)). This parameter makes it possible to compare the head loss properties for different filters under various conditions. Figure 3 compares SS removal efficiency (SS%) and specific head loss (dH/SS$_a$) between the K1L+K2H filter and the K1L/K2L filter. The raw wastewater during this period had an SS concentration in the influent of 264 ± 46 mg/l. The filtration rates varied from 10 to 30 m/h and the dosage of polymer varied from 0 to 4 mg/l.

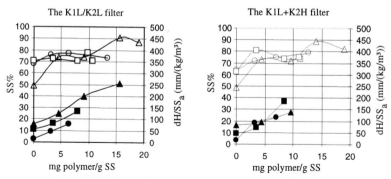

Figure 3. Comparison of SS removal efficiency and specific head loss as a function of dosage ratio (mg polymer/g SS) in (left) the K1L/K2L filter and (right) the K1L+K2H filter (SS% 10 m/h, dH/SS$_a$ 10 m/h; SS% 20 m/h, dH/SS$_a$ 20 m/h; SS% 30 m/h, dH/SS$_a$ 30 m/h)

SS removal efficiency was similar in both filters with a critical dosage ratio of about 5 mg polymer/g SS at which the SS removal was 70-80%. When the dose was increased to 15 mg polymer/g SS, the SS removal could be increased up to 90% even at the high filtration rate of 30 m/h. This caused, however, a dramatic increase in head loss, reducing the length of the filter run significantly. When no polymer dose was used, the filtration rate of 20 m/h seemed to be a critical one, above which the SS removal was significantly lower. At the filtration rates of 10 and 20 m/h, the SS removal was found to be as high as 65-70% without polymer dosage.

It was found that the specific head loss in the K1L/K2L filter was related (but not exactly proportional) to the filtration rate at the same dosage. In the K1L+K2H filter this filtration-rate related difference in specific head loss was insignificant. The sieve immediately underneath the K2H media layer in the K1L+K2H filter may have taken out part of the flocculated aggregates, masking the difference. However, in this filter, the K2H layer had a tendency to move up during the experiments when high polymer dosages and high filtration rates were used. Compared with the single K1L filter (Liao and Ødegaard, 2001), the specific head loss in the two media filters was nearly halved, while the SS removal efficiencies were nearly the same. This indeed reveals the benefit of combining K1 and K2 media into dual media filters.

Performance of the K1H/K2H filter

The performance of the K1H/K2H filter was to some extent different from the two filters discussed above. The SS in the influent varied quite a bit over this experimental period, being on average 371 ± 182 mg SS/l when 10 m/h was applied and on average 162 ± 75 mg SS/l when 20 m/h was applied.

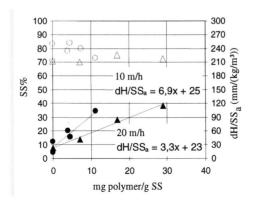

Figure 4. SS removal efficiency and specific head loss in the K1H/K2H filter (SS% 10 m/h, dH/SS$_a$ 10 m/h; SS% 20 m/h, dH/SS$_a$ 20 m/h)

When no polymer was applied, the SS removal efficiency at 10 m/h was found to be 70-85% and it was about 70% at the filtration rate of 20 m/h. Dosing polymer actually did not improve the SS removal very much. Since the SS in the influent at 20 m/h was lower, more than 70% SS removal or less than 50 mg SS/l in effluent in the K1H/K2H filter was considered to be acceptable.

The head loss property of the K1H/K2H filter is definitely different from that of the other two filter configurations. Lower specific head loss at a higher filtration rate (20 m/h) demonstrates that more even distribution of SS accumulation throughout the filter bed depth was obtained, as a result of more SS deposition in the deeper layers.

An analysis of how the two media K1H and K2H were packed after each backwash procedure is helpful for the explanation of the head loss property. Although the material density of the two media, K1H and K2H, is the same (1450 kg/m^3), the drag force on the K2H media was larger during the backwash process so that more K2H media were brought up by the simultaneous air and water backwash. Therefore, a grading effect through the depth of the filter bed occurred, with more K2H media in the top layer and more K1H media in the bottom layer. This resulted in an ideal bed configuration with more bigger pores and higher porosity in the top layer.

Comparison of specific head loss in the K1L+K2H and the K1H/K2H filter

The specific head loss in the K1L+K2H and the K1H/K2H filters is plotted as a function of dosage ratio (mg polymer/g SS) for the filtration rates of 10 and 20 m/h, respectively, in Figure 5. Trend lines of specific head loss demonstrate that the specific head loss in the K1H/K2H filter developed slower and had lower initial specific head loss at the same filtration rates than the K1L+K2H filter. These results suggest that a longer filter run time can be expected in the K1H/K2H filter.

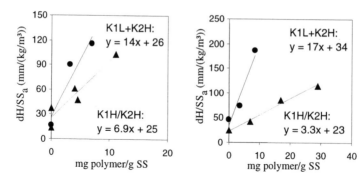

Figure 5. Comparison of specific head loss in two dual media filters at filtration rates of (left) 10 m/h and (right) 20 m/h (■ the K1L+K2H filter; ▲ the K1H/K2H filter)

Based on the filter performances (SS removal efficiency and head loss properties), all three dual media filters may be used for enhanced primary treatment. We conclude, however, that the K1H/K2H filter is the best filter bed configuration based on the lowest specific head loss and similar SS removal efficiency under similar SS concentration in influent.

Optimisation of backwash procedure for the K1H/K2H filter

The backwash procedure for the K1H/K2H filter was 3-5 min air scouring followed by simultaneous air and water washing with continuous draining from the top of the filter. The air flow rate (A) varied from 60 to 120 $m^3/m^2_{filterbed}$/h and the water flow rate (W) varied from 30 to 40 $m^3/m^2_{filterbed}$/h. The variable rates ensured that the K1H/K2H media bed could be fluidised by air scour and that the sludge accumulated in the filter bed could be detached from filter media and cleaned completely. Three combinations of air and water flow rates were A60W30 (air flow rate of 60 $m^3/m^2_{filterbed}$/h and water flow rate of 30 $m^3/m^2_{filterbed}$/h), A60W40 and A120W40, respectively.

Samples of sludge water were taken at intervals of 5 min from the start of water supply for backwash and the SS concentration was measured. Then the water consumption was calculated by multiplying water flow rate by the elapsed time of water flow. SS concentration of each sample in the sludge water could then be correlated with water consumption ($m^3/m^2_{filterbed}$) (Figure 6). The denotation of "3F+A60W30" indicates that the corresponding filtration cycle was operated at dosage of 3 mg of the polymer FO4440SH/l. All experiments on backwash were carried out for filter cycles at a filtration rate of 10 m/h.

First of all, Figure 6 shows that SS and water consumption for backwash are related exponentially. The exponential constants illustrate how fast the SS concentration is reduced in the sludge water.

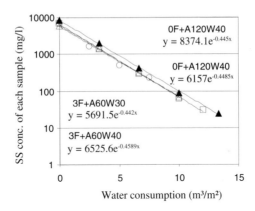

Figure 6. Correlation between SS concentration and water consumption for backwash in the K1H/K2H filter (○ SS (3F+A60W30), □ SS (3F+A60W40), △ SS (0F+A120W40(1)), ▲ SS (0F+A120W40(2)))

Various combinations of air and water flow rates from A60W30 to A120W40 do not seem to have a significant influence on the backwash performance in the K1H/K2H filter. An exponential constant of −0.45 can be assumed for the optimisation evaluation. It is recommended from the backwash performance that the K1H/K2H filter should be backwashed at an airflow rate of 60 $m^3/m^2_{filterbed}$/h and a water flow rate of 30 $m^3/m^2_{filterbed}$/h.

The time needed for the total backwash procedure also depends on the total SS accumulation. For example, if the total SS accumulation is 20 kg $SS/m^3_{filterbed}$, the SS concentration in the filter column (including the space of 1 m height between the surface of the filter bed and the effluent/draining) after 3-5 min air scour is calculated to be 10 $kg/m^3_{filtervolume}$ or 10000 mg/l. If the exponential constant is set at −0.45 and the final SS concentration in the spent water after backwash is assumed to be 100 mg/l (SS concentration in sludge water is reduced by 99 %), the water consumption is calculated to be 10 m^3/m^2. With a water flow rate of 30 $m^3/m^2_{filterbed}$/h, the time for water washing is 0.33 hour or 20 min. The total time for backwash is then 25 min.

Optimal performance of the K1H/K2H filter—a prediction

In order to predict the optimal performance of the K1H/K2H filter, both the filtration performance and the backwash procedure must be described by comprehensive parameters. For example, it is first assumed that the influent SS is 200 mg/l, the optimal dosage ratio is 5 mg polymer/g SS (or dosage 1 mg/l) and the filtration rate is in the range of 10-30 m/h. At an SS removal of 75 %, the specific

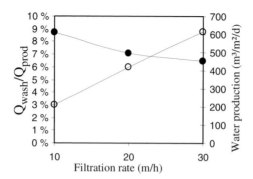

Figure 7. The ratio of wash water consumption to water production (—●—) and water (filtrate) production (—○—) as a function of filtration rate for the K1H/K2H filter

head loss is estimated as 59, 45, and 40 mm/(kg/m³) for filtration rates of 10, 20 and 30 m/h, respectively. Finally the exponential constant of the backwash procedures is set at –0.45 and the final SS concentration after backwash is assumed to be 100 mg/l.

The length of filter run time to 1 m head loss is calculated to be 11.5, 7.5 and 5.5 hours for filtration rates of 10, 20, and 30 m/h, respectively. The ratio of water consumption for backwash relative to the amount of treated water (filtrate) in each filter cycle and the daily water (filtrate) production are plotted as a function of the filtration rate in Figure 7.

The ratio of water consumption for backwash to water production decreases from 8.7 % for 10 m/h to 7 % for 20 m/h and to 6.5 % for 30 m/h. The water (filtrate) production per day reaches 210, 417, and 616 m³/m²/d at filtration rates of 10, 20, and 30 m/h, respectively. Even though a filtration rate of 30 m/h yields higher water production, we believe that a filtration rate of 20 m/h will be more typical in practice.

Conclusions

1. Three different dual media configurations were tested for coarse media filtration. All three dual media filters were found to be feasible for enhanced primary treatment under high filtration rates (10-30 m/h) and low dosages (1-2 mg/l) of the high MW, medium charge density, cationic polymer FO4440SH.

2. A dosage ratio higher than 5 mg polymer/g $SS_{influent}$ did not improve SS removal significantly, but may have a negative effect on specific head loss. A dosage of 1-2 mg FO4440SH at an SS concentration in the influent between 200 and 400 mg/l was found to be optimal.

3. Among the three filter configurations for dual media coarse filters that were

tested, the downflow K1H/K2H filter gave the best results when all factors (removal efficiency, head-loss development, filter run length) are taken into consideration.

4. At the optimal dosage, the K1H/K2H filter had a SS removal of 70-80 %, and a filter run time (to reach 1 m head loss) of more than 6 hrs at filtration rate of 10-30 m/h.

5. The optimal backwash procedure for the downflow K1H/K2H filter was 3-5 min air scouring followed by continuous air (60 m/h) and water (30 m/h) washing and draining from the top. The water consumption for backwash can be expected to be between 9 % - 6 % for filtration rates of 10-30 m/h.

6. Further improvement of the SS removal could only be obtained by combining the K1H/K2H dual media with other finer granular media in a multi-media filter. A Kaldnes-Filtralite-Sand (KFS) filter (Ødegaard, *et al.*, 2002) is under investigation for advanced particle separation from wastewater.

Acknowledgement

Financial support from the Norwegian State Education Funds and Kaldnes Environmental Technology AS is greatly appreciated. We thank the staff at the Ladehammeren Wastewater Treatment Plant in Trondheim, Norway for their support and cooperation.

References

Levine, A. D., Tchobanoglous, G., and Asano, T. (1985) Characterisation of the size distribution of contaminants in wastewater: treatment and reuse implications. *J. WPCF* **57**, 805-816

Levine, A. D., Tchobanoglous, G., and Asano, T. (1991) Size distribution of particulate contaminants in wastewater and their impact on treatability. *Wat. Res.* **25**(8), 911-922

Liao, Z., and Ødegaard, H. (2001) Coarse media filtration for enhanced primary treatment of municipal wastewater. In *Proceedings of the 2nd World Water Congress*, Oct. 15-19, 2001, Berlin, Germany

Liao, Z. (2002) Coarse media filtration for enhanced primary treatment of municipal wastewater. PhD thesis (in preparation), Norwegian University of Science and Technology

Nieuwenhuijzen, A. F. van (2002) Scenario studies into advanced particle removal in the physical-chemical pre-treatment of wastewater. PhD thesis, Technische Universiteit Delft, The Netherlands

Ødegaard, H., Ulgenes, Y., Brevik, D., and Liao, Z. (1998) Enhanced primary treatment in floating filters. In *Chemical Water and Wastewater Treatment V* (Hahn, H.H., Hoffmann, E. and Ødegaard, H., eds.) Springer-Verlag, Berlin, Heidelberg, New York, pp. 149-162

Ødegaard, H. (1999) The influence of wastewater characteristics on choice of wastewater treatment method. In *Pre-print for Proceedings of the Nordic Conference on Nitrogen Removal and Biological Phosphorous Removal*. Oslo, Norway

Ødegaard, H. (2000) Advanced compact wastewater treatment based on coagulation and moving bed biofilm processes. *Wat. Sci. Tech.* **42**(12), 33-48

Ødegaard, H., Liao, Z., and Hansen, A. T. (2002) Coarse media filtration—An alternative to settling in wastewater treatment. In *Proceedings of the 3rd World Water Congress*, April 7 - 12, 2002, Melbourne Australia

H.H. Hahn, E. Hoffmann, H. Ødegaard (Eds.)
Chemical Water and Wastewater Treatment VII, pp. 233-242
© IWA Publishing, London
ISBN: 1 84339 009 4

Granulated Iron-Hydroxide (GEH) for the Retention of Copper from Roof Runoff

M. Steiner and *M. Boller*

*Environmental Engineering Water Resources and Supply, Dübendorf, Switzerland
email: michele.steiner@eawag.ch

Introduction

In Switzerland, copper in roof runoff originates to a significant amount from corroding copper surfaces on or attached to roofs. Nowadays, up to 50% of the copper load in a combined sewer system can be related to roof runoff (Boller and Steiner, 2001). According to the Swiss water protection law, on-site roof runoff infiltration has to be prioritised to direct discharge into receiving waters or into combined sewer systems. In future urban water systems, copper will therefore be distributed in a diffuse way to soil infiltration layers, to receiving waters and sediments. The consequences are the formation of hazardous wastes in infiltration sites and sediments and elevated copper concentrations above critical concentration levels for many aquatic organisms especially in small rivers. Until source control measures such as the use of alternative materials will be implemented, on site filter systems for the removal of copper from roof runoff are considered to be a feasible solution.

In view of the large roof areas which are already or will be covered with copper, a special adsorber system was developed which removes copper from runoff and retains it in a controlled smallest possible volume. In contrast to present infiltration techniques, exhaustion can be predicted which enables the exchange of the adsorber material before severe breakthrough into the environment occurs. In order to achieve a feasible technical solution for the treatment of metal containing runoff water, sufficient hydraulic conductivity, high copper removal efficiency, long life cycle time of several years and reasonable costs are important characteristics of the new adsorber. After exhaustion, the adsorber has to be properly disposed of or the copper can be regained from the adsorber.

[233]

Materials and methods

Because of the well documented ability of iron-hydroxide to adsorb copper (Benjamin *et al.*, 1996, Benjamin and Leckie, 1981), granulated iron-hydroxide (GEH) was chosen as adsorber material for the new type of infiltration system. GEH was obtained from Wasserchemie GmbH, Osnabrück, Germany. Because of its granular structure, GEH may be directly applied as a filter layer in infiltration facilities or deep bed filter systems. Internal porosity and BET surface are 75% and 280 m^2 g^{-1}, filterbed porosity and density are 0.22 - 0.28 and 1.3 kg GEH l^{-1}, respectively, and grain size is 0.32 - 2 mm.

Mass transport in porous media can be described as (1) transport from the solute phase to the sorption sites and (2) the adsorption process itself. While the latter process is assumed to be not limiting, transport is regarded as the crucial kinetic process. Hence, a steady state model was used to describe the adsorption process which was combined with a dynamic transport model. In order to gain data for model calibration and verification, batch as well as column experiments were performed.

Batch experiments were carried out in 50 ml centrifugation tubes (TPP, γ sterilised), stored as all other equipment in a 0.1 M HNO_3 solution for at least 24 hours. HEPES buffer, adjusted with NaOH to the desired pH, ionic strength (KNO_3), solution of suspended GEH and finally the required amount of copper were added subsequently, before filling up to 40 ml with nanopure water (NANOpure, ultrapure water systems) or stored roof runoff. Although adsorption was, after a very fast initial step, almost complete after 1 hour, 22 hours at 23°C were allowed for equilibration in the overhead shaker. In order to detect eventual adsorption on tube walls, a blank sample without iron was prepared with each series. Initial copper concentrations and pH were chosen in a range where no precipitation is expected. After equilibration time, the suspension was filled in a syringe and filtered with 0.45 µm PES membrane (GyroDisc 25, Orange Scientific). The filtrate was filled in 10 ml analyse tube, acidified with 100 ml 69 % HNO_3 and analysed either with ICP or AAS, depending on the copper concentration. Filters and centrifugation tubes were used once only.

The columns, made of transparent PVC are 1m long with a diameter of 5 cm. A 0.3 mm sieve was inserted to retain the GEH grains 15 cm above the column outlet. Data for pH, flow rate and temperature were monitored and recorded automatically. In order to avoid algae growth, the columns and all tubes were wrapped in lightproof. In contrast to real process conditions, the columns were operated under fully water saturated conditions. Samples can be taken at the column outlet after the flow regulation valve as well as in the filterbed in steps of 5 cm along the column height. For this purpose, tubes with an inner diameter of 5 mm were mounted perpendicular to the column reaching into the middle of the column where they were sealed with a 0.3 mm sieve. Sampling was performed at a flow rate of 10 ml h^{-1}, similar or smaller than the pore water velocity in order not to disturb the flow pattern in the filterbed. Before sampling, the profile tube as well as the annexed short tube were flushed for at least twice their volume at sampling velocity. Only when flushed with the appropriate velocity, samples were taken. Samples from several profiles were taken sequentially.

Adsorption capacity

Batch experiments

Adsorption of copper on GEH shows, as expected, a strong pH dependency (Dzombak and Morel, 1990). Adsorption in nanopure matrix starts at pH 5 and reaches its maximum at pH 6.5. With increasing initial concentration, saturation can be observed, and even at higher pH, increasing adsorption is not possible. The maximum calculated adsorption capacity from experimental data is 33.9 mg Cu g^{-1} GEH, corresponding to $5.3 \cdot 10^{-4}$ mol Cu g^{-1} GEH, which is less than the GEH site density of $7.6 \cdot 10^{-4}$ mol Cu g^{-1} GEH measured by Teermann (2000). Adsorption behaviour in the roof runoff matrix is similar. The maximum capacity according to the experiments is 27.1 mg Cu g^{-1} GEH. With lowest initial concentrations, adsorption is enhanced at low pH but reduced above pH 5. With increasing initial copper concentration, adsorption in the roof runoff matrix is more and more enhanced, but is still slightly less at pH higher than 6.2 compared to the nanopure matrix (Figure 1).

Several processes can be considered to explain the different adsorption pattern. Complexation of copper with organic ligands in solution, existent in roof runoff at 1 mg DOC l^{-1}, could reduce adsorption at lowest initial copper concentration (Xue

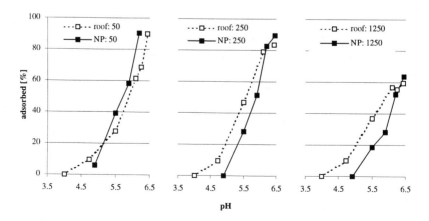

Figure 1. Adsorption of copper on GEH with roof runoff and nanopure matrix. Initial concentrations from left to right 50, 250 and 1250 µg Cu l^{-1}

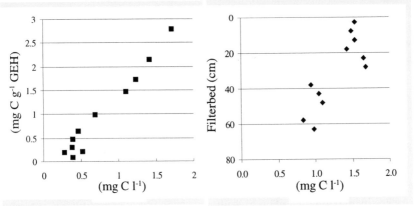

Figure 2. Adsorption of roof runoff DOC on GEH

Figure 3. DOC profile samples in a GEH column

and Sunda, 1997; Kunz and Jardim, 2000). As illustrated in Figure 2 and 3, DOC adsorbs in batch and column experiments on GEH (Holliger, 1999; Korshin et al., 1997). Here, first-flush roof runoff with 2.7 mg C l⁻¹ was used. Therefore, competition for surface sites may reduce copper adsorption. On the other hand, formation of ternary complexes with DOC bound copper in solution can enhance adsorption. Further, at a pH below the point of zero charge, which is 8 for GEH, the surface is positively charged and adsorption of negatively charged ligands results in charge neutralisation or reversal (Teermann, 2000; Weirich, 2000). This weakens or even changes electrostatic repulsion into attraction of Cu^{2+} and adsorption increases.

In different runoff waters pH values from 6.5-7.5 corresponding to pH values in the rain from 5.3-6.2 could be observed (Zobrist et al, 2000). Obviously, adsorption of copper in such roof runoff is not critical. But, if the pH is lower, e.g. if the runoff water is not sufficiently buffered, a reduction of the adsorption capacity may occur. As a consequence, the addition of granulated carbonate is a possibility to improve GEH performance by increasing the pH of the runoff water.

Modelling

Since pH may be regarded as crucial parameter, a model must be able to depict its influence on adsorption. A simple but sufficient approach to describe adsorption in a roof runoff matrix is a modified Langmuir Model according to equation 1.

$$q = \frac{q_{max} K_L C}{H^+ + K_L C} \qquad \text{(Eq. 1)}$$

C, H^+: copper and proton concentrations
q: adsorption capacity
q_{max}: maximum adsorption capacity
K_L: Langmuir constant

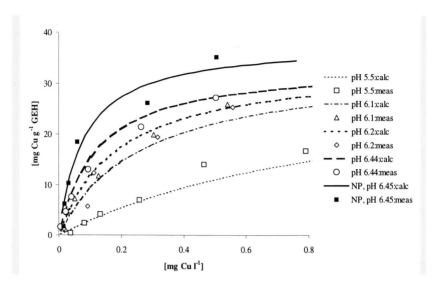

Figure 4. Copper adsorption in roof runoff at different pH and fitted isotherms. For comparison, data and fitted isotherm for pH 6.45 in nanopure matrix are also indicated

However, application of the same model type in a nanopure matrix shows unsatisfactory results. In this case, more sophisticated models, e.g. surface complexation models with correction of the electrostatic influence and allowing for multi-site adsorption, are able to describe adsorption properly. Consideration of these additional processes allows to describe adsorption in both matrices and to explain the observed differences (Robertson and Leckie, 1998).

In Figure 4, the measured data at pH 5.51, 6.1, 6.25 and 6.45 from Wimander (2001) and Langbein (2000) were used to fit model parameters according to the least square method, resulting in a maximum capacity of 33.7 mg Cu g^{-1} GEH in the roof runoff matrix. With a filterbed density of 1.3 kg GEH l^{-1} adsorption capacity of 43 g Cu l^{-1} can be achieved. The calculated single maximum capacity in the nanopure matrix at pH 6.45 is 38 mg Cu g^{-1} GEH.

Column experiments

The aim of this serie of column experiments running continuously for two years was to verify the above calculated loading capacity. In order to avoid a possible pH decrease resulting in a lower adsorption capacity, granulated carbonate ($CaCO_3$) was added and mixed on a one to one weight ratio, resulting in a filter bed depth of 30.6 cm. The filter velocity ranged between 0.1- 0.18 m•h^{-1}. The inlet concentration of 300 µg Cu l^{-1} in the roof runoff water is typical for a roof with new copper gutters. At this concentration, copper precipitation can be avoided.

After 23 months of incessant operation, copper removal was still almost complete (> 99 %) at the column outlet. In a parallel operated column filled with carbonate only, the removal was 39 % only with the same operation time and similar loading. Profile samples revealed a decrease of 90 % and 99 % at a filter bed depth of 4.5 cm and 14.5 cm, respectively. From a mass balance calculated over the first 4.5 cm of the filter bed, it can be concluded that a total amount of 1.1 g Cu was adsorbed, leading to a capacity of 21 mg Cu g^{-1} GEH which is close to the value of 23.9 mg Cu g^{-1} GEH calculated from the isotherm. However, because the roof runoff was stored in a concrete tank, the pH was 7.6 and particulate copper, 0.45 μm filtrated was 10 % in average. Samples taken from the profile at 4.5 and 14.5 cm did not show a significant difference in total and dissolved copper, indicating that filtration or adsorption of particles may be an additional removal process. Finally, the influence of additional processes on adsorption, such as biological growth or ageing of the adsorber material could not be observed under the prevailing conditions.

Full-scale applications

Adsorber design example

Based on the adsorption capacity of GEH with 43 g Cu l^{-1} and hydraulic loading tests (not shown here), confirming no limitation within the filter velocity range in infiltration systems, an adsorber layer can be roughly designed. In addition to removal efficiency and adsorption capacity, several criteria such as maintenance, realistic life cycle periods and grain size of the adsorber have to be considered. In order to guarantee satisfactory operation, clogging must be avoided. Therefore, the removal of coarse material and fine particles preceding the adsorber system is important. Additionally, particulate copper can be removed by filtration (Ochs and Sigg, 1995; Nadler and Meissner, 1999). Assuming an average copper corrosion rate of 1.8 g $m^{-2} a^{-1}$, as suggested from KBOB/IPB (2001), a copper roof area of 100 m^2, an infiltration pit with an adsorber layer depth of 0.4 m with 50 % $CaCO_3$, and an application time of 10 years, the diameter should be 52 cm.

Infiltration of façade runoff

The Swiss Federal Office of Metrology and Accreditation (METAS) in Berne constructed a building with a pre-oxidised copper façade of 2300 m^2. Being aware of environmental relevance and in cooperation with the Swiss Federal Office of Buildings and Logistics (BBL) and EAWAG, an infiltration trench, situated directly under the façade and filled with a 1:1 mixture of GEH and carbonate was built to immobilise copper in the façade runoff (Figure 5).

Two sampling units were installed at two sides of the building exposed to different meteorological conditions. Each unit consists of two sampling stations for the façade

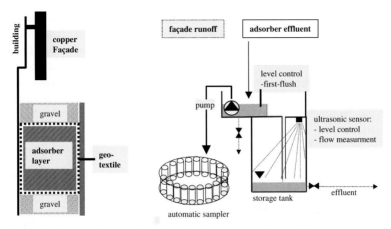

Figure 5. Construction details of the adsorber layer and a sampling station

and for the adsorber effluent, with a sampling length below the façade of 2 m each. Samples were taken automatically and dynamically to follow runoff dynamics.

The removal efficiency of the adsorbing layer was calculated from corresponding event averaged concentrations of the façade runoff and the adsorption ditch effluent. Overall, 28 runoff events in 16 months with more than 2 samples per event were analysed. As a result, runoff concentrations varied over almost two orders of magnitude in single samples and within the rainfall events, concentration variations were remarkable. Among the many influences causing these fluctuations, runoff intensity and duration of the dry period preceding rainfall are most significant and could surpass first-flush effects. Maximum concentration values of up to 100 mg Cu l^{-1} were measured. Typical event averaged concentrations were mostly between 1-5 mg Cu l^{-1} with some values above 10 mg Cu l^{-1}. Despite the dynamic behaviour, the removal efficiency reached from 95-99.7 %, predominantly above 98 %, with exception of two heavy storm events when the efficiency dropped to 92 % and 93 %, (Steiner and Boller, 2001). Whenever the efficiency was reduced, it was due to the washout of fine GEH particles. Additional lab experiments as well as recent data show that this phenomenon occurs only during the initial operation phases.

Reuse of roofwater from a copper roof

In Zurich, Switzerland, the roof of a building of the university will be reconstructed. As many historical buildings in this area of the town have copper roofs, a green patinated copper sheet will be applied. The roof area is 940 m² and the runoff will be collected in a 50 m³ storage tank to be reused for toilet flushing. The management for public buildings of the Kanton of Zurich considered environmental aspects and decided to remove the copper from the roof runoff. The runoff reuse plant is presently under design.

Because a filtration facility and pump will anyway be installed to supply the toilets, the best choice to place the adsorber was after the fine particle filter and after the pump. In contrast to an adsorber in an infiltration site, the adsorber will be operated under saturated flow pattern in a pressure vessel. Thus, headloss is not crucial since the pumps will provide sufficient pressure head. Samples can be taken before and after the particle filter as well as after the adsorber and headloss will be monitored. Exhaustion of the adsorber is expected to occur clearly after 5 years.

Adsorber layer in an infiltration gallery

As a more typical application, an adsorber layer will be added to a conventionally designed subsurface infiltration trench between the percolating soil layer and the inlet pipe to the infiltration gallery, connected to a copper roof of 1500 m^2 of a municipal multipurpose building in Hirschthal, Switzerland. The gallery length and width are 12 m and 1.5 m, respectively, and the adsorber layer is 0.3 m in depth. The calculated exposure time is estimated to reach between 15 and 20 years.

Conclusions

Granulated iron hydroxide (GEH) has been tested in order to investigate its ability to adsorb copper from roof runoff. The adsorption capacity was evaluated with equilibrium batch experiments using nanopure and roof runoff matrices and different copper concentration levels and pH. Adsorption has shown to be strongly pH dependent starting at pH 5 in the nanopure system and at pH 4.5 in the roof runoff system. With increasing initial copper concentrations, adsorption in the roof runoff system is enhanced, but is slightly diminished at pH above 6.2. In both cases, the adsorption capacity for technical applications revealed to be sufficient. In contrast to the nanopur system, modelling of the adsorption in real roof runoff water is possible using an adapted Langmuir model including pH effects. The equilibrium model can be implemented in a transport model for the simulation of fixed bed adsorbers.

With the help of longterm column experiments, it could be shown that extrapolated adsorption capacities from model calculations can actually be reached in the top layers of an adsorber bed provided that the contact time is sufficient. In this case, almost complete copper removal was observed. As shown in the design example conditions, implementable volumes of GEH are required and varying typical flow rates in infiltration sites should not be process limiting.

Full scale application of a mixed GEH/carbonate layer in the form of a trench for the removal of copper from a copper façade runoff shows excellent performance after an operation time of almost two years. Other applications such as the treatment of recuperated roof runoff water from a copper roof used for toilet flushing and the addition of an adsorption layer to a standard infiltration gallery design reveal the broad applicability of GEH for copper removal.

In practice, it is good to see the increasing awareness of decision makers to use barrier systems in cases where large copper surfaces are unavoidable. But it has to be stressed that the technical solution presented with the GEH adsorber system should not be used as an argument to promote the use of copper on roofs. The ultimate goal of modern urban water management is to reach higher levels of sustainability which may be achieved by proper source control only. For new buildings, the use of less corrosive metals than copper or zinc or the use of alternative non-metal materials should be studied and evaluated with highest priority.

References

Benjamin, M.M., Sletten, R.S., Bailey, R.P., and Bennet, T. (1996) Sorption and filtration of metals using iron-oxide-coated sand. *Wat. Res.* **30**(11), 2609-2620

Benjamin, M.M., Leckie, J.O. (1981) Multiple-site adsorption of Cd, Cu, Zn and Pb on amorphous iron oxyhydroxide. *Journal of Colloid and Interface Science* **79**(1), 209-221

Boller, M. Steiner, M. (2001) Diffuse emission and control of copper in urban surface runoff. *Proceedings of the 2nd IWA World Water conference, Berlin*

Dzombak, D.A., Morel, F.F.M. (1990) Surface complexation modeling. Hydrous ferric oxide. *John Wiley & Sons, New York*

Holliger, U. (1999) DOC Adsorption an Eisenhydroxid. *Diplomarbeit, ETH/EAWAG*

KBOB/IPB (2001) Coordination of the Federal Construction and Properties Services, Association of professional Corporate Building Owners, Metals for roofs and façades. *Recommendation 2001/1, KBOB/IPB*

Korshin, G.V., Benjamin, M.M., Sletten, R.S. (1997) Adsorption of natural organic matter (NOM) on iron oxide: effects on NOM composition and formation of organo-halide compounds during chlorination. *Wat. Res.* **31**(7), 1643-1650

Kunz, A. Jardim, W.F. (2000) Complexation and adsorption of copper in raw sewage. *Wat. Res.* **34**(7), 2061-2068

Langbein, S. (2000) Schwermetallproblematik bei der Meteorwasserversickerung, *Diplomarbeit, in Zusammenarbeit mit der EAWAG, Institut für Siedlungs- und Industriewasserwirtschaft, TU Dresden.*

Nadler, A., Meissner, E. (1999) Zwischenbericht, Entwicklungsvorhaben: Versickerung des Niederschlagswassers von befestigten Verkehrsflächen. *Report, Bayrisches Landesamt für Wasserwirtschft.*

Ochs, M., Sigg, L. (1999) Speziierung von Schwermetallen im Dachwasser. *Report, EAWAG*

Robertson, A.P., Leckie, J.O. (1998) Acid/Base, Copper Binding, and Cu^{2+}/H^+ Exchange Properties of Goethite, an Experimental and Modeling Study. *Environ. Sci. Technol.* **32**(17), 2519-2530

Steiner, M., Boller, M. (2001) Untersuchungen zum Kupferabtrag einer Kupferfassade und zur Wirksamkeit einer Eisenhydroxid-Sickerschicht zur Abtrennung von Kupfer aus dem Fassadenwasser. *Report, EAWAG, Duebendorf, Switzerland*

Teermann, I. (2000) Untersuchungen zur Huminstoffadsorption an b Eisenoxidhydrat. *Thesis, TU Berlin*

Weirich, D. (2000) Influence of organic ligands on the adsorption of copper, cadmium and nickel on goethite. *Thesis, ETH Zürich*

Wimander, H. (2001) Modelling of copper adsorption on Ironhydroxide. *Diploma work, in cooperation with EAWAG, Master Science Program, Departement of Environmental Engineering, Technical University of Lulea, Sweden*

Xue, H., Sunda W.G. (1997) Comparison of [Cu^{2+}] measurements in lake water determined
    by ligand exchange and cathodic stripping voltammetry and by ion-selective electrode.
    Env. Sci. Technol. **31**(9), 1902-1909
Zobrist, J., Müller, S.R., Ammann, A., Bucheli, T.D., Mottier, V., Ochs, M., Schoenenberger,
    R., Eugster, J., and Boller, M. (2000) Quality of roof runoff for groundwater infiltration.
    Wat. Res. **34**(5), 1455-1462

H.H. Hahn, E. Hoffmann, H. Ødegaard (Eds.)
Chemical Water and Wastewater Treatment VII, pp. 243-252
© IWA Publishing, London
ISBN: 1 84339 009 4

Innovative Sewerage Management Using a High Rate Physico-Chemical Clarification Process

A.J. Davey*, S.R. Gray and R.A. Jago

*CDS Technologies, Mornington, Australia
email: alexd@cdstech.com.au

Abstract

An alternative primary treatment process utilising continuous deflective separation (CDS) technology combined with a flocculation process has been developed at the pilot plant level and is now undergoing field evaluations. In the first field trial, the process treated sewage to a quality that allowed effective disinfection. The effluent was stored in an open basin within an urbanised area without any notable odour or aesthetic issues. This strategy will enable the water authority involved to store sewage during high flow periods saving significant capital outlay for upgraded sewerage or enclosed storage.

Introduction

The removal of solids is a critical process in the treatment of domestic wastewater, as the majority of pollutants that are harmful to humans and the environment exist as solids or are adsorbed onto the surface of the solids. The majority of these pollutants are associated with fine particles ($< 50\ \mu m$) and colloidal solids (Levine et al., 1985).

Primary treatment processes that can remove high levels of fine and colloidal solids in a compact and cost-effective operation are increasingly finding application in the management of sewage infrastructure. This is because such processes are rapid to start up, compact, low in capital cost, able to achieve good removal of solids and associated pollutants, and the effluent may be highly amenable to UV disinfection. These characteristics make such processes flexible and ideal for use where intermittent operation is required.

[243]

Such processes include chemically assisted vortex separators, ballasted flocculation processes (Actiflo® & Sirofloc™), enhanced sedimentation processes (Densadeg®), induced air flotation processes (Jetflote) and high rate filtration processes. There are several cases now of high rate clarification processes being used for sewer overflow remediation, treatment of sewage bypass flows and primary treatment prior to biological treatment (Keller, 2001; Brewer *et al.*, 1997).

One application for these technologies proposed by Booker (2000) is the use of high rate clarification for *peak load lopping*. In this scenario, the high-rate clarification process is used to treat periodic peaks in the hydrograph allowing the conventional infrastructure to deal with the base load. This limits any over designing of the conventional infrastructure, namely the pipes, pump stations, and treatment plants. Both Booker (2000) and Davey *et al.* (2001) have presented modelling data showing the significant capital savings that may be achieved. In addition, the life of existing infrastructure may be extended utilising this approach.

High rate clarification processes do not remove all the pollutants of a conventional biological treatment plant (namely the soluble fraction of the pollutants). In a peak load lopping application they should only treat a small percentage of the flow, and so overall treatment efficiencies would not be significantly affected.

The CDS screening technology

The CDS screening technology is not difficult to understand. Utilising indirect screening along with hydraulic balancing and appropriate geometric design, a non-mechanical, non-blocking screening process is achieved. This technology as been applied across a wide range of solid/liquid separation applications. The most common of these is stormwater and sewer overflow remediation (CSO & SSO). As the screens are non-blinding, they maintain a constant throughput operation, and are capable of operating at very high loading rates (up to 450 $m^3/m^2 \cdot h$ for the largest units).

Combining flocculation and screening

Unfortunately the CDS technology itself is not suitable for removal of fine neutral density solids (< 0.25mm). Although finer screens can be employed for the removal of fine particles in small-scale operations, their installation is not straightforward and they are not suitable for sewage and stormwater applications. Flocculation of the fine solids, creating aggregates large enough to be captured in a CDS unit, produces a clarification process that is not limited by settling or flotation velocities.

This combination of flocculation and the patented screening technology has been demonstrated to achieve high levels of solids removal from raw sewage in a compact treatment process (Gray *et al.*, 2000; Davey *et al.*, 2001). This process,

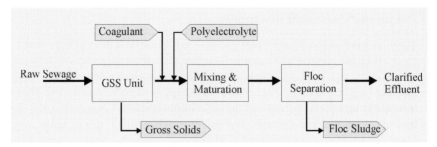

Figure 1. Fine solids separation (FSS) process flow sheet

termed FSS (Fine Solids Separation) was developed recently in Australia by CDS with input from the CSIRO. The process, shown schematically in Figure 1, consists of:

- a CDS Gross Solids Separator (GSS), (a proprietary non-mechanical screening unit that removes all larger material as well as grits and sediments by physical separation)
- coagulant and polyelectrolyte addition
- rapid mixing and maturation of flocs
- separation of flocs in a second screening operation.

Previously reported work (Gray *et al.*, 2000) focussed on developing the chemistry to produce flocs suitable for separation via a rapid screening operation. This involved a combination of laboratory work and verification of this work in pilot plant trials.

Pilot plant work

The pilot plant was set up at Mornington Sewage Treatment Plant, Victoria, Australia, while the supporting experimental laboratory work carried out on floc size and floc shear resistance or strength was being undertaken. The plant's first task was a preliminary trial to determine the feasibility of the floc screening concept.

The process is now proven at loading rates up to 110 m³/m²•h (up from 57 m³/m²•h as reported by Gray *et al.*, 2000), and only limitations in plant setup prevent this figure from being increased further. Recent pilot plant work has focussed on engineering refinements including solids removal options, and a simplified flocculation system. Demonstration of process performance over extended periods of operation, and data collection for a wide range of water quality parameters was also undertaken.

Pilot plant results and discussion

Typical results from the pilot plant, averaged from several days of operation at up to 3 MLD, are presented in Table 1. The chemical dosing regime was 18 mg/L alum (as Al^{III}) and 9-10 mg/L polyelectrolyte C (low charge density, very high molecular weight, see Gray et al., 2000 for description of polyelectrolytes). These results are by no means optimal, and the table includes results taken during periods where operation was sub-optimal. For example, when high levels of algae were present due to decanting of the sludge drying basins into the plant influent. This significantly reduced the effectiveness of the flocculation process. A point of interest is that a high proportion of both the BOD_5 and COD are present as solids in this sewage, resulting in unusually high levels of removal for these parameters.

It has been assumed that the size and strength of the flocs are the controlling process issues in determining whether a high level of solids removal is achieved. The flocs have to be quite large (> 1mm) to be effectively separated in a CDS unit. In addition, the shear forces experienced by the floc in the CDS unit are large due to the high velocities in the units and the high mechanical shear experienced by the flocs when they impact the screen.

The combination of both large and strong, shear resistant flocs is not easily achieved without high doses of polyelectrolyte because size and strength require significantly different flocculation conditions. Low shear flocculation enables large flocs to be formed (Vlaski et al., 1997), but flocculation under low shear conditions generally leads to weak flocs of low density (Gregory, 1997). Flocculation under high shear conditions generally leads to smaller flocs of higher density and strength, for which the floc size may be increased by raising the polyelectrolyte concentration (Arld, 1998).

The initial laboratory work attempted to identify which type of polyelectrolytes gave the best results in terms of size and strength (Gray et al., ,2000). Recently, trials have been undertaken to verify this laboratory work. Polyelectrolyte A (low

Table 1. Average water quality data from the FSS pilot plant system

Parameter	Units	Influent	Effluent	Reduction %
Turbidity	NTU	240	7.6	97
TSS	mg/l	259	13.5	95
BOD_5	mg/l	302	38	87
COD	mg/l	531	82	80
FC	CFU/100ml	$12.5 \cdot 10^6$	$58.6 \cdot 10^3$	99.5
TP	mg/l P	12	0.6	95
TN	mg/l N	71	55	16
NH_4-N	mg/l NH_4-N	40	37	8

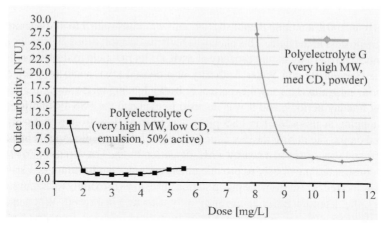

Figure 2. Effluent turbidity results achieved with different polyelectrolytes

charge density, very high molecular weight) required high doses to achieve the results outlined in Table 1. Figure 2 is a plot of outlet turbidity as a function of polyelectrolyte dose for the pilot plant. It shows that doses in excess of 9 mg/l of polyelectrolyte C (the emulsion equivalent of polyelectrolyte A) were required to get good removal of solids. This dose is well above that required for good flocculation, demonstrating the importance of achieving floc size and strength for this process. However, polyelectrolyte G (medium charge density, very high molecular weight) was able to operate in the pilot plant at doses of 2 mg/l or less. This corresponds well with laboratory work, which indicated that doses of 2.5 mg/l were required for effective treatment (Gray *et al.*, 2000).

Process improvements are likely to be achieved in the mixing and maturation process where the formation of the flocs occurs. Initially the pilot plant utilised a tanks-in-series process with a residence time of a few minutes. However, over time this was reduced to a system fed by a pump, which is used to mix in the alum. This is followed immediately by polyelectrolyte dosing, a static mixer, and a short length of pipe before entering the second screening unit for floc removal. Total residence times for this flocculation system can be less than 20 seconds, while still achieving results as in Table 1. Little has been done, however, to optimise this system. Significant process changes may be needed in order to further reduce dose requirements, or to deal with lower concentrations of solids that may be experienced in treating sewer overflows.

Field trial program

Early in 2001, a mobile demonstration unit was constructed to allow testing of the FSS process in the field. This plant was built inside a soft-sided 20 foot shipping container and had a hydraulic capacity of 3 MLD. Its first field trial was conducted for the Barwon Region Water Authority at a coastal town south of Geelong in Victoria, Australia. This town experiences heavy loading of the sewerage system during holiday periods, and it was decided to evaluate the FSS system for peak load lopping at this site. The FSS plant was to treat flow from a pump station located on the site during the day at periods of high flow. Following disinfection the effluent was stored in a nearby 7 ML detention basin. The detention basin had been constructed previously to provide emergency storage for untreated sewage, and was located 500 m from the pump station. Due to its proximity to neighbouring houses, the detention basin is currently only used as a last resort. When used to store the FSS effluent during periods of high flow, effluent from the basin is discharged back to the sewer during periods of lower flow, such as during the night. It should be noted that in Australia, open storage basins are typically around 5 % of the cost of equivalent below-ground storage facilities.

Two branch sewers were accessible on the site. One supplied the pump station from which the sewage was drawn for the trial, the other by-passed the pump station and provided a convenient means for disposal of the sludge from the FSS. In this way, no solids were stored at the pump station site, thus conforming with the Environment Protection Authority's (EPA) permit requirements for the trial. Both sewers receive only domestic sewage.

Barwon Water set two objectives for the trial – an assessment of the effluent from the FSS for storage in an open basin as above, and evaluation of the effluent for possible reuse applications.

For the trial, raw sewage was pumped from the pump station to the FSS for treatment (see Figure 3). Gross solids and grits, separated from the sewage by the GSS, were returned periodically to the pump station for disposal when each day's trial was completed and the pump station resumed normal operation. During startup, and while the appropriate dose rate was being established, effluent from the FSS was also returned to the pump station. Once satisfactory, steady-state operation was achieved, effluent was then discharged to the detention basin. A side stream from this discharge was passed through a UV disinfection facility to assess effluent suitability for disinfection.

One limitation on the trial was the availability of sufficient sewage to operate the system at the maximum treatment rate (30 l/s). The hydrograph for the sewage at the pump station showed a peak from about 08.00 to 11.30 am and another early in the evening. For this reason, the system could only be run at the high treatment rates for short periods, although longer runs at the lower flow rates were possible.

Operation of the FSS system was carried out at flow rates of 15, 20, 25 and 30 l/s and samples were taken from influent, effluent, storage basin and UV

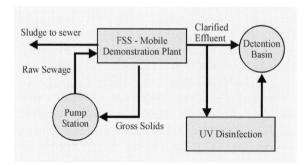

Figure 3. Schematic layout of site showing major functional components for field trial

disinfection facility. All samples were analysed by Barwon Water's NATA (National Association of Testing Authorities, Australia) registered laboratory, and were taken by Barwon Water staff in accordance with procedures set by the laboratory.

Field trial results and discussion

Results and observations reported here are taken from an internal report (Teng and Williams, 2001) prepared by staff of Barwon Water. Analytical results are summarised in Table 2 below. System flow rates in the range 15-30 l/s appeared to have no influence on system performance. Dosage rates were approximately 12 mg/L alum (as Al^{III}) and 12 - 16 mg/L of polyelectrolyte C.

The UV disinfection unit achieved significant kills, with the concentration of coliforms below 10 orgs/100 ml in some cases.

The performance of the system, while satisfactory for the purpose, was inferior to that recorded for the pilot plant (Table 1). In addition, achieving the above results required a higher polyelectrolyte dose than had been used on the pilot plant. The variation in BOD_5 and COD removals may be largely explained by a different split between the soluble and insoluble fraction of these components. The solids removal

Table 2. Water quality data from field trials of the FSS system

Parameter	Units	Influent	Effluent	Reduction%
Turbidity	NTU	131	8	94
TSS	mg/l	137	19	86
BOD_5	mg/l	151	48	68
COD	mg/l	365	166	55
FC	CFU/100ml	5.8×10^6	62×10^3	98.9
TP	mg/l P	9.5	1.2	87
NH_4-N	mg/l NH_4-N	46	38	17
Oil/Grease	mg/l	40	10	75

performance variation between the two plants is not so easily explained but the most probable reason for this was variation in the wastewater chemistry.

Much higher levels of salinity were encountered during the field trial at the Barwon Water site (1500 - 2000 mg/l NaCl) compared to the pilot plant trials at the Mornington STP (350 mg/l NaCl). It is known that adsorption of polyelectrolytes onto surfaces decreases with increasing electrolyte concentration when electrostatic interactions are important in the adsorption process (van de Steeg *et al.,* 1992). The importance of the electrostatic interactions for the flocculation process here has not been established. However, the better performance of cationic polyelectrolytes compared to anionic polyelectrolytes (Gray *et al.,* 2000) would suggest that electrostatic interactions play a significant role in building of the flocs. Therefore, an increase in salinity might lead to poorer flocculation of sewage.

The effect of higher salinity can also be seen in the reduced alum dose for the field trial (12 mg/l) compared to the pilot plant trial (18 mg/l). The increased salinity would lower the surface potential of the solids present, and therefore a reduction in the amount of coagulant required for destabilisation of the solids is observed.

Time on the site was limited, preventing measures to improve the process performance being tested. Such measures included use of higher charge density polyelectrolyte to increase the electrostatic attraction between the particles and the polyelectrolyte or raising the surface charge on the solids by adjusting the pH.

While such measures may assist in optimising the flocculation process and therefore solids removal, it should be noted that the performance of the system was still satisfactory for the purpose of the application. The clarified effluent was stored for periods up to 2 months in the open storage basin with no odour issues. The installation of an FSS on the site will allow the basin to regularly store flow during periods of high utilisation of the sewer. This will have the same net effect as increasing the downstream sewer capacity by around 25 %, but for only 10 % of the capital outlay required for a conventional upgrade of the sewer.

Further monitoring of the fluid in the basin was undertaken, and the soluble BOD was shown to be consumed within only a few hours in the basin. This brought the water within the parameters required for reuse (EPA, 2001), and enables Barwon Water to move towards greater water reuse within its jurisdiction. Water reuse is currently being actively promoted by the Victorian EPA.

Conclusions

The CDS FSS process utilises a combination of flocculation and screening to achieve high rate clarification of sewage. While there are many improvements still to be made, the process has been proven to operate well in both pilot plant and field trials.

The CDS FSS pilot plant results confirmed previous laboratory investigations of the performance of different polyelectrolyte types. High removals of turbidity, TSS and associated pollutants were demonstrated at separator loading rates up to

110 m^3/m^2•h. The importance of optimising the flocculation process was demonstrated, as selection of appropriate polyelectrolyte was able to reduce the polyelectrolyte dose from 9 mg/l to 2 mg/l.

The field trial demonstrated the ability of the process to produce an easily disinfected effluent that could be stored in an open basin in an urban environment with no odour, noise, or aesthetic issues. The process would be suitable for application in a peak load lopping operation on this site, allowing significant capital savings in the management of sewerage infrastructure and providing reuse opportunities.

Field trial results were inferior to those recorded for the pilot plant. Sewage chemistry, and in particular salinity, was also thought to affect flocculation and process performance, highlighting the possible variability in performance that might result from changes in sewage chemistry.

Future pilot plant development will focus on optimisation of the flocculation process and engineering improvements. The mobile plants will be used to develop the process for use in new applications such as sewer overflows, and to explore the limits in hydraulic loading rates for the units.

References

Arld, R. (1998) Effect of Floc Strength on High Rate Filtration. Diploma thesis, Department of Process Engineering, Georg-Simon-Ohm Fachhochschule Nürnburg

Booker, N. A. (2000) Peak Load Management at WWTPs, *Water* **20** (5), 20-21

Brewer, P., Martin, J. C., Bedard, P. (1997) Lamella Plate Separators and Biological Aerated Filters at Pool STW, *Proceedings of International Conference on Advanced Wastewater Treatment Processes*, Leeds, UK, September 1997

Davey, A., Booker, N., Gray, S. (2001) High rate sewage treatment for peak flow management, *Proceedings of the AWA 19th Federal Convention*, Canberra, Australia.

EPA (2001) Draft Environmental Guidelines for the Use of Reclaimed Water, *Best Practice Management Series*, Environmental Protection Authority of Victoria, Melbourne, Australia.

Gray, S.R., Harbour, P.J., Dixon, D.R. (1997) Effect of polyelectrolyte charge density and molecular weight on the flotation of oil in water emulsions, *Colloids and Surfaces A: Physiochem. Eng.* Aspects 126, 85-95

Gray, S.R., Booker, N.A., Arld, R. (1998) Effect of floc characteristics on high rate filtration of sewage. *Chemical Water and Wastewater Treatment* V, H.H. Hahn, E. Hoffmann and H. Ødegaard (Eds), Springer-Verlag, Berlin Heidelberg New York, 1998, 205-217

Gray, S.R., Becker, N.S.C., Booker, N.A., Davey, A., Jago, R.A., and Ritchie, C. (2000) The Role of Organic Polyelectrolytes in High Rate Alternatives to Primary Clarification, *Chemical Water and Wastewater Treatment*-VI, H.H. Hahn, E. Hoffmann and H. Ødegaard (Eds), Springer-Verlag, Berlin Heidelberg New York, 2000, 223-233

Gregory, J. (1997) The density of particle aggregates, Wat. Sci. Tech. **36** (4) ,1-13

Keller, J. (2001) Lawrence, Kansas: Detailed Design Issues for a Ballasted Flocculation System, *Proceedings of the 74th Weftec Conference,* Atlanta, GA, October 2001

Levine, A.D., Tchobanoglous, G., Asano, T. (1985) Characterisation of the Size Distribution of Contaminants in Wastewater: Treatment and Reuse implications, *Journ. Water Pollution*, Control Federation **57**, 805-816

Lurie, M., Rebhun, M. (1997) Effect of properties of polyelectrolytes on their interaction with particulates and soluble organics. *Wat. Sci. Tech.* **36**, (4), 93-101

Teng, M. L. and Williams, P. G. (2001) Trial of the CDS Fine Solids Separation System at Ocean Grove, Victoria, Barwon Region Water Authority report, Geelong, Victoria.

Van de Steeg, H.G.M., Cohen Stuart, M.A., de Keizer, A., Bijsterbosch, B.H., (1992) Polyelectrolyte adsorption: a subtle balance of forces, *Langmuir* **8** (10) 2338-2546

Vlaški, A., van Breemen, A.N., Alaerts, G.J. (1997) The role of particle size and density in dissolved air flotation and sedimentation, *Wat. Sci. Tech.* **36** (4), 177-189

H.H. Hahn, E. Hoffmann, H. Ødegaard (Eds.)
Chemical Water and Wastewater Treatment VII, pp.253-260
© IWA Publishing, London
ISBN: 1 84339 009 4

Simulation Program for Wastewater Coagulation

L. Lei, H. Ratnaweera and O. Lindholm*

*Norwegian Institute for Water Research, Oslo, Norway
email: harsha.ratnaweera@niva.no

Introduction

Simulation programs for wastewater treatment processes offer a number of advantages for a broad user group. They provide a basis for process design alternatives for designers, considerably reducing the need for pilot scale tests and avoiding costly mistakes in full-scale. Plant operators may simulate operational alternatives for process optimisation and as a learning tool for operational activities while researchers and consultants get a wider range of opportunities to experiment with more economical and efficient process combinations and operational conditions. Simulation programs may also provide guidance for operational conditions under extreme situations like accidents, where previous experience may be absent or inadequate. In certain situations, these programs may even be employed for on-line process control to secure efficient and economical operation.

A number of models and advanced simulation programs for mechanical and biological treatment processes have been on the market for more than a decade. The dominance of these treatment processes in wastewater treatment resulted in the need and the interest to develop simulation programs by a number of research groups, leading to excellent and well-proven products on the market. However, simulation programs for chemical treatment are far from that situation. While only a handful of simulation programs available on the market claims to work for coagulation, functionality is really only limited to simultaneous precipitation or chemically enhanced precipitation processes.

Chemical coagulation has gained popularity as a leading wastewater treatment concept in many countries. Its efficiency in particle and phosphate removal, flexibility and the robustness for climatic and shock loads are among the major reasons for this popularity. For example, over 75 % of all treated wastewater in Norway goes through a coagulation stage. Thus the need for simulation programs for chemical treatment, both on-line and off-line, is as obvious as for the biological processes.

At present, there are no comprehensive conceptual models for the wastewater coagulation process due to its complicated physical-chemical behaviour. Unlike many other industrial processes, the coagulant dosing is difficult to control using feedback concepts, primarily due to the 2-6 hours of sedimentation time and the rapid influent quality fluctuations, e.g. every 15 minutes. On the other hand, there is no simple method to determine the optimum coagulant dosage even if the influent quality is well defined. This situation forces most of the treatment plants to run either with an overdose or an under dosage of coagulants, which results in many adverse effects. With the developments in water quality measuring technologies, and using the process knowledge and the process control facilities, radical improvements in process optimisation of coagulation are expected. With the availability of a simulation program for coagulation available on-line, considerable improvements in process optimisation are thus anticipated.

This paper presents a concept for a model for coagulation and a simulation program demonstrating its applicability at two full-scale plants.

Concept

First it is necessary to identify the most important factors relevant for coagulation. It is well documented that coagulation efficiency is dependent on influent quality, coagulant properties and the process conditions. pH, particle and phosphate content, alkalinity and temperature are the most important factors influencing coagulation. Particle removal and phosphate removal are the two major processes that take place during wastewater coagulation. Fettig *et al.* (1990) determined that when an inorganic coagulant is added to wastewater, both the hydrolysis and the chemical reaction with phosphates takes place in parallel, contrary to the results of a number of earlier studies. Ratnaweera (1991) showed that these two processes compete with each other depending on the coagulant type - and thereby also on the pH - and the concentration of particles and phosphates. It was shown that pre-polymerisation ratio also has a major influence on the preferred reaction between the two processes. Fettig *et al.* (1990) found that at a given pH range, the proportion of coagulant used for precipitation of phosphates decreases from 57 % to 25 % when the pre-polymerisation ratio increases from 0 to 2. These conclusions were later confirmed by a number of wastewater treatment plant (WWTP) owners, as coagulants with high pre-polymerisation ratios were reported to remove less phosphates compared with non-pre-polymerised ones like aluminium sulphate, under otherwise identical conditions.

Hydrolysis products of the conventional coagulant salts are dependent on pH (Baes and Mesmer, 1976). The hydrolysis products with high, positive charges are most favourable for particle separation and the presence of these favourable species are dominant at the so-called "optimum coagulation pH range," which is, for aluminium salts, from pH 6.5 to 6.7. Thus, there could be a significant portion of non-contributing coagulant dependent on the pH which does not participate in the above mentioned competition between particle and phosphate removal mechanisms. A significant amount of coagulant is also consumed just to bring pH down to more optimal conditions, which is also a part of this non-active portion. One assumption used in the concept presented herein is that the coagulant added to wastewater can be divided into three portions depending on pH, coagulant type and particle to phosphate ratio in wastewater (Figure 1). Al(PA) represents the amount of coagulant available for particle removal, Al(PO) phosphate removal and Al(IN) the inert portion. The equations with the necessary constants were determined in lab and pilot scale experiments, which are reported elsewhere (Fettig *et.al,* 1990; Ratnaweera, 1991). These equations also consider the influence of other mechanisms, like sweep floc, and make it possible to identify a representative distribution among the three fractions.

The hydrolysis process reduces the pH of the wastewater, which has an impact on the distribution among the above three components. The change of pH during hydrolysis can be estimated using equations adopted from Baes and Mesmer (1976).

The next assumption is that only the soluble phosphates react chemically with the relevant coagulant portion. The particulate phosphates are removed in a similar manner to the particles, although particulate phosphate removal may not be as comprehensive as that of other particles or colloids. Proper coefficients must be selected to reflect any deviations from the above assumptions.

The particles are also removed by two methods. The percentage of the first portion that is removed is based on the amount of coagulant that is available for particle removal compared to that which is necessary. The latter is an experimental value, which is often a constant, but may also change to reflect the process conditions. The second portion of particles is removed by conventional sedimentation. It must

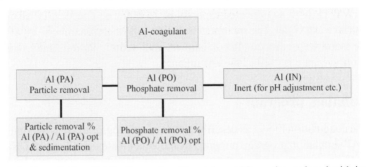

Figure 1. Coagulation distribution paths. Aluminium (Al) can be replaced with iron

be noted that it is not possible to give any physical meaning to these two portions, as they are only divided in abstract form.

The number of particles to be removed during the first stage is based on the ratio Al(PA)/Al(PA)opt, where Al(PA) is the portion of available coagulant for particle removal while Al(PA)opt is the optimum amount of coagulant required for particle removal. The latter is only a function of the particle concentration and pH. The final particle removal efficiency depends on the sedimentation characteristics of the plant, which are considered in the second stage, under conditions similar to a primary settling tank. The phosphate removal efficiency is directly given as the ratio Al(PO)/Al(PO)opt, where Al(PO)opt is the coagulant amount required to precipitate all phosphates in a given sample, where Al(PO) is the fraction of coagulant available for phosphate removal. The Al(PO)opt is experimentally found to be 1.1 to 2.0 times the molar concentration of the phosphates, in a pH range between 4 and 8. In the common coagulation pH range of 6 - 7, this value is found to be around 1.4 (Fettig *et al.,* 1990).

The organic matter and nitrogen removal during chemical coagulation is estimated using a simplified validation. These values are mainly estimated from the particulate organic matter and particulate nitrogen removal efficiencies, which are proportional to the particle removal. Since a small, but variable, fraction of soluble organic matter and nitrogen may also be removed during coagulation, the total removal efficiencies are calibrated using a coefficient that is proportional to the particle removal efficiencies.

The coagulation model was then integrated as a specific module into the leading wastewater simulation software, STOAT, with the assistance from the Water Research Centre, UK. The module is as flexible as other modules in the software, enabling the simulation of any treatment process that combines various single processes.

The model was calibrated and tested at the Toensberg WWTP and the Fredrikstad WWTP. Toensberg WWTP is a medium scale WWTP with a design and connected capacity of 60 000 p.e. The treatment plant has grit chambers, input of septic waste, sand traps, flocculation chambers and sedimentation tanks. Toensberg WWTP are using iron chloride sulphate as the coagulant. Fredrikstad WWTP is a medium scale WWTP with a connected capacity of 75 000 p.e. The WWTP also receives industrial wastewater. The treatment train consists of grit chambers, input of septic waste, sand traps, flocculation chambers and sedimentation tanks, where iron chloride sulphate is used as the coagulant.

Simulation program

The coagulation model as described above was programmed as a module for the STOAT simulation program, provided by the Water Research Centre of UK. STOAT has a library of process units that can be used to build a flow sheet using the drag and drop method.

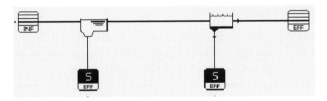

Figure 2. Flow sheet of a basic coagulation plant

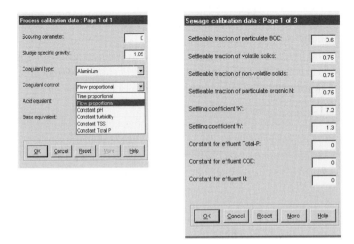

Figure 3. Menu examples to select the operational conditions and to specify process calibration data

Figure 2 illustrates the basic flowsheet of a coagulation plant. As described earlier, to represent the coagulation stage it is necessary to have two units, to represent the coagulation and separation stages. Using drop down menus it is possible to register the influent profiles and dimensions of the units. It is possible to simulate the coagulation process in flow proportional, time proportional or constant pH mode, which are the most commonly used operational conditions in practice. The operational conditions and related parameters can be selected from a drop down menu as illustrated in Figure 3.

After defining the necessary coefficients, or leaving default values unchanged, the simulation process can start for the specified period. The results can be presented in graphical and/or tabular form.

A comprehensive description of STOAT and its extensive process and simulation capabilities is beyond the scope of this paper. Further information on STOAT is found elsewhere (WRC, 2002).

Results and discussion

The simulation program was calibrated for the Toensberg WWTP and the results are shown in Figures 4 and 5, for turbidity and pH, respectively. A period with a considerable variation in the influent turbidity as well as the effluent turbidity due to insufficient coagulant dosage is presented. The initial calibration results at Toensberg show reasonable agreement between the measured and calculated values. Figure 5 demonstrates the good agreement between the effluent pH calculated by the simulation program and the measured pH. The slight deviation between the measured and calculated pH probably results from the inaccuracy of measuring pH under open conditions, while the simulation program calculates the resulting pH under ideal conditions.

The calibrated model of STOAT was then used to predict the coagulant dosage for a data series to result in an effluent having certain desired treatment efficiencies. The variation in the estimated dosages is illustrated in Figure 6.

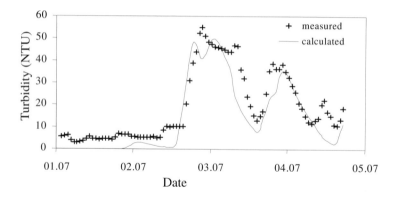

Figure 4. Calibration of effluent turbidity with STOAT at Toensberg WWTP

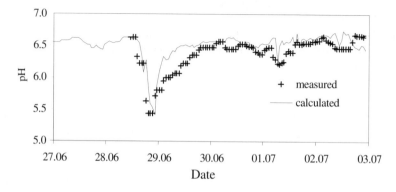

Figure 5. Calibration of effluent pH with STOAT at Toensberg WWTP

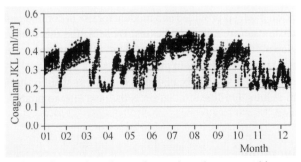

Figure 6. Estimated coagulant dosage for a selected year, to achieve a certain desired treatment efficiency

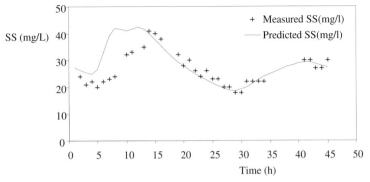

Figure 7. Measured and calculated effluent suspended solids values at the Fredriksatd WWTP

The simulation program was then calibrated for the Fredrikstad WWTP. The results are presented in Figure 7 for suspended solids. The calibration results show that the measured and calculated values are in considerable agreement.

The results confirm the validity of the simulation program concept, which was found to be valid for turbidity, suspended solids, pH and phosphates. In addition to the possible offline simulation activities to identify optimisation options, the concept will also be tested on-line to predict the optimal coagulant dosage, aiming to reduce the operational costs while retaining the required treatment efficiencies.

Summary

A simulation program for wastewater coagulation was presented. The concept is based on defining three fractions of the added coagulant, which consist of those involved in particle and phosphate removal, and an inert fraction. The proportion between these fractions varies depending upon the wastewater composition, pH and coagulant type, making it necessary to build coefficients into the program using previous experimental results.

The coagulation model is provided as a module for use in commercially available software for wastewater treatment processes. By selecting this option rather than developing independent software, it is possible to utilise the advanced and well-proven powerful features of the main program.

The simulation model was calibrated at two medium-sized coagulation plants with successful results. On-line simulation is the next step to predict the optimal coagulant dosage.

Acknowledgement

The authors wish to express their gratitude to the management of the Toensberg and Fredrikstad WWTPs, particularly to Mr Dagnfinn Fremstad and Mr Knut Lilleng who provided practical assistance and valuable advice during the experimental procedures. The Water Research Centre, UK, and Mr Jeremy Dudley is thanked for his assistance in programming the integration of the coagulant module into the main program.

References

Fettig, J., Ratnaweera, H., and Ødegaard, H. (1990) Simultaneous phosphate precipitation and particle destabilisation using aluminium coagulants of different basicity. In Hahn. H.H. and Klute. R. (Eds.) *Chemical water and wastewater treatment,* Springer-Verlag, Berlin. pp 221-242

Ratnaweera, H. (1991) Influence of the degree of coagulant prepolymerization on wastewater coagulation mechanisms. PhD dissertation, The University of Trondheim, Norwegian Institute of Technology, Norway

Tenny, M.W., and Stumm, W. (1965) Chemical flocculation of microorganisms in biological waste treatment. J. WPCF 37, p 1370

Baes, C.F., and Mesmer, R.E. (1976) The hydrolysis of cation. John Wiley and Sons, New York

WRC (2000) User manual for STOAT - a simulation program for wastewater processes. Water Resecarch Centre, Swindon, UK

H.H. Hahn, E. Hoffmann, H. Ødegaard (Eds.)
Chemical Water and Wastewater Treatment VII, pp. 261-272
© IWA Publishing, London
ISBN: 1 84339 009 4

Dissolved Air Flotation of Bioreactor Effluent Using Low Dosages of Polymer and Iron

*E. Melin, H. Helness and H. Ødegaard**

*Norwegian University of Science and Technology, Department of Hydraulic and Environmental Engineering, Trondheim, Norway.
email: Hallvard.Odegaard@bygg.ntnu.no

Introduction

Many cities still discharge large amounts of inadequately treated wastewater to marine waters. In many of these cities, the effluent standard that is to be met is secondary treatment. Secondary treatment is, however, not implemented because of excessive costs, lack of available land, and costly sludge disposal. These cities need a treatment method that is compact, reaches secondary treatment effluent standard, and has minimal sludge production. A competitive solution would be a treatment plant consisting of fine screening/sieving for primary treatment, a highly loaded biofilm reactor, and a highly loaded separation reactor (Ødegaard et al., 2000).

Most of the organic loading from municipal wastewater is in particulate matter (Levine et al., 1985; Ødegaard 1998, 1999). Therefore, with a good separation process, most of the organic loading can be removed. However, concentration of soluble organic matter is often too high for treatment plants to meet secondary standards with particle removal only. In the high-rate treatment concept, the intention is to operate the bioreactor at such high loading rates that it removes soluble matter but hydrolysis of particulate organic matter does not take place. The removal of particulate organic matter is left for the separation reactor after the bioreactor.

Compact biological treatment systems require biofilm reactors. Biofilters clog easily at high loads of particulate matter, which results in too frequent filter washing. For high-rate treatment concepts, a moving bed biofilm reactor (MBBR) has been shown to be a good alternative since the process can accept high particulate and soluble organic loading rates (Ødegaard et al., 2000).

One alternative for a high-rate separation process is flotation. The flotation system can be used with higher loading rates than sedimentation and has been shown to be effective in secondary wastewater treatment (Ødegaard, 2000; Filho and Brandao, 2000). With flotation, sludge settleability is not an issue. This can be a problem in a MBBR at high soluble COD loading rates (Ødegaard et al., 2000). Flotation has become more attractive after recent developments of a very highly loaded flotation process (so called turbulent flotation) where the surface loading can be as high as 25-40 m/h (Kiuru, 2000).

Optimal separation processes require addition of a coagulant. Inorganic metal salts are often used. However, with high metal dosages, the sludge production becomes high because of chemical precipitation. With use of cationic polymer the sludge production can be reduced but the dosages required can be relatively high (Fettig et al., 1990). When an inorganic metal salt is combined with low dosages of polymer, the metal dosage and sludge production can be significantly reduced without compromising treatment efficiency (Ødegaard, 1998).

This paper reports the results from preliminary screening tests of different cationic polyacrylamide (PAM) and poly-diallyl-N,N-dimethylammonium chloride (polyDADMAC) polymers. The purpose was to investigate what kind of polymer is best for flotation of the MBBR effluent and to find optimal dosage for coagulation with combination of metal salt and polymer.

Experimental set-up

Flotation experiments

The principle of the treatment process is presented in Figure 1. A laboratory-scale MBBR was used to treat domestic wastewater, which was pumped into a buffer tank from a nearby residential area. The MBBR loading was very high with a detention time of only 15 min. An Aztec flotation jar tester (Severn Trent Services, Capital Controls Ltd, England) was used for flotation tests. One-litre samples collected from the MBBR outlet were used in each jar.

The iron and polymer were dosed with syringes under rapid mixing (400 rpm) which was continued for 0.5-1 min. The water was then flocculated for 20 min while mixing with 80 rpm. In the flotation step, 150 ml of dispersion water (15% recycle rate) was used, saturated with air under 5 bar pressure. Distilled water was used as dispersion water. The dilution effect of dispersion water was taken into account when calculating the results. The samples from clarified water were taken 10 min after dispersion water was applied.

Iron was dosed as JKL ($FeCl_2SO_4$), which is a product of Kemira Chemicals. PAMs were manufactured by Kemira and polyDADMACs were from Cytec.

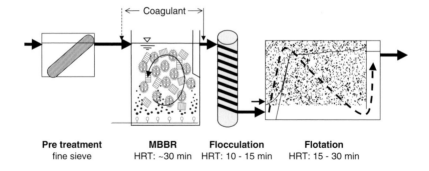

Pre treatment   **MBBR**   **Flocculation**   **Flotation**
fine sieve   HRT: ~30 min   HRT: 10 - 15 min   HRT: 15 - 30 min

Figure 1. Schematic of the treatment process

Experimental design and data analysis

The effect of polymer properties and polymer and iron dosages on the treatment results were evaluated. Figure 2 shows the design regions for molecular weight and charge density. Two different designs were used for PAMs. In addition, three different low molecular weight, high charge density polyDADMACs were tested.

The low iron dosages varied from 0 to 0.2 mmol Fe/l and polymer dosages from 0.5 to 3 mg/l. Some polyDADMAC tests were done with polymer dosages up to 3.4 mg/l and 0.3 mmol Fe/l. Since a real wastewater was used in the experiments, the wastewater quality could not be used as a design variable.

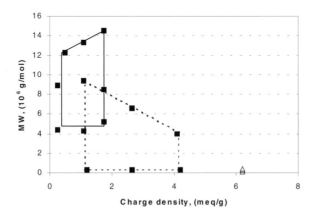

Figure 2. Design regions for charge density and molecular weight (■ PAMs, ▲ polyDADMACs)

The results were evaluated using Partial Least Squares Regression (PLSR), a multivariate analysis method based on analysis of the variation in the data (Martens and Næs, 1989). In the PLSR, a new set of x-variables called PLS components (PC) are computed in such a way that the first PC lies in the direction of the largest variation of the data. The second PC lies in the direction of second largest variation and so on. The logic is that the largest variation in the data is likely to be caused by important or real effects while small variations in the data can be caused by less important effects or noise. One advantage of PLSR is that one avoids focusing on large variations in the x-data that have little importance for the variation in the response variables. Another advantage is that the PCs are orthogonal, i.e. linearly independent. Using PCs in regression can therefore overcome problems caused by collinear x-variables. However, one must be aware of the danger of over-fitting and meaningless results and put heavy emphasis on validation of the models.

The multivariate regression model has so far been developed for suspended solids (SS) removal. The conclusions from the model are presented and the average results from experiments are used to illustrate the observed effects.

Wastewater

The effluent from the MBBR was used in the flotation tests. The water quality is presented in Table 1 for the tests with PAMs and polyDADMACs. The temperature of the water was 9-11°C.

Table 1. Raw water quality during the experiments

	PAMs			PolyDADMACs		
	Average	Min	Max	Average	Min	Max
SS (mg/l)	143	98	187	111	53	163
COD (mg/l)	249	161	316	207	115	306
SCOD (mg/l)	67	51	92	61	39	125
pH	7.71	7.54	7.92	7.70	7.43	7.89

Results and discussion

Effect of polymer and iron dosage

The multivariate regression model shows that the treatment results are primarily governed by the polymer and iron dosages. The effect of polymer dosage was linear for both types of polymer. The response for iron dosage was non-linear, i.e. the effect of iron on the treatment results became smaller with increasing iron dosage. Figure 3 shows the average SS removals in all the tests with different polymers.

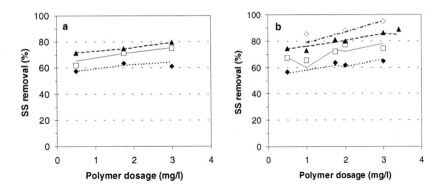

Figure 3. Effect of iron and PAM (a) or polyDADMAC (b) dosages on suspended solids removal (◆ 0 mmol Fe/l, ☐ 0.1 mmol Fe/l, ▲ 0.2 mmol Fe/l, ◇ 0.3 mmol Fe/l). Lines show model predictions

Without metal coagulant, only moderate removal of SS was achieved. At PAM dosages of 1.75 and 3 mg/l, increasing iron dosage from 0.1 to 0.2 mmol Fe/l did not increase removal efficiency as much as demonstrated between no iron and 0.1 mmol Fe/l, explaining the non-linearity in the model. The non-linear response is difficult to see in the polyDADMAC results, probably due to variation in the wastewater quality as discussed below.

The lines in Figure 3 show the model predictions for SS removal. The model fits the experimental results well and shows that the variations in the removal efficiency are a result of varying raw water quality rather than experimental error. Generally, the removal efficiency is similar between PAMs and polyDADMACs but polyDADMACs give slightly higher SS removal with 0.2 mmol Fe/l. With 3 mg/l

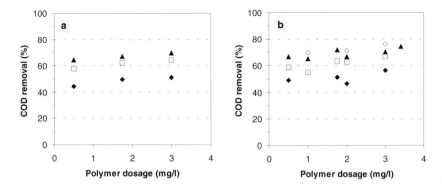

Figure 4. Effect of iron and PAM (a) or polyDADMAC (b) dosages on COD removal (0 mmol Fe/l, 0.1 mmol Fe/l, 0.2 mmol Fe/l, 0.3 mmol Fe/l)

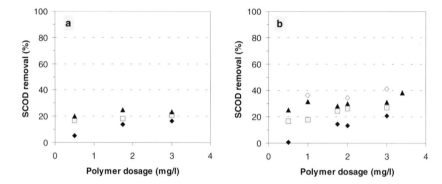

Figure 5. Effect of iron and PAM (a) or polyDADMAC (b) dosages on SCOD removal
(◆ 0 mmol Fe/l, ☐ 0.1 mmol Fe/l, ▲ 0.2 mmol Fe/l, ◇ 0.3 mmol Fe/l)

of polymer and 0.2 mmol Fe/l the SS removal efficiency varied with PAMs from
72 to 89%, resulting in 16-46 mgSS/l (average 29 mgSS/l) in the treated water.
With polyDADMACs, the removal efficiency varied from 79 to 91% with residual
SS of 6-21 mg/l (average 14 mg/l). However, because of the variation in the
wastewater quality (Table 1), it cannot be concluded that polyDADMACs in general
give better results than PAMs.

Figure 4 shows the removal of chemical oxygen demand (COD) in the flotation
tests. The removal patterns are the same as with SS although overall removal
efficiencies are lower. Figure 5 shows the removal of soluble COD (SCOD). Only
minor removal is observed in tests without metal coagulant and iron improves
SCOD removal. Soluble COD is normally defined as the COD measured after
filtering sample through GF/C filter, which has a nominal pore size of about 1 μm.
This fraction still includes some colloidal material and the truly soluble fraction is
below 0.1 μm. Therefore, the removal of SCOD can actually be coagulation of
the colloidal fraction.

Effect of raw water quality

The multivariate analysis indicates that the raw water SCOD and pH influenced
SS concentration in clarified water with PAMs. In the case of polyDADMACs, the
raw water COD and SS affected the treatment efficiency while pH and SCOD did
not have a significant effect. Figure 6 shows the effect of raw water SCOD on SS
removal for all the tests. With PAMs, removal efficiency decreased when SCOD
was over 65-70 mg/l while the polyDADMAC results were unaffected by raw water
SCOD. However, the results need to be verified by further tests because the PAM
results with high SCOD concentration are from experiments without iron dosage
while the polyDADMAC results with high SCOD concentration are from
experiments with 0.2 mmol Fe/l.

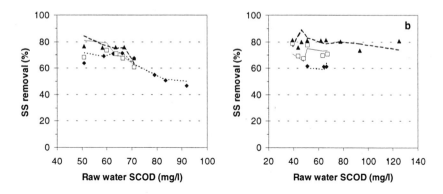

Figure 6. Effect of raw water SCOD on SS removal in flotation test using PAMs (a) and polyDADMACs (b) (◆ 0 mmol Fe/l, ◻ 0.1 mmol Fe/l, ▲ 0.2 mmol Fe/l). Lines show model predictions

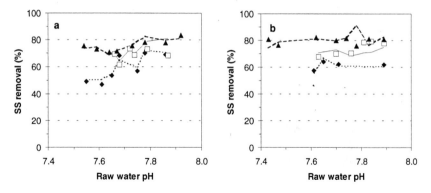

Figure 7. Effect of raw water pH on SS removal in flotation test using PAMs (a) and polyDADMACs (b) (◆ 0 mmol Fe/l, ◻ 0.1 mmol Fe/l, ▲ 0.2 mmol Fe/l). Lines show model predictions

Figure 7 shows the results for both polymers. While they seem to confirm the model, the raw water pH varied only in a very narrow range (from 7.4 to 7.9) and therefore the results are not very conclusive. However, the results indicate that PAMs are more sensitive to raw water quality properties like SCOD and pH than polyDADMACs.

Effect of polymer properties

The model indicates that with PAMs there is an effect of the molecular weight that depends on the iron dose. The model predicts that when iron is not used or the dosage is low, it is a benefit to have a high molecular weight polymer. When the iron dosage is increased, slightly better results are predicted with a low molecular weight polymer. Figure 8a shows the average results from all the tests done with different polymers. It should be noted that the wastewater quality and average polymer dosages are not the same for the different data points. The average trends in the data for the different iron dosages support the results from the model. Without

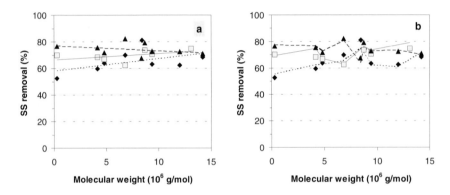

Figure 8. The effect of molecular weight on SS removal efficiency with different iron dosages for all the tested PAMs (a) and the same results together with the model predictions (b) (◆ 0 mmol Fe/l, ◻ 0.1 mmol Fe/l, ▲ 0.2 mmol Fe/l)

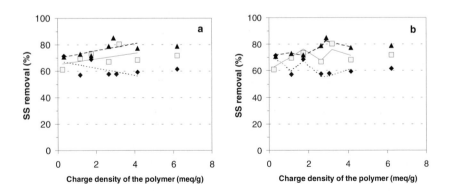

Figure 9. Effect of charge density of the polymer on the SS removal with all the tested polymers (a) and the same results together with the model predictions (b) (◆ 0 mmol Fe/l, ◻ 0.1 mmol Fe/l, ▲ 0.2 mmol Fe/l)

iron, the removal efficiency is lowest with low molecular weight PAMs. If the polymer is going to be used alone without metal coagulant, low molecular weight PAMs do not seem to be the best alternative as also observed by Pilipenko and Ødegaard (2002). In Figure 8b, the model prediction of the different data points is included. The good agreement between the model and the experimental results shows that the scatter in the results can largely be explained by the variation in the wastewater quality and polymer dosage.

The model for PAM indicates that a high charge density is a benefit. Figure 9a shows the removal efficiency as a function of the charge density of the polymer with all the tested polymers. In these figures, polyDADMAC is also included for comparison (charge density 6.2 meq/g). The average trend in the PAM data with iron doses of 0.1 and 0.2 mmol Fe/l supports the model, while the average trend for the PAM data with no iron shows the opposite. However, this is due to two data points with high removal at charge densities of 0.3 and 1.8 meq/g. In Figure 9b, the model prediction of the different data points is included. The good agreement between the model and the experimental results indicates that the scatter in the results can largely be explained by the variation in the wastewater quality and polymer dosage. The results for both molecular weight and charge density demonstrate that although these polymer properties probably have an effect on the removal efficiency, they are small compared to the effect of dosage and wastewater quality.

There were no significant differences between the tested polyDADMACs. Also, the best PAMs performed equally with polyDADMACs. When a PAM is used together with iron, the best choice seems to be a medium molecular weight, high charge density polymer.

COD fractions in the flotated water

In some tests, COD was analysed from water samples that were filtered through filters having different pore size. Figure 10 shows the COD fractions in raw water and flotated water. In the experiment, polyDADMAC was used at variable dosages (0.6-3.4 mg/l) and the iron dosage was 0.2 mmol Fe/l. The results show that particles above 11 μm are effectively removed by flotation, which is consistent with general understanding that flotation is effective in removing particles down to 10 μm in size (Kiuru, 1990). The truly soluble COD fraction (<0.1 μm) is the largest fraction in the clarified water and is not removed very well. Since the aim is removal of particulate organic matter, this size fraction is of no interest. The preceding biological process should be operated so that the truly soluble fraction is removed to desired levels. The results show that the observed SCOD removal is mostly the removal of colloidal material (size fraction 0.1-1 μm). The 1-11 μm size fraction is, however, critical for successful particle removal and further process optimisation should concentrate on good flocculation of this particle size range.

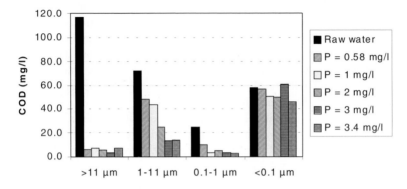

Figure 10. Particulate COD fractions in raw water and flotated water with different polyDADMAC dosages and 0.2 mmol Fe/l iron

Conclusions

1. Good SS removal could be obtained by flotation when low dosages of iron and polymer were combined.
2. The dosages of iron and polymer together with raw water properties like SCOD are more important than molecular weight or charge density of the polymers.
3. There were no significant differences between polyDADMACs and the best PAMs. With metal coagulant, the best PAM was a medium weight, high charge density polymer. Without metal, high molecular weight PAMs gave the best result.
4. Multivariate analysis is a good tool when analysing results with variable water quality. Although the regression model developed for SS removal was able to predict the results well, it needs additional tests to fill the gaps in the data set and an independent verification test.

Acknowledgement

This study was financed by Kemira OYj, Finland, National Technology Agency, Finland and Kaldnes Miljøteknologi AS, Norway.

References

Fettig, J., Ratnaweera, H. ,and Ødegaard, H. (1990) Synthetic organic polymers as primary coagulants in wastewater treatment. *Water Supply* **8**, 19-26
Filho, A. C. T. P., and Brandao, C. C. S. (2000) Evaluation of flocculation and dissolved air flotation as an advanced wastewater treatment. *Wat. Sci. Tech.* **43**(8), 83-90

Kiuru, H. (1990) Unit operation for the removal of solids and their combinations in water treatment. In Hahn, H.H. and Klute, R. (Eds.) *Chemical Water and Wastewater Treatment*, Springer-Verlag, Berlin, pp. 169-186

Kiuru, H. J. (2000) Development of dissolved air flotation technology from the 1st generation to the newest or 3rd one (very thick micro-bubble bed) with high flow-rates (DAF in turbulent flow conditions). *Wat. Sci. Tech.* **43**(8), 1-7

Levine, A.D., Tchobanoglous, G. and Asano, T. (1985) Characterization of the size distribution of contaminants in wastewater; treatment and reuse implications. *J. WPCF* **57**(2), 805-816

Martens, H., and Næs, T. (1989) *Multivariate Calibration*, John Wiley & Sons Ltd.,Chichester

Ødegaard, H. (1998) Optimized particle separation in the primary step of wastewater treatment. *Wat. Sci. Tech.* **37**(10), 43-53

Ødegaard, H. (1999) The influence of wastewater characteristics on choice of wastewater treatment method. In *Proc. Nordic Conf. on Nitrogen removal and biological phosphate removal*, Oslo, 2-4 February

Ødegaard, H. (2000) The use of dissolved air flotation in municipal wastewater treatment. *Wat. Sci. Tech.* **43**(8), 75-81

Ødegaard, H., Gisvold, B., Helness, H., Sjøvold, F., and Liao, Z. (2000). High rate biological/chemical treatment based on the moving bed biofilm process combined with coagulation. In Hahn, H.H., Hoffmann, E. and Ødegaard, H. (Eds.) *Chemical Water and Wastewater Treatment VI*, Springer-Verlag, Berlin, pp. 245-255

Pilipenko, P., and Ødegaard, H. (2002) Removal of suspended solids and sludge production in coagulation of municipal wastewater with cationic polyelectrolytes. In *Proc. of the International Congress and Trade Fair Water: Ecology and Technology,* ECWATECH, June 4-7, Moscow

H.H. Hahn, E. Hoffmann, H. Ødegaard (Eds.)
Chemical Water and Wastewater Treatment VII, pp. 273-284
© IWA Publishing, London
ISBN: 1 84339 009 4

Effects of Chemical Agents on Filamentous Growth and Activated Sludge Properties

J. Kegebein, *E. Hoffmann and H.H. Hahn*

*Institute for Aquatic Environmental Engineering, University of Karlsruhe, Germany
email: kegebein@isww.uka.de

Abstract

Sludge bulking and foaming caused by filamentous organisms are well known phenomena of activated sludge systems. The effluent may deteriorate due to poor settling properties, thus the effluent standards cannot be met. The addition of aluminium containing chemicals, such as $AlCl_3$ or prepolymerized aluminium chloride, is seen as a means of controlling and improving sludge properties. $AlCl_3$ and PAX-18 were tested in a low loaded nutrient removal sequencing batch system. A second reactor was operated under the same conditions except for the chemicals addition, and served as the control. A single pulse dose of 30 mg PAX-Al/gMLSS damaged the filamentous organisms and none of them could be found after 10 days, whereas the filamentous growth continued in the control reactor. PAX-18 was effective for neutralizing the surface charge (zeta potential) of activated sludge flocs at pH 6.5, whereas $AlCl_3$ was effective in a lower pH range. At low pH, Al^{3+} may inhibit the activity of enzymes that are involved in the growth of filamentous bacteria. Thus a strategy for dosing aluminium, which is directed towards optimising the concentration of free aluminium rather than total dose, may be more effective for depressing filament growth.

Introduction

Waste water treatment plants for advanced nitrogen and phosphorus removal are operated under low-load conditions with sludge ages of 10 days or more. The nutrient removal activated sludge process utilizes the metabolic properties of several types of microorganisms, which can be controlled by substrate levelling and

[273]

provision of different electron acceptors. The subdivision of the bioreactor into anaerobic, anoxic and aerobic zones and a proper design of flow rates and tank volumes are necessary preconditions. However, even a perfect biochemical performance results in a poor effluent quality if effective sludge separation cannot be achieved.

Sludge sedimentation has turned out to be the crucial step in the treatment process. The floc properties (size, density and shape) are the governing factors for the sedimentation process and can hardly be predicted.

The basic aim of an aluminum dosage in the treatment process is the removal of phosphorus by simultaneous precipitation. In addition a floc structure improvement can be expected. Two mechanisms are known for aggregation of colloids and creation of stronger flocs:

1.   Positively charged ions, such as Al^{3+}, lower the colloidal surface charge thus enabling aggregation. This effect occurs at low pH values, when the ions are soluble and do not form neutral hydroxides.
2.   At higher pH values hydroxides are formed and act as sweeping flocs.

These processes are instantly effective, because of their physicalchemical nature. Recent research and operational experience indicate a drawback of filamentous growth when aluminium is added to the activated sludge system (Eikelboom *at al.,* 1998). When filamentous bacteria profilerate, they grow out of the floc, slowing the settling due to increased drag forces, and may interconnect with other flocs, thus creating a network where a single floc cannot move independently. Moreover, gas bubbles entangle in the web, leading to buoyancy and foaming sludge. The filamentous decrease due to aluminium dosing is not well understood, but it may turn out to be a strong instrument for improving the floc structure.

The most common filament is *Microthrix parvicella* with the following selection factors known for mixed culture systems (Knoop and Kunst, 1998):

1.   Low food to microorganism ratio or long sludge age
2.   Alternating availability of oxygen and nitrate as electron acceptors
3.   Low temperature
4.   Availability of long chain fatty acids (sole carbon source)

Although these selection factors are well known, they can hardly be altered as a means of filament control. Selection factors 1. and 2. are preconditions for nutrient removal systems and raising the temperature consumes to much energy. Blocking the access of the carbon source is the most promising approach, but surprisingly, no literature on the removal efficiency of grease traps is available.

The objective of this study was to investigate the performance of $AlCl_3$ and PAX 18 regarding their instant physicalchemical and delayed physiological effects on the floc structure.

Materials and methods

General

Origin: Sludge and wastewater was taken from a municipal nutrient removing waste water treatment plant (WWTP) in Blankenloch, Germany.

Waste water parameters: Suspended solids were measured according to standard methods. COD, N and P were measured with test kits from Dr. Lange and Merck respectively.

Floc and settling properties

Settling velocity was measured in a column with 0.25 m diameter and a height of 0.8 m. $AlCl_3$ and PAX 18 was added to 1 L of activated sludge. After 15 minutes of gently stirring, the suspension was diluted with 20 L effluent from the WWTP and allowed to settle in the column. Samples were taken for solids content analyses 40 cm below the surface at the beginning and after 2, 5, 10, 20 and 40 minutes. The calculated settling velocity distributions represent single floc settling (Type I settling).

Zeta potential was measured with Pem Kem 501, Pem Kem Inc., US.

Filamentous levels: Filamentous growth was classified into 4 classes using a dark field microscope at 480 x magnification (Figure 1).

  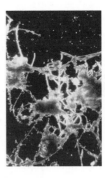 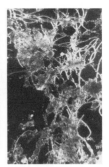

Figure 1. Classification of filamentous levels. (Left to right) Class 0: no filamentous growth. Class 1: Filaments grow out of the floc. Class 2: Filaments connect flocs. Class 3: Filamentous web contains flocs

Lab scale SBR treatment system

The sequencing batch reactor consisted of a 10 L column serving as aerobic, anoxic and phase separation reactor. A second reactor of 0.6 L Volume was placed before the main reactor to produce anaerobic conditions prior to the aerobic/anoxic treatment. A flow scheme of the reactor setup is shown in Figure 2. The raw wastewater mixed with settled sludge was fed to the anaerobic reactor then transferred to the main reactor in the subsequent filling. The retention time within the anaerobic reactor therefore equals the time lag between two transfers to the reactor and was expected to be sufficient for biological phosphorus removal. The entire system was cooled down to 10°C – 12°C and the F/M ratio was kept below 0.2 g COD/gMLSS • d in order to provide good growth conditions for filamentous bacteria. The operational data are displayed in Table 1.

Two lab-scale treatment plants, as displayed in Figure 2, were operated in parallel, one for experiments (EXP) , the other served as control (REF). Both reactors operated for 20 days in parallel before the experiments.

Table 1. Data during operation in parallel; Mean values; N = 13

	INFLUENT	EXP	REF
MLSS [g L^{-1}]		1,6	1,8
F/M [gCOD gMLSS^{-1} d^{-1}]		0,18	0,15
HRT [d]		1,8	1,9
COD [mg/L]	321	49	36
NH$_4$-N [mg/L]	28	<0,1	<0,1
Turbidity [NTU]		18	10

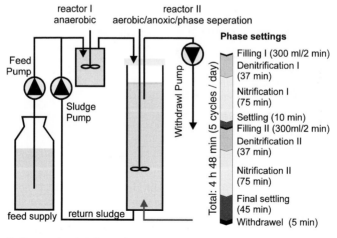

Figure 2. Sequencing batch reactor set up

COD, ammonia, nitrate and phosphorus were measured of both, influent and effluent. Sludge samples were taken from the reactors for determination of MLSS. Sludge volume was read directly at the reactor columns after the 45 minute settling phase. Thus the calculated sludge volume index does not match with the determination according to standard methods and must be considered as a plant specific relative value.

Results

PAX-18 long term performance

PAX-18 was dosed at day 0 only. The initial Al/MLSS ratio was set to 30 mg PAX-Al per gMLSS. Due to the dosage the pH decreased to 5.5 but recovered with the next filling. Initially the effluent quality deteriorated and the COD reached 160 mg/L and was accompanied by a high turbidity of 200 NTU. The system recovered within three days and no more significant variations of the effluent quality were recorded with respect to the control reactor.

Figure 3 displays the filamentous levels identified by microscopic observation. No filaments were detected within the experimental reactor ten days after dosage, whereas filamentous growth proceeded in the control reactor. These levels were maintained for further 20 days without additional dosage. The depression of filamentous growth was reflected by the SVI measurements to some extent, although the decreasing tendency in the control reactor did not reflect the increase of filament growth. When the experiment was terminated, the descent of the sludge blanket was recorded in a 100 ml cylinder. The results, shown in Figure 5, indicate a faster and more intense thickening even though the Al-application dated back 30 days.

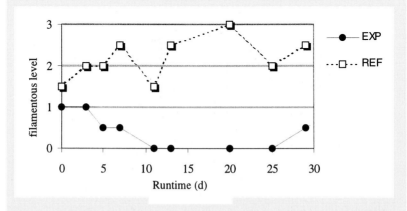

Figure 3. Filamentous levels within the reactors with single PAX dosage at day 0 (EXP) and without PAX dosage (REF)

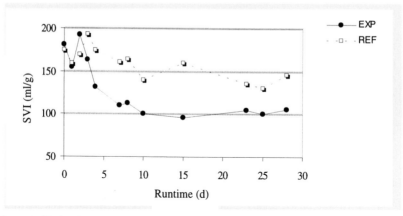

Figure 4. Sludge Volume Index after PAX dosage

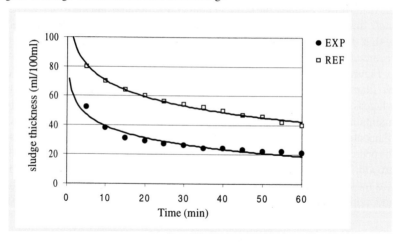

Figure 5: Sludge thickening curve within a 100 ml cylinder
MLSS content: EXP = 2,4 g/L; REF = 2,5 g/L

$AlCl_3$ long term performance

In the preliminary stages of the $AlCl_3$ experiment the filamentous growth was very poor. Only a few short filaments grew out the flocs, representing a filament level somewhere below 1. The Al dosage was therefore set to 6 mgAl/gMLSS, which led to a pH drop to 6. In contrast to PAX-18, $AlCl_3$ had no significant effect on effluent quality. In fact, there was no significant variation in the filamentous levels in either the control or the experimental reactor during the 15-day experiment. Although the filament levels within the experimental reactor were found to be slightly below the reference reactor, the difference was marginal and was not reflected by the sludge volume index measurements.

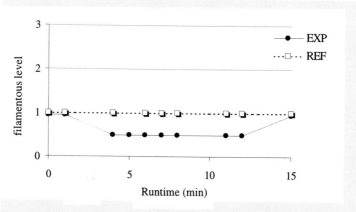

Figure 6. Filamentous levels within the reactors after AlCl₃ dosage

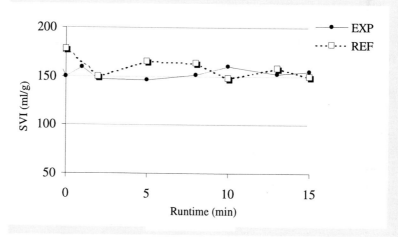

Figure 7. Sludge Volume Index after AlCl₃ dosage

Zeta potential and settling velocity

Zeta potential measurements were carried out with an activated sludge suspension (MLSS: 500 mg/L) at pH 4.5, pH 5.5 and pH 6.5. PH values were adjusted by adding HCl prior to the addition of $AlCl_3$ and PAX-18. $AlCl_3$ was least effective for colloid destabilization at pH 6.5. At lower pH values the destabilization with $AlCl_3$ was faster and 2 mgAl/L were sufficient to reach zeta 0. PAX-18 turned out to be effective even at pH 6.5.

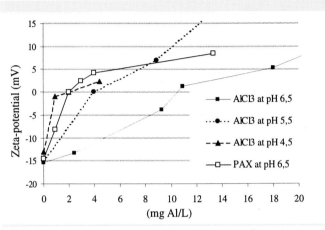

Figure 8. Zeta Potential vs. Al-addition

Figures 9 and 10 display the settling velocity distributions of activated sludge flocs treated with $AlCl_3$ and PAX-18, respectively. Generally, both precipitants effected a settling velocity increase, indicated by the right-shift of the graphs. The settling velocity increase can be attributed to the floc destabilization and agglomeration. As shown in Figure 8, the application of PAX led to a sharp rise of the zeta potential, as compared to the addition of $AlCl_3$ at pH 6.5 (which was the pH during settling). The lower PAX dosage (6.4 gAl/gMLSS) led to a complete destabilization and maximum agglomeration, therefore no further increase was achieved by adding a higher dose. The effects of $AlCl_3$ on zeta potential and settling velocity correspond in almost the same manner.

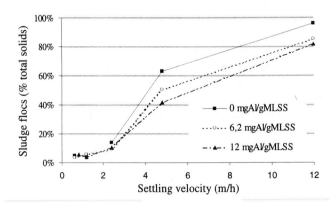

Figure 9. Settling velocity distribution of activated sludge flocs treated with $AlCl_3$

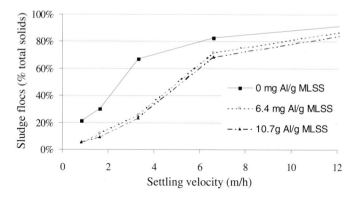

Figure 10. Settling velocity distribution of activated sludge flocs treated with PAX-18

Discussion

The zeta potential measurements demonstrate that the destabilization of biocolloids by PAX-18 occurred in an almost neutral pH range, whereas a destabilization with $AlCl_3$ requires either a higher dosage or an additional pH adjustment. The settling velocity increase corresponded with the degree of floc destabilization and agglomeration, although sweeping floc effects must have contributed. Aluminium exhibits a complex speciation chemistry which is highly dependent on pH. At pH values greater than 6.0 only 0.03 % of Al in solution is present as the Al^{3+} ion with the rest in different forms of $Al(OH)_x^y$. Edzwald *et al.* (2000) described that prepolymerized aluminium exhibits a better solubility at pH values greater than 6.0 than non-prepolymerized forms, thus the zeta potential response is somehow correlated to the presence of soluble charged Al species. Moreover, the presence of soluble species is probably the governing factor for biological inhibition within the activated sludge process. It is well known that free aluminium ions inhibit the growth of plant roots (Bertram *et al.*, 1996). Jones and Kochian (1997) investigated the impact of Al^{3+} on enzyme activities, and ascertained a *phospholipase A* activity decrease. These findings are reproduced in Figure 11 and may be of particular importance for depression of filament growth, since the physiology of many filaments, particular *Microthrix parvicella* is linked to the uptake of fatty acids (Andreason and Nilsen, 1997; Slijkhuis, 1983) and phospholipases are known as key enzymes for the degradation of certain fats (Gaudy and Gaudy, 1981). It was Eikelboom (1994), who entitled one of his publications "The *microthrix parvicella* puzzle". Several new pieces of the puzzle have been published within the last years, and the results presented here may contribute one more piece.

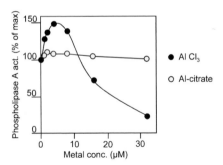

Figure 11. Effect of AlCl$_3$ and Al-citrate on *phospholipase A* activity (adapted from Jones and Kochian, 1997). Citrate is known as an Al complexing ligand, thus it becomes not effective even at low pH

If it turned out to be true, the dosing strategy of aluminium for depression of filament growth should target on the concentration of free aluminium rather than the total dose. Only a few reports are available on aluminium based filament depression and most of them neglect to mention the relevant pH value at the dosing spot. Engl (1999) gives a very detailed report of a full scale experiment in Austria and indicated that dosing VTA, a poly-aluminium-magnesium-silicate, failed to depress *Microthrix parvicella*. VTA was dosed directly into the activated sludge tank and, furthermore, lime was dosed into the same tank for keeping the pH at 7.1, which probably counteracted the VTA .

The results of the presented long-term experiments show that a single dosage of PAX-18 led to long lasting filament depression. As for the AlCl$_3$ application, the result is not so clear. However, a dosing strategy, which brings together the presented results and assumptions could be realized with a low-pH-bypass-dosing. A small portion of the return sludge flow could be diverted to a dosing reactor, where the pH is reduced and either PAX-18 or AlCl$_3$ is added. The treated bypass sludge is released back to elsewhere in the aerobic/anoxic system. It is assumed, that the treated portion would not suffer from filamentous growth subsequently as long as one sludge age. Furthermore, the proposed dosing strategy may reduce the chemicals demand as compared to the treatment of the entire return sludge, since the percentage of active (free) aluminium would be greater due to the lower pH value.

The proposed dosing strategy will be the subject of further research. The optimum flow rate, retention time and pH/Al - ratio shall be investigated in order to achieve a sustainable filament depression and to minimize the chemicals demand.

Acknowledgments

We would like to thank the students Christiane Bories and Andreas Blank for their ambitious thesis work on the subject.

References

Andreason, K. and Nilsen, H. (1997) Application of microautoradiography to the study of substrate uptake by filamentous microorganisms in activated sludge, *App. and envir. microbiology* **63** (9), 3662-3668

Bertram, R., Stiebert, E., Geßner, W. (1996) Toxizität von Aluminium - Al Spezies in protolysierten Aluminiumchloridlösungen, Umweltwissenschaften und Schadstoffforschung, *UWSF* **8** (2), 78-82 *(in German)*

Edzwald, J.K., Pernitsky, D.J. and Parmenter, W.L.(2000) Polyaluminium coagulants for drinking water treatment: Chemistry and selection, in Hahn, H.H., Hoffmann, E., Ødegaard, H. (Eds.) Chemical water and wastewater treatment VI, Springer Berlin, Heidelberg, New York, 3-14

Eikelboom, D., Andreadakis, A, and Andreasen, K. (1998) Survey of filamentous populations in nutrient removal plants in four European countries. *Wat.Sci.Tech.* **37**(4/5), 281-289

Eikelboom, D.H. (1994) The *Microthrix parvicella* puzzle, *Wat.Sci.Tech.* **29** (7), 271 - 279

Engl, K.(1999) Gegenüberstellung von VTA und $FeSO_4$ in einem Langzeitversuch auf der Kläranlage Mittleres Pustertal, available on http://www.aratobl.com/dt/wissen/wissen1.html *(in German)*

Gaudy, A.F. and Gaudy, E.T. (1981) Microbiology for environmental scientists and engineers, *Mc Graw Hill International Book Company,* 425

Jones, D.L. and Kochian, L.V. (1997) Aluminium interaction with plasma membrane lipids and enzyme metal binding sites and its potential role in Al cytotoxicity, *FEBS Letters* **400**, 51-57

Knoop, S. and Kunst, S. (1998) Influence of temperature and sludge loading on activated sludge settling, especially on Microthrix parvicella. *Wat. Sci.Tech.* **37** (4/5), 27-35

Slijkhuis, H. (1983) *Microthrix parvicella*: a filamentous bacterium isolated from activated sludge: cultivation in a chemicially defined medium, *Appl. Environ. Microbiol.* **23** (4), 77-80

Sludge Treatment

H.H. Hahn, E. Hoffmann, H. Ødegaard (Eds.)
Chemical Water and Wastewater Treatment VII, pp. 287-296
© IWA Publishing, London
ISBN: 1 84339 009 4

Polymers in Alum Sludge Dewatering: Developments and Issues

*D.H. Bache**

*University of Strathclyde, Dept. of Civil Engineering, UK
email: d.bache@strath.ac.uk

Abstract

Issues surrounding the use of polymers for conditioning alum sludge derived from the coagulation of low turbidity coloured waters are discussed. The factors that control the optimum dosage *vis a vis* individual test methods such as SRF and CST are reviewed. Consistency amongst the test methods was considered to emanate from the underlying pattern of polymer adsorption. It was demonstrated that good estimates of the solids feed could be obtained via the suspended solids concentration in the flocculators. An algorithm based on water quality parameters was recommended for planning purposes. Operational problems linked to the delivery of the optimum dose were also described. Difficulties arising from variations in the solids delivery to the thickeners were shown to influence the separation efficiency. A survey of polymer routing within a wastewater works, in which there was multiple polymer use and extensive recycling, helped identify a number of quality control issues and demonstrated the value of polymer measurement.

Introduction

Polymers have long been recognised for their utility in sludge dewatering. As a result, they are used extensively in water treatment. For both economic and environmental reasons, there is a continuing need to optimise their usage and to avoid waste. Although many decisions concerning their use are based on their efficacy at full scale, there are many laboratory based test methods, such as the Capillary Suction Time (CST) to assist with the task of identifying the optimum dose. Until comparatively recently, there has been a lack of a straightforward and sufficiently sensitive method for measuring polymer concentration. Size exclusion chromatography (SEC) described in Keenan *et al.* (1998) provides a means of solving

[287]

this problem, and represents a tool that can be used to gain a better understanding of the influence of polymers on conditioning and dewatering. It can also be used for assessing the efficiency of polymer use at full scale. The goal of this paper is to consider some of these factors and to identify some of the operational difficulties and design issues surrounding the specification and control of the optimum dosage in the form kg/tonne DS (*viz* kg/t). The focus is on alum sludge gained from the treatment of low turbidity coloured waters.

Concept of optimum

The optimum is often assessed using laboratory scale test methods such as the specific resistance to filtration (SRF) or CST. In the case of an alum sludge gained from the treatment of low turbidity coloured waters, the optimum dose is usually in the range of 1-3 kg/t. The 'optimum' is usually based on a behavioural feature on the dose-response curve, such as the existence of a minimum, or the lowest dose to generate an acceptable rate of water release. As such, the optimum is tied to specific equipment and specific behavioural features. In the case of CST, the limitations are well known. For example, the CST measures the rate of water release at a low pressure head, whereas real systems operate at much greater pressures at which compressibility may be significant. The CST is particularly sensitive to the shear history of the sludge, a facet that can render the test meaningless unless the shear history can be mimicked. The SRF-dose curve also passes through a minimum, this being caused by an interplay between the cake resistance and the resistance of the filter support medium (Zhao *et al.*, 1998). Dentel and Abu-Orf (1995)

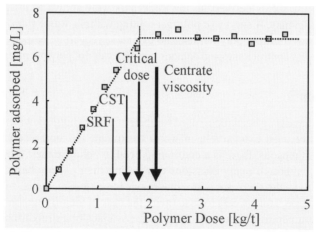

Figure 1. Comparison of estimates of the optimal dosage of Flocmiser 50 (a non-ionic polymer) for an alum sludge of 4250 mg/l solids concentration using SRF, CST and centrate viscosity measurements. The critical dose corresponds to the onset of adsorption saturation (from Papavasilopoulos, 1998)

demonstrated the use of centrate viscosity as a means of identifying the optimum dose. Subsequent analysis by Bache and Papavasilopoulos (2000) showed that a minimum on the viscosity-dose curve was controlled by the interplay of the effects of turbidity on viscosity and the viscous nature of polymer residuals. In this case, the factors controlling the minimum were different from those that apply to the SRF-dose curve. In spite of different modes of behaviour amongst the test methods, there is reasonable consistency in their estimates of the optimum dosage (Figure 1). This arises because the individual traits of behaviour are controlled by the underlying polymer adsorption. The adsorption scales on a mass/mass basis (see Keenan *et al.*, 1998) and represents the *raison d' etre* for the form of the dosage ratio. When a residual exists, it is either caused by incomplete mixing, or arises when the adsorption capacity is exceeded. The advantage of the viscosity measurement is that it picks up this feature without the need for sophisticated measuring equipment. Experience with an alum sludge shows that the optima associated with the SRF and CST are below the dose at which saturation polymer adsorption occurs.

Specification of solids feed

As noted above the target dosage is based on the control of the ratio kg/t. In principle, this is straightforward, but in practice it is influenced by many operational factors. Amongst these is the need to specify the solids feed. In the UK, guidance is provided by the WRc formula (Warden, 1983):

Within this formula, the specification of the colour is ambiguous, because it is

$$SS\ (mg/L) = 2.9 \cdot Al\ (mg/L) + 0.2 \cdot Colour\ (^{\circ}Hazen) + 2 \cdot Turbidity\ (NTU) \qquad \text{(Eq. 1)}$$

not clear whether 'colour' refers to the 'true' colour or the 'apparent' colour. In strict terms, it should also refer to the true colour *removed*, a factor which also applies to other components contributing to the suspended solids. To assess the viability of Eq. (1), the solids feed (L_s; kg/d) was assessed by a number of different methods.

a) By direct measurement of the suspended solids in the flocculators (SS_f) and calculation via $L_s = Q_w \cdot SS_f$ in which Q_w is the works flow. Samples were taken on a weekly basis from three wastewater works (WTWs) over a 7-month period.

b) From the mass of solids (M_w) disposed to landfill (recorded for the purposes of landfill tax in the UK) in conjunction with the moisture content (θ) of the sludge ($L_s = M_w \cdot (1 - \theta)$).

Comparisons between methods (a) and (b) at two works over a 4-month period showed agreement to within 8 %. Errors in (b) arise from many causes such as accumulations in the washwater recovery tanks; spillages; gains of mass due to trapped rain water in the storage skips, and also from inhomogeneities and variations during measurement. In spite of these factors, the agreement between the methods supported the use of the floc solids as a means of estimating L_s. The second stage was to compare the measured SS_f , with estimates based on water quality data. These were:

c)   The WRc algorithm shown by Eq. (1)
d)   A revised algorithm shown below:

$$SS\ (mg/L) = 2.9 \cdot Al\ (mg/L) + 0.1 \cdot True\ Colour\ (°H) + 2 \cdot Turbidity\ (NTU) \qquad \text{(Eq. 2)}$$

Table 1 summarises the estimates of SS by different methods in which SS_f is regarded as a datum. It is seen that column (4) is in reasonable agreement with column (2), whereas with 'colour' interpreted as 'apparent colour' in Eq. (1), the latter overestimates SS by ~50 %.

Table 1. Comparison of measured values of SS and estimates based on water quality data

WTW	SS_f (mg/l)	Eq. (1) (mg/l)	Eq. (2) (mg/l)
Muirdykes	15.1	21.5	15.2
S. Moorhouse	21.4	33.9	24.8
Neilston	11.2	17.0	11.5

Dosage Control

Solids variations

Maintenance of the target dosage prior to thickening depends on the ability to track the solids in the instantaneous feed, and from this to deliver the required polymer dose. Solids variations will emanate from changes in raw water quality, but of greater consequence are those arising from the internal flow regime. This is illustrated in Figure 2, which refers to output from a balance tank receiving sludge (at about 1% DS) from DAF units, and sludge of more variable solids content from the wash water recovery tanks. When the balance tank received sludge from DAF, the output solids were in the range 0.55-0.9 %. However, when the tank received wash water sludge, the output solids was lower (range 0.35 - 0.55 %). As is common in many WTWs, the polymer feed was flow proportional, but there was no hardware for tracking the continuously changing solids concentration. Under this regime with a constant flow and constant feed, a target dose of say 1.5 kg/t manifests itself as about 0.8 to 2.1 kg/t as a result of the solids variation.

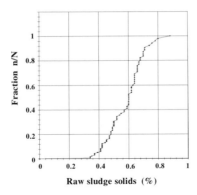

Figure 2. Cumulative fraction of samples with solids less than stated value

Polymer delivery

As part of a quality control exercise at a WTW, the concentration of polymer stock solutions was checked periodically using SEC. At the WTW (see Figure 5), three polymers were in use: a cationic polymer (A) serving as a floc aid; an anionic polymer (B) used to assist thickening and a further anionic polymer (C) added prior to dewatering in a belt press. Table 2 shows the target concentration and measured values together with other pertinent information. It is evident that there are major departures from the target levels, a factor which can have significant impact on plant operation. An example is considered below.

Table 2. Stock solution data

	A	B	C
Molecular weight	6×10^6	$14\text{-}15 \times 10^6$	$14\text{-}15 \times 10^6$
Target concentration (g/l)	0.56	2.45	0.92
Measured concentration (g/l)	0.50, 0.52 1.31, 1.41	2.83, 3.85 3.99, 2.47	0.99, 0.57 0.94, 0.96

Dosage and thickener performance

The thickeners associated with the data shown in Figure 1 were found to be particularly sensitive to overdosing. For example, a dose of 2.9 kg/t caused a solids overflow over the weirs. The symptoms of this behaviour were evident in a CML30 test, performed on samples from the thickener. The CML30 test is an *ad hoc* test described in Bache and Zhao (2001) which provides insight into the settleability of a dosed sample in a 100 ml measuring cylinder. For example, Figure 3 shows that optimal settlement occurs at a polymer concentration of 10 mg/l ($\equiv 2.2$ kg/t dosage). Below the optimum, settlement is enhanced by increasing

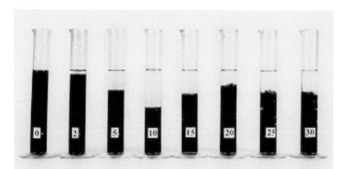

Figure 3. CML30 test showing state of settlement of polymer-treated sludge after 30 min in a 100 ml cylinder. The solids concentration was 4595 mg/l. Polymer dose is given on each cylinder

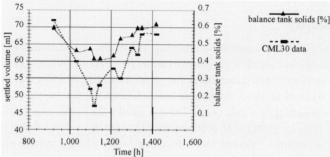

Figure 4. Relationship between solids concentration in the tank and settleability in CML30 test over time

dose. In the overdose regime, the separation process is dominated by the rapid formation of a poorly consolidated solids network in the settled layer (the likely cause of the solids overflow noted above). The settlement is largely insensitive to dose because it lies in the regime of adsorption saturation. With changing solids concentration at a fixed polymer feed rate, the dosage ratio is altered. The potential effects on settlement are illustrated in Figure 4 on the basis of CML30 tests performed on the balance tank output during a period in which wash water tanks were being emptied.

Polymer routing

The advantages of polymers for supporting flocculation and dewatering are well known. In practice, one must try to avoid residuals because of their effects on sand filters and filter cloths (e.g. Hall and Hyde, 1992; Zhao and Bache, 2002). There is also concern about residuals in the discharge streams. Aspects of polymer routing within a WTW were considered in Warden (1983) who assumed that they behaved as a conservative tracer. However, polymers are reactive and do not necessarily behave as an inert material. For example, if cationic and anionic residuals come into contact, there is a degree of precipitation following the charge interaction. In these circumstances residuals would be 'lost' from the aqueous phase. Residuals are also subject to adsorption during transport. Issues of this nature arose during a study of polymer routing at a direct filtration plant treating a coloured water with low turbidity. At the WTW under scrutiny (see Figure 5), it was the practice to recycle all aqueous streams. There was also extensive use of polymer (see Table 2). The polymers in the aqueous phase could be distinguished on the basis of their individual retention times in size exclusion columns at a fixed wavelength (see Keenan *et al.*, 1998). Polymer separation was facilitated by using two size exclusion columns in series (type PL aqualgel-OH 40; 8 μm pore size; Polymer Laboratories, UK). The eluent was nanopure water. Samples were taken at different points in works over three non-consecutive days. The solids transport (summarised in Table 3) was estimated via the inlet water quality and coagulant demand (Eq. 2), and the load in the flocculators and in the thickener feed.

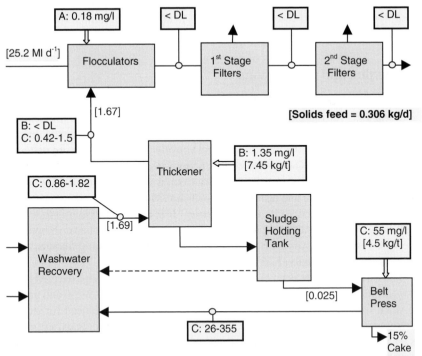

Figure 5. Schematic of WTW summarising flow (Ml d^{-1}), target dosages for polymers A, B and C (kg/t), and residual polymer concentration (mg/l). DL refers to detection limit of about 0.2 mg/l

Table 3. Estimates of solids load on sampling days with alum as coagulant

	Via Eq (2)	Flocculator	Thickener feed
Load (kg d^{-1})	299	306	~312

The estimates of the solids feed were in good agreement. Owing to large variations in the SS, the estimate for the thickener feed is only approximate. Aside from factors of accuracy, the solids feed in the flocculator was expected to be marginally greater than the estimate via Eq. (2), because some solids were recycled (> 15 mg/l) via the thickener supernatant.

Figure 5 shows that it was feasible to track excess residual from polymer C (from the belt press) back through a series of units to the head of works. The concentration values were roughly in line with mass balances based on the local flows. The target concentration of polymer C in the belt press was 55 mg/l, though one sample showed a concentration of 355 mg/l in the aqueous stream. Generally the wastage of C was ~ 1kg/d. The 7.5 kg/t dosage for polymer B appears high, but no residual was detected.

An in-line meter monitoring the suspended solids in the thickener feed showed readings that bore little resemblance with SS in co-adjacent samples measured by standard methods.

Discussion and Conclusions

The beguiling simplicity of the dosage mass ratio disguises a host of issues and problems surrounding its specification, control and operational implications.

The final choice of the target dosage must be guided by experience, but one cannot get away from the fact that this is controlled by the underlying pattern of polymer adsorption. Knowledge of this pattern is of great assistance in dosage control, but it is impractical to generate such information for day-to-day management. It is here that test methods such as CST have merit, their behavioural features being determined by the adsorption. It is of no surprise that individual test methods provide differing views of 'the optimum'. This is because optima in the dose-response curves are tied to the interaction of behavioural features which are specific to each individual test method.

The CML30 test has been closely scrutinised for an alum sludge derived from a low turbidity coloured water and solids content in the range 0.2-1% and appears to work well in this context. It is a simple test, which can be used for trouble-shooting purposes with respect to thickener performance. Analysis reported in Bache and Zhao (2001) showed that optima identified by the CML30 test were virtually identical to those indicated by SRF analysis; this emphasises that the patterns of water release in both sedimentation and pressure filtration are intimately connected.

It hardly needs stating that the specification of the solids feed is central to the design of the sludge handling plant. Calculations based on water quality parameters are useful for first order estimates. Eq. (2) avoids some of the ambiguities contained in Eq. (1) and is well-supported for a range of soft low turbidity coloured waters. For an operational plant, it is far better to estimate the solids feed via the suspended solids in the flocculator.

The data shown in Figure 2 and 4 illustrate the pitfalls of poor balance tank design. Fluctuations in the solids feed are not a problem if they can be accurately tracked. To date, the author has not encountered an in-line solids meter that is up to this task.

The polymer feed arrangements warrant close scrutiny. In the absence of polymer measurements, good records of polymer use can greatly assist this task. Failure to control the target dosage will lead to both underdosing and overdosing, with their own class of problems. The high concentration of polymer appearing in the belt press aqueous phase probably arises from inadequacies in mixing rather than being caused by overdosing; this highlights yet another problem.

The data shown in Figure 5 illustrates the potential of using SEC for tracking polymers; this is believed to be the first study of its kind. SEC is a valuable tool for supporting the polymer use in a WTW, but is sensitive to interferences (e.g. from algal presence) which can mask the polymer signal and preclude its use.

Acknowledgements

This paper springs from pioneering work by Drs EN Papavasilopoulos, H Keenan and YQ Zhao on methods and techniques, and major contributions from Dr E Rasool and FJ McGilligan. The support of West of Scotland Water and EPSRC under Grant No GR/L 61026 is gratefully acknowledged.

References

Bache, D.H. and Papavasilopoulos, E.N. (2000) Viscous behaviour of sludge centrate in response to polymer conditioning. *Wat. Res.* **34** (1), 354-358

Bache, D.H. and Zhao, Y.Q. (2001) Optimising polymer use in alum sludge conditioning: an *ad hoc* test. *J. Wat. SRT-Aqua* **50**(1), 29-38

Dentel, S.K. and Abu-Orf, M.M. (1995) Laboratory and full scale studies of liquid stream viscosity and streaming current for characterization and monitoring of dewaterability. *Wat. Res.* **29**(12), 2663-2672

Hall, T. and Hyde, R.A. (1992) *Water Treatment Processes and Practices*. WRc Swindon, UK

Keenan, H., Papavasilopoulos, E.N. and Bache, D.H. (1998) On the measurement of polymer residuals following the conditioning of an alum sludge. *Wat. Res.* **32**(10), 3173-3176

Papavasilopoulos, E.N. (1998) On the role of aluminium hydroxide in the conditioning of an alum sludge. PhD thesis. University of Strathclyde, UK

Warden, J.H. (1983) *Sludge treatment plant for waterworks*. Tech. Rep. TR189, WRc Swindon, UK

Zhao, Y.Q., Papavasilopoulos, E.N. and Bache, D.H. (1998) Clogging of filter medium by excess polymer during alum sludge filtration. *Filtr. & Sep.* Volume 35 No.10, 945-950

Zhao, Y.Q. and Bache, D.H. (2002) Polymer impact on filter blinding during alum sludge filtration. *Wat. Res.* (in press)

H.H. Hahn, E. Hoffmann, H. Ødegaard (Eds.)
Chemical Water and Wastewater Treatment VII, pp. 297-308
© IWA Publishing, London
ISBN: 1 84339 009 4

Optimisation of Floc-Stability by Mechanical Pre- and Post-Stressing

J. A. Müller* and S. K. Dentel

*Technical University of Braunschweig, Inst. of Sanitary Engineering, Braunschweig, Germany
email: jo.mueller@tu-bs.de

Abstract

The influence of mechanical shear stress on the structure and dewatering properties of flocculated sewage sludge was investigated. Shear forces were applied before, during and after floc formation. Linear and x-linked polymers were used for conditioning.

Introduction

Sedimentation and dewatering properties as well as the degree of separation are dependent upon the size and stability of activated sludge floc. For a given biological process operation and its primary floc characteristics, these floc properties are controlled by the type of flocculant, the imposed mixing conditions, and the stressing conditions after floc formation. It is thus useful to classify the overall process of floc formation and stressing into 4 phases:

1. The pre-stressing phase, where the existing primary floc is formed
2. The rapid mixing phase, where flocculant and sludge are mixed
3. The floc formation phase, where breakage and (re)flocculation take place
4. The post-stressing phase, where final floc size and stability are determined

This step-model is similar to ones suggested by others (O´Melia, 1972; Langer *et al.*, 1995). The pre-stressing phase is included in order to describe the influence of the primary floc strength. Usually only Phases 2 and 3 are considered in research and practice, even though Phase 4 is crucial in such processes as mechanical thickening and dewatering.

[297]

Influence of shear forces on flocculation

Flocs that have been created during the conditioning process are stressed in many different ways until their separation. Mechanical compression and shear stress appear within pumps and during the flow to the solid-liquid separation process. Especially large strain is created within dewatering devices like centrifuges and filter presses (Novak, 2001).

Two influences of shear forces on the dewatering properties have to be distinguished:

1. Shear forces lead to a destruction of flocs and a decrease in particle size. The dewatering result after filtration or centrifugation may be inferior (Mühle and Domasch, 1991).
2. Shear forces may create especially firm flocs that are found to be advantageous for the subsequent dewatering process (Kleine, 1992).

These stable flocs can be obtained by providing appropriate polymer dose and mixing during conditioning (Hemme *et al.*, 1995; Langer, 1995). Residual flocculant in solution or partially adsorbed polymer on floc surfaces will cause flocculation or reflocculation. The ongoing process of breakage and reflocculation (pelletisation) decreases the excess or reserve flocculant concentration while it increases the strength of the sludge flocs. The size, stability and density of these flocs are high, which is beneficial for the subsequent dewatering process. But stronger flocs may not be desirable for maximum dewatering results, since they tend to hold more of the trapped water within the floc assemblies.

Prior to conditioning, the primary floc can show very different structure and strength (Bruus, 1992), which partially depends on its stress history. The influence of pre-stressing on sludge conditioning and dewatering has not been widely investigated. Friedrich *et al.* (1993) tried to improve the dewatering result by disintegrating sewage sludge. Mixtures of treated and non-treated sludge were produced in order to achieve an optimum particle distribution. Only marginal improvements could be achieved (Müller, 1996).

Materials and Methods

Samples of digested sludge were taken from two municipal wastewater treatment plants in Wilmington, Delaware, USA (TS = 25-27 g/kg) and in Hildesheim, Germany (TS = 16-21 g/kg).

Three different cationic polyelectrolytes were used, all supplied by SNF Floerger and indicated as Chemtall 640 CT (linear), 640 TRM (branched) and 640 BD (x-linked). All products were 60 mol% cationic. The molecular weight of the 640 CT is 10^6 daltons while no information was available on the other formulations. The emulsions were made up to 0.3 % active using tap water. The water was stirred at 300 rpm and after injecting the emulsion, mixing was continued for 10 min. Then the solution aged for another 30 min. The mixing of sludge and polymer (500ml) was done using a blade stirrer of 2.5 • 7.5 cm in a 1000 ml jar with cap.

Capillary suction time (CST) measurements were conducted using two different types of cylinders, with an inner diameter of 18 mm (CST_L) and 10 mm (CST_S), respectively. A beaker-centrifuge was used for solid-liquid-separation at 1000 g for 5 min. Zeta potential (Zetamaster, Malvern), turbidity (Turb 350 ID, WTW), and total solids (TS) were measured in the supernatant. The specific resistance to filtration (SRF) was measured using a pressure cell having a volume of 300 ml at 8 • 10^5 Pa.

Approaches to quantify mixing in terms of the mean velocity gradient (G), the time of mixing (t) and the product G•t have generally shown that the optimum polymer dose increases as G, t, or G•t is increased (Novak, 1988). The problem with using G is the spatial distribution of the velocity gradient in a mixing chamber. Peak gradients and the gradient distribution are dependent on many parameters and Christensen *et al.* (1995) showed that different mixing devices lead to quite different sludge conditioning results at the same G value. Therefore the rotational speed was used as a relative parameter to describe the mixing intensity.

Influence of polymer dosage and mixing time

The optimal polymer quantity for the dewatering of sludge is usually determined in the laboratory by running jar tests. Published instructions for performing this test (APHA, 1995; Dentel, 1993) do not provide precise specifications for mixing conditions, which may be site or process-specific. However, dewatering results are highly dependent on the mixing conditions. Thus variations in polymer dose and mixing time in jar tests result in different conditioning behaviour (Figure 1).

It can be seen that for all selected polymer doses an optimal mixing time does exist where a minimal CST can be measured. With a shorter duration of mixing, the polymers and the sludge particles have not been completely mixed yet. However, during long mixing times increasing numbers of small flocs are formed, because sludge-polymer-flocs are torn apart by the shear forces of the stirrer. For higher polymer dosage a longer duration of mixing is needed to achieve the best CST values. Minimal CST values are slightly lower with increasing polymer dose. The relevance of CST should not be overrated, as it merely presents a measurement of the ability to release water from the sludge. Therefore the CST values do not have to correspond to the solids content that is reached by dewatering devices. However, it has to be emphasized that there is not one optimal polymer dosage; rather, only optimal combinations of shear energy and dose exist. Consequently, the determination of an optimal polymer dose is only accurate for a certain energy input applied during the conditioning and the subsequent process steps, respectively, up to the separation of the floc from the liquid. Therefore the application of laboratory experiments in full scale dewatering devices requires energy inputs that are equal to each other.

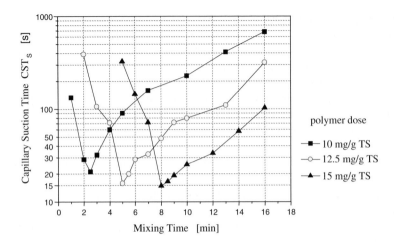

Figure 1. Influence of the duration of mixing on the floc properties measured by CST. Linear polymer was dosed at 300 rpm mixing speed

Influence of polymer structure and mixing velocity

By using controlled degrees of cross-linkage in polymer formulation, highly branched or structured polymers are produced. It is known that a greater dosage of the modified polymers is required to reach optimum performance when compared to the analogous linear product without cross-linkage (Dentel, 2001). On the other hand, the flocs attained are larger and more shear-resistant.

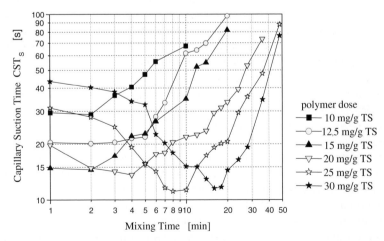

Figure 2. Influence of mixing time on floc properties as measured by CST. An x-linked (branched) polymer was dosed at a mixing speed 300 rpm

The conditioning behaviour of branched polymers is fairly similar to that of the linear type (Figure 2). However, in order to get good dewatering results, noticeably higher quantities of polymer had to be added. With the lowest chosen dosage no optimal CST values were achieved. At a mixing time of 3 minutes or more, the formed flocs were broken up and the CST values decreased. With increasing polymer quantities, the duration of mixing had to be extended in order to get the minimal CST value. The minimal values declined until a polymer quantity of 25 mg/g TS was used, but required shear stress that is equivalent to a mixing of several minutes. The measured CST values also showed that this polymer type appears to have a broader optimum concerning the mixing time, thus more shear resistant flocs were attained. Nevertheless, there were optimal mixing times in this case as well, but exceeding or falling below these optima merely resulted in slightly deteriorated values.

With the highly cross-linked or structured polymer, good CST values were only achieved when using a certain polymer quantity (30 mg/g TS, results not shown). Over- or under-conditioning does not lead to an acceptable dewatering behaviour. According to the producer this polymer is only employed in combination with the linear polymer, because only in this combination good dewatering results can be achieved.

Influence of shear forces before, during, and after floc formation

Shear forces during and after the flocculation were varied with different rotational speeds of the stirrer (300 and 600 rpm). If necessary, the rotational speed was changed one minute after the flocculant had been added.

In Figure 3 the measured CST values are presented as a function of the mixing time. More stable flocs were formed if the sludge was exposed to higher shear forces in the first minute after adding the polymer. This can be seen from the minimal values of the CST, because the sludge that has been conditioned with higher rpm can endure a longer duration of mixing until the optimal CST is reached. Hardly any difference between the minimal values exists here. Therefore a sludge that has been exposed to a short mixing time at high shear forces will provide better dewatering results, if also considerable shear forces will affect the flocs between conditioning and dewatering.

The mixing intensity in the first mixing period seems to influence the dewatering result. Local overdosing, resulting in an uneven distribution of the polymer, was observed for 300 rpm mixing. In case of a subsequent mixing at 600 rpm, better results could be obtained because of the better dispersion of polymer. The initial mixing intensity has a particular influence on the flocs' shear resistance. High initial mixing will lead to flocs that can withstand high shear forces in subsequent processes. But if these high shear forces do not occur, the dewatering results will deteriorate.

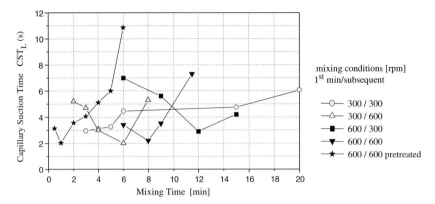

Figure 3. Variation of mixing intensity during conditioning on floc properties as measured by CST. Linear polymer (13 mg polymer/g TS) was used

Thus the past history of a sludge is of considerable importance. Apart from many other factors, shear forces affect the particle size distribution before conditioning, which is of major importance for conditioning and dewatering. Flocs that have already been exposed to variable shear stresses by stirrers, pumps, and other devices or processes during the previous sludge treatment appear to be shear resistant. Pre-shearing in a jar with the stirrer at 600 rpm for 5 minutes did not make any significant difference in the sludge qualities. In order to clearly show the influence of the shear history, a shear-gap-homogeniser (disperser) was employed, with which the digested sludge was treated at 8000 rpm for 1 minute. The conditioning and post-shearing was carried out with a blade type stirrer at 600 rpm.

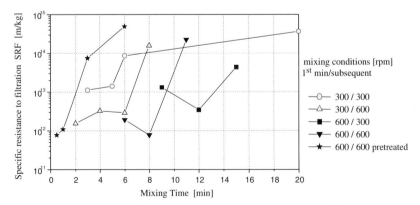

Figure 4. Influence of mixing conditions on floc properties as measured by the specific resistance to filtration. Linear polymer (13 mg polymer/g TS) was used

Preliminary pre- and post-stressing experiments (Müller, 1996) showed that extreme shear stress (disintegration) applied during Phase 1 reduced the floc size and increased the demand of flocculants needed to reach a complete neutralisation of the surface charge. The pre-stressing of the sludge in the shear-gap-homogeniser resulted in an additional decrease in the optimal mixing time (Figure 3). Thus pre-stressing does not influence the dewatering behaviour as long as the optimal mixing time is maintained. An improvement of the dewatering result by pre-stressing could not be demonstrated.

Use of a branched polymer showed results similar to those with a linear polymer. The minima of the CST values were reached in the same chronological order for the 4 combinations of rotational speeds. However, again there was a broader optimum with respect to mixing time, as shown in Figure 2.

Further experiments with the conditioned sludge were carried out using a filtration cell. The specific resistance to filtration (SRF) values obtained from these experiments (Figure 4) correspond to the CST results in terms of mixing time. The duration of mixing for the maximum degree of dewatering are equal to the ones with the minimum SRF (results not shown).

Figure 5 shows the total solids content achieved in the pellet after centrifugation in a beaker centrifuge. Again the outcome corresponds very well to both the CST and SRF results. The positions of the optima, in regard to the mixing time, is of particular interest here. The peak values for TS in the sediment should not be overrated, because the conditions during separation in a beaker centrifuge can only be applied to the dewatering in a solid-bowl-centrifuge in an approximate sense.

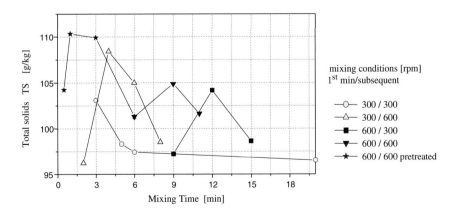

Figure 5. Solids content in the sediment after centrifugation in a beaker centrifuge. Linear polymer (13 mg polymer/g TS) was used

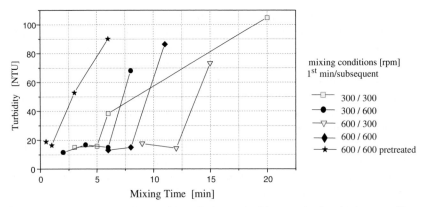

Figure 6. Turbidity in the supernatant after solid-liquid separation in a beaker centrifuge. Linear polymer (13 mg polymer/g TS) was used

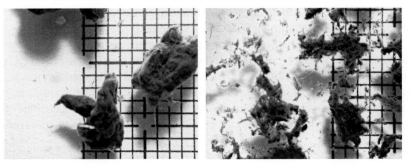

Figure 7. Effect of mixing at 600 rpm for 1 min (left) and 3 min (right) on pre-sheared sludge (shear-gap-homogeniser)

Measurements of supernatant particle concentrations provide additional insight into the centrifugal dewatering results. As one such measure, supernatant turbidity increased significantly with a longer mixing time (Figure 6). Solids capture percentages, though not presented here, ranged from 93 % to 96 % in exhibiting similar trends to the turbidity results. More intense mixing not only led to detachment of some particles from the sludge flocs, but also to their far-reaching destruction (Figure 7). Furthermore, with low mixing time the turbidity was nearly constant between 15 and 20 NTU (Figure 6). This was the case with the selected mixing times regardless of how much the sludge was stressed before, during or after the flocculation. Photographs (Figure 7) showed that there were only large, well-developed flocs in all these experiments.

When the mixing duration was inadequate, dewatering results were also suboptimal, as indicated by both the cake-TS and the low supernatant turbidities. Conversely, the mixing times that led to maximum dewatering also produced low turbidity values. In other words, increased turbidity always involves a deterioration of the flocs during dewatering. If the flocculation takes place at a low rotational speed (300 rpm), a deterioration of the dewatering results can already be observed even when the turbidity has not yet increased. For flocculation at a higher rotational speed (600 rpm), the increase in turbidity and the decrease of the dewatering effect occur at the same mixing time. Finally, for mechanically pre-sheared sludge, a distinctive turbidity can be measured already after a mixing time of 3 minutes. However, the dewatering results are still very good.

The determination of the zeta potential provided surprisingly little information regarding the optimal flocculant dosage. Experiments with different dosages and mixing durations demonstrated that when charge neutralisation was reached, the minimal CST values had not yet been achieved. These were only reached after a slightly longer mixing time, when the zeta potential had gone down to a value of –7.5 mV. For all samples that were close to the optimal dewatering behaviour with respect to dosage and mixing time, a negative zeta potential between –8 and –12 mV could be measured. The transition through a charge neutralisation state indicates that all excess polymer was already taken up by flocs or floc fragments. These results agree with the results of other authors, who found that, especially in the case of high molecular weight polymers, optimal dewatering results will already be achieved when the zeta potential is clearly negative. Previous results indicating optimal dewatering near a charge neutralisation dose (Abu-Orf and Dentel, 1998) have generally been reported when using a streaming current detector, which may indicate polymer depletion in solution as well as particle charge. Such results have been demonstrated in full scale dewatering processes and it is also possible that laboratory experiments do not simulate all shear or charge-related phenomena operative in actual treatment.

Treatment of filamentous sludge

A remarkable improvement of the settling properties was observed when filamentous flocs were subjected to mechanical pre-stressing (Phase 1; Figure 8). The sludge volume index of bulking activated sludge could be reduced from 180 down to 80 ml/g. After the treatment the sludge lost its voluminous filamentous floc structure, resulting in settling properties similar to those of sludge with non-filamentous floc structure. Sludge flotation could be suppressed completely.

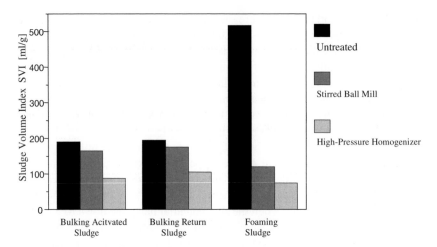

Figure 8. Improvement of settling properties of bulking activated sludge by disintegration

Summary

Best dewatering results were obtained only if polymer dose, mixing intensity and mixing time were matched. Sludge-stressing prior to conditioning led to more shear-sensitive flocs. Stronger flocs that can withstand extended shear stress in the subsequent dewatering process were formed by increased shear during conditioning. A higher dosage of flocculants was required to achieve the same results when low shear forces were used during conditioning as in shear intensive dewatering processes. Shear resistant flocs could be obtained by applying x-linked polymers, but a higher polymer dosage was still required. Filamentous flocs destroyed by shear forces had better settling properties.

Acknowledgements

We thank the German Research Council for their financial support.

References

APHA-AWWA-WEF (1995) Standard methods for the examination of water and wastewater, 19th ed., A.D. Eaton, L.S. Clesceri, A.E. Greenberg (Eds.)

Abu-Orf, M.M., and Dentel, S.K. (1999) Rheology as tool for polymer dose assessment and control. *J. Env. Engng. ASCE* 125(12), 1133-1141

Abu-Orf, M.M., and Dentel, S.K. (1998) Automatic control of polymer dose using the streaming current detector, *Wat. Envt. Res.* **70**, 1005-1018

Bruus, J.H., Nielsen, P.H., Keiding, K. (1992) On the stability of activated sludge flocs with implications to dewatering. Wat. Res. 26, 1597-1604

Christensen, J.R., Christensen, G.L., and Hansen, J.A. (1995) Improved characterisation of mixing for sludge conditioning. *J. Env. Engng*. 121(3), 236-244, and 122(9), 880-881

Dentel, S.K. (2001) Conditioning. Chapter 16 in *Sludge into Biosolids*, L. Spinosa and P.A. Vesilind (Eds.) IWA Publishing

Dentel, S.K., Abu-Orf, M.M., and Griskowitz, N.J. (1993) Guidance manual for polymer selection in wastewater treatment plants. *Water Environment Research Foundation*, Alexandria, VA: WERF Publication D0013

Friedrich, E., Friedrich, H., Heinze, W., Jobst, K., Richter, H.-J., and Hermel, W. (1993) Progress in Characterisation of Sludge Particles. *Wat. Sci. Techn.* **28**(1)145-148

Hemme, A., Polte, R., Ay, P. (1995) Pelletierungsflockung - Die Alternative zur herkömmlichen Schlammkonditionierung. *Aufarbeitungs-Technik* 36(5), 226-235

Klute, R. (1990) Destabilization and Aggregation in Turbulent Pipe Flow. In: *Chemical Water and Wastewaster Treatment,* Hahn, H.H., Klute, R. (Eds.), Springer Verlag Berlin Heidelberg, pp. 33-54

Kleine, U. (1992) Der Einfluß der Flockenbildungsbeanspruchung auf die Festigkeit und das Sedimentationsverhalten von Flocken bei der Zentrifugalabscheidung, Dissertation TH Karlsruhe, Germany

Langer, S.J., Klute, R. and Hahn, H.H. (1995) Mechanisms of floc formation in sludge conditioning with polymers. *Wat. Sci. Tech.* 30(8), 129 –138

Langer, S.J., and Klute, R. (1993) Rapid mixing in sludge conditioning with polymers. *Wat. Sci. Tech.* 28(1), 233-242

Mühle, K., Domasch, K. (1991) Stability of Particle Aggregates in Flocculation with Polymers. *Chem. Eng. Process*. 29, 1-8

Müller, J. (1996) Mechanischer Klärschlammaufschluß. Dissertation, TU Braunschweig, Germany

Novak, J.T. (2001) Dewatering. Chapter 18 in *Sludge into Biosolids*, L. Spinosa and P.A. Vesiling (Eds.) IWA Publishing

Novak, J.T., Prendevill, J.F. and Sherrard, J.H. (1988) Mixing intensity and polymer performance in sludge dewatering. *J. Envir. Engrg.* ASCE 114(1), 190-198

O´Melia, C.R. (1972) Coagulation and Flocculation. Chapter 2 in *Physicochemical Processes for Water Quality Control* (W.J. Weber, Jr., Ed.), Wiley Interscience, New York

H.H. Hahn, E. Hoffmann, H. Ødegaard (Eds.)
Chemical Water and Wastewater Treatment VII, pp. 309-316
© IWA Publishing, London
ISBN: 1 84339 009 4

Effect of PH and Oxidation on Sludge Dewatering

K. Jansson* and H. Palonen

*Kemira Chemicals OY, Vaasa, Finland
kaj.jansson@kemira.com

Introduction

The pulp and paper industry in Finland normally purifies its waste water by using sedimentation and activated sludge treatment. Discharges to the environment decreased continuously in the 1990s, although production increased. This implies a very effective waste water treatment as well as the production of increasing amounts of solid substances in the form of sludge. At the same time the amount of sludge disposed in landfills decreased by 8 % in the year 2000.

Sludge consists of primary sludge from the sedimentation basin and biosludge from biological treatment. The amount of primary sludge has been decreasing due to more efficient fibre recovery and less use of fresh water. This gives a bigger portion of biosludge, which makes it more difficult to dewater. This in turn leads to larger volume and lower dryness making it more difficult to find an end use for the sludge. The most common way to dispose of sludge is to incinerate it in a boiler. The low dry matter content assumes a use of a support fuel, as bark.

Mainly polyelectrolytes have been used for sludge dewatering. Their effect is based on the flocculation of small particles into compact agglomerates, which enable a better dewatering from outside of microbial cells and other particles. But in the near future there will be an increased demand for even smaller sludge volume and higher dryness content. Manufacturers of chemicals must develop and produce more effective chemicals and chemical systems for better dewatering in sludge conditioning.

This study focuses on the enhancement of sludge dewatering by adjusting the pH, adding an inorganic coagulant, and oxidative treatment ahead of the flocculant.

[309]

Experimental

Sludge samples were taken from an integrated pulp and paper mill, located in Western Finland. Samples were either from mixed sludge or separated primary and biosludge, all before press filter. Biosludge comprised 50-75 % of total sludge dry substance in these laboratory studies.

All chemicals used in this study were produced by Kemira Chemicals.

The polyelectrolytes used were acrylic amide-based polymers, with different cationicity and molecular weight, sold under the trade name Fennopol.

An iron-based coagulant consisting of ferric sulphate in acidic solution (trade name Fennoplus 105) was used. The product consisted of 10 % Fe^{3+}. 35 % hydrogen peroxide was used in the test and the dosages were calculated as 100 % H_2O_2.

The PolyTest method was used for the dewatering tests. The sludge (500 mL) was poured into a filter, which was connected to a funnel with a wire at the bottom. The filtrate was collected in a beaker, which was set on a balance. The filtration rate during the first 30 seconds was measured.

The sludge sample was centrifuged, and both the supernatant and the sedimented solid were titrated with a PC-Titrate™ device using 0.1 mol/L H_2SO_4.

Results

PH titrations

Primary sludge consists mainly of solids originating from fibrous material, bark, pigments and other impurities separated in the mill's screening processes. If a paper mill uses calcium carbonate as a filler or for coating colour, pH is at a slightly higher level than without carbonate, and this kind of sludge also possesses a higher buffering capacity against pH variation.

Slowly sedimenting solid and colloidal particles run to the biological stage, where they are mixed with microbiological mass to form biosludge. This means that the pH character of this fraction first of all depends on the properties of the biological matter and the by-product formed in the activated sludge process.

One way to characterize sludge is to measure the amount of acid titrant required to decrease the sludge pH to some convenient value. Sludge samples were titrated separately and blended alike (Figures 1 and 2). Biosludge consumes 3 times more acid to reach pH 4 than primary sludge, 6 meq/g versus 2 meq/g. Mixtures of bio and primary sludges had an acid consumption corresponding to the proportion of each sludge.

Supernatants from bio and primary sludges consumed about 10 times less acid than uncentrifuged samples containing solids, which indicates that the main buffering capacity originates from the solid particles (fibres, pigments, microbes, and polysaccharide slime).

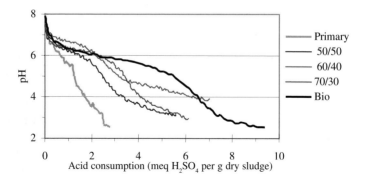

Figure 1. Amount of sulphuric acid required to titrate different kinds of sludge

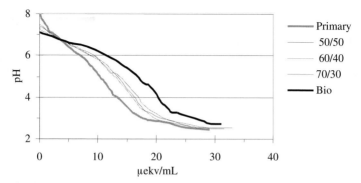

Figure 2. Sludge supernatant (after centrifugation) titrated with sulphuric acid

The choice of flocculant

The mixture of primary and biosludge is thickened by filter press or screw. To maximize the dry solids content, a flocculant or a combination of coagulant + flocculant is commonly added to the sludge. Typical flocculants are polyelectrolytes with different ionic character and chain length.

The flocculant was chosen using the Polytest-method. The proportion of bio/primary sludge in the test sample was 50/50. Figure 3 shows how much filtrate was obtained during the first 30 seconds of dewatering. In this case, the most effective flocculant was Fennopol K1384, which is a low cationic, long chain polymer. The other polymers are either more cationic or have shorter chain length. Fennopol K1384 was used in all the following experiments, always at a dosage of 4 kg/t.

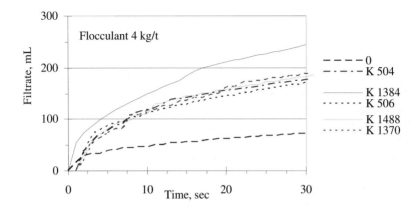

Figure 3. Dewatering results for different Fennopol products

PH

The effect of pH was studied on the same sludge as in the flocculant test. PH was adjusted with sulphuric acid. The filtrate volumes at dewatering times of 10 and 30 seconds were almost constant between pH 4 to 7 (Figure 4). Below pH 4, the dewatering speed was decreased. One possible reason for this is that the amount of negative surface charge diminishes around this pH value and adsorption of flocculant decreases, which leads to poorer flocculation.

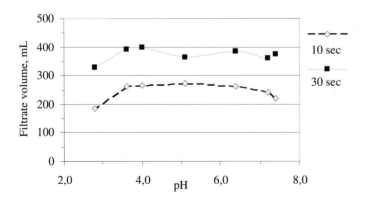

Figure 4. Dewatering speed at different pH

Iron salt

One way to intensify the effect of the flocculant is to add inorganic coagulant before flocculant. The commercial iron coagulant (Fennoplus), which is in acidic solution, decreases the pH with increasing dosages of coagulant. When Fennoplus was added to sludge having a bio/primary proportion of 75/25 in the Polytest series, the filtrate volume decreased from 350 mL to around 100 mL at 30 seconds (compare Figures 4 and 5).

When iron coagulant was first added, dewatering speed decreased, but then increased considerably with higher dosages of iron. At low iron concentrations, new particles (Fe-precipitates) are formed from soluble materials. This phenomenon is clearly demonstrated in Figure 6, where turbidity increases significantly around 100 kg/t iron coagulant. When dosage exceeds 150 kg/t, dewatering speed accelerates and turbidity decreases dramatically. By 200 kg/t dewatering is about 2 times faster than without iron coagulant and turbidity has already reached its minimum level.

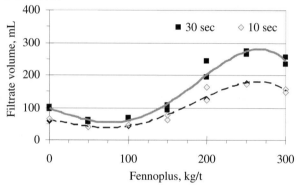

Figure 5. The effect of adding an iron coagulant on dewatering of 75/25 bio/primary sludge

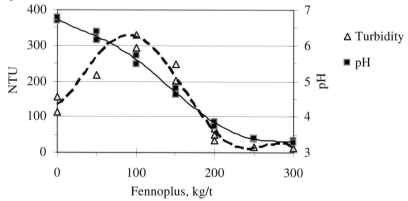

Figure 6. The effect of iron dosage on turbidity and pH of filtrate

Peroxide

The Fenton reaction is based on the oxidation of hydrogen peroxide catalyzed by Fe^{2+} ions. In this reaction peroxide is decomposed to a highly reactive hydroxyl radical, which will decompose organic molecules and polymers. The optimum pH for this reaction is between 3 and 5. The temperature should be below 40 °C, because the efficiency declines as a result of the decomposition of hydrogen peroxide.

Different concentrations of hydrogen peroxide were added to the same sludge that was used in the studies for iron salts (Figure 7). The iron concentration of 50 kg/t was chosen because it is sufficient for catalysis, but, on the other hand, it has no positive effect on dewatering. Reaction time was 1 hour at 40 °C.

Increasing the dosage of peroxide had a positive effect on dewatering (Figure 7). The first 10 kg/t had the greatest effect in that the dewatering speed almost doubled from 100 mL to 200 mL in 30 seconds. The same result was obtained when 200 kg/t iron was added. Therefore the Fenton reaction seems to work in these circumstances. The mechanism is probably depolymerisation and oxidation of extracellular polysaccharides, by means of the well known Fenton reactions, in which hydroxyl radicals are created. (These are capable to oxidice most hydrocarbons).

By decreasing the pH and adding an iron coagulant, it is possible to dewater sludges with a high biosludge component more effectively. A disadvantage of this method is that large amounts of iron are needed, which increases sludge volume and ash content if the sludge is incinerated.

The amount of iron coagulant can be cut by 75 % by using oxidative treatment. Utilisation of hydrogen peroxide seems to improve the dewatering properties of these kinds of sludges even more, while maintaining the current sludge treatment costs.

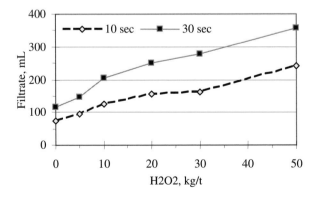

Figure 7. The effect of hydrogen peroxide concentration on dewatering

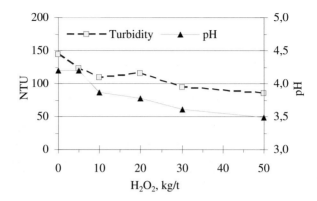

Figure 8. The effect of hydrogen peroxide concentration on turbidity and pH of the filtrate

Conclusion

By decreasing the pH and adding an iron coagulant, it is possible to dewater sludges with a high biosludge component more effectively. A disadvantage of this method is that large amounts of iron are needed, which increases sludge volume and ash content if the sludge is incinerated.

The amount of iron coagulant can be cut by 75 % by using oxidative treatment. Utilisation of hydrogen peroxide seems to improve the dewatering properties of these kinds of sludges even more, while maintaining the current sludge treatment costs.

Enhanced Removal and Reuse

H.H. Hahn, E. Hoffmann, H. Ødegaard (Eds.)
Chemical Water and Wastewater Treatment VII, pp. 319-330
© IWA Publishing, London
ISBN: 1 84339 009 4

Trace Contaminant Removal Using Hybrid Membrane Processes in Water Recycling

A.I. Schäfer and *T.D. Waite*

*University of New South Wales, Centre for Water and Waste Technology, Sydney, Australia
email: A.Schaefer@unsw.edu.au

Introduction

Water recycling plays an essential role in integrated water management, especially in an arid country like Australia but also worldwide (Seckler *et al.,* 1999). Water recycling, however, has suffered extensive constraints due to "toilet to tap" media campaigns and "yuck factor" attitudes in the community. The support of the community for water recycling projects generally decreases as the personal contact with the recycled water increases (SWC, 1999). Some of the very valid concerns of the community stem from uncertainties involved in water recycling, such as the issue of persistent organic pollutants (POPs) potentially present in recycled waters or the ever growing group of endocrine disrupting chemicals.

Endocrine disrupters have the potential to interfere with our normal growth, development and reproduction. Modulation of that system could cause severe adverse health effects. Industrial chemicals, consumer chemicals and chemicals in the environment can be endocrine disrupters that mimic, enhance or inhibit the action of hormones (Fawell *et al.*, 2000; EU, 1999). Sewage disposal to water sources may be a major exposure pathway for pharmaceuticals, synthetic and natural hormones, and industrial chemicals to reach humans and wildlife, directly and via the food chain. This issue of trace contaminants concerns disposal of treated effluents and applications of recycled water.

This paper aims to address some of the uncertainties and risks involved in recycling technology and aims to stress caution and the need for well designed recycling projects. This risk expands to water treatment in situations where contaminated waters are treated.

[319]

Selection of High Priority Trace Contaminants

The list of trace contaminants or endocrine disrupters stemming from human activity found in wastewaters is long (Council, 1999; Harland, 1999; Lucier, 1997; Stahlschmidt-Allner *et al.,* 1997; Crisp *et al.,* 1998; U. EPA and US EPA, 1997; U. EPA, 1997; Arcand-Hoy *et al.,* 1998; Manning, 2000; Colborn *et al.,* 1997; Keith, 1998) and the problem has been apparent since the 1970s (Kirchner *et al.,* 1973).

The most relevant compounds were selected for our research based on four criteria - abundance in waters and wastewaters, high persistence in the environment, high potency as endocrine disrupters and analysis to below ng/L levels.

Compounds that meet those criteria best are natural and synthetic estrogens excreted by men and women in urine and faeces and with increased levels during pregnancy and hormone replacement therapy. Those compounds are excreted in conjugated form and are reactivated during biological treatment (Jurgens *et al.,* 1999; Ternes *et al.,* 1999). While those compounds have an average persistence, synthetic hormones and chemicals have a much higher persistence, but lower potency. Natural hormones can be expected to be present in all municipal wastewaters and hence have a global relevance.

Removal of Trace Contaminants in Treatment

Conventional wastewater treatment is not an effective barrier to trace contaminants. While removal rates published in the literature vary greatly, this appears to depend on local conditions and the nature of the contaminant (Ono *et al.,* 1996; Tyler and Routledge, 1998; Johnson *et al.,* 2000; Johnson and Sumpter, 2002).

The main characteristic that determines the fate of such contaminants in the water cycle is their ability to interact with particulates. These particulates can occur naturally (clays, sediments, colloids coated with natural organics, microorganisms) or be added during treatment (activated sludge, powdered activated carbon, ion exchange resin, coagulants). The transition of trace contaminants to the solid phase will greatly enhance chances of removal. In contrast, the interaction of trace contaminants with dissolved organics can increase their mobility in the environment and through treatment (Ohlenbusch *et al.,* 2000).

The fate of natural and synthetic estrogens in wastewater treatment plants is uncertain. It is estimated that less than 10 % of the compounds are removed via biodegradation; the majority of the compounds remain in the water phase while a considerable amount is adsorbed to the sludge (Mastrup *et al.,* 2001).

The results reported here represent a component of a multi-faceted investigation into the mechanisms of removal of natural and synthetic estrogens in water and wastewater treatment. This involves adsorption to common particulates and membrane processes. An understanding of those mechanisms is essential to predict the likelihood of removal and the reliability of treatment.

Materials and methods

Chemicals and analysis

All chemicals were of analytical grade. Radiolabelled estrone-2,4,6,7-^{3}H(N) and estradiol-2,4,6,7-^{3}H(N) (Figure 1) were purchased from Sigma Aldrich (Saint Louis, Missouri, USA). The background electrolyte consisted of 1 mM $NaHCO_3$, and 20 mM NaCl unless otherwise stated.

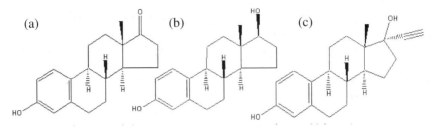

Figure 1. Structure of the contaminants of concern A: Estrone, B: 17β estradiol and C: 17α ethinylestradiol

Adsorbents

Adsorbents and other compounds that interact with pollutants used in this study were natural organic matter and IHSS FA and HA as previously characterised (Schäfer, 2001). Cellulose with an average particle size of 20 μm, kaolin and bentonite, and $FeCl_3$ were purchased (Sigma Aldrich (Saint Louis, Missouri, USA)). Monodispersed spherical hematite particles with an average diametre of 75 nm were synthesized (Schäfer, 2001) and activated sludge was taken from Brendale sewage treatment plant (Pine Rivers Shire, QLD, Australia). Magnetic ion exchange resin MIEX$^\oplus$ was supplied by Orica Watercare (Melbourne, Australia) and powdered activated carbon type NORIT SAM 52 was purchased from Norit (The Netherlands). The PAC had an average diametre of 20 μm and a surface area of 72.5 m^2/g.

Membranes and filtration protocols

Membrane filtration was performed in stirred cell experiments as described elsewhere (Schäfer, 2001). Ultrafiltration (UF) and microfiltration (MF) flat sheet membranes were supplied by Millipore (Australia). UF submerged modules were supplied by Zenon Environmental (Burlington, Canada) and MF submerged membranes by Memcor (Windsor, Australia). Nanofiltration (NF) and reverse osmosis membranes (RO) were supplied by Koch Membrane Systems (San Diego, USA) and Trisep Corporation (Goleta, USA).

Results and discussion

Trace contaminants - natural particle interactions

The interactions of estrogens with natural particles were studied at particle concentrations to be expected in the environment, or in the case of activated sludge, during treatment. A summary of results is presented in Table 1.

The adsorbed amount of contaminants depends on particle size and roughness and hence the available particle surface besides material characteristics. While results here are shown as a function of particle mass, one should bear in mind that at identical particle concentration and smooth surfaces, the available area decreases with the square of particle diameter. This is reflected in the results in Table 2.1 where activated sludge, with the largest particle size of about 100 μm shows the lowest adsorption despite a very high octanol water distribution coefficient of estrone. Corrected for particle surface area the adsorption on activated sludge is the highest of all the compounds studied. These adsorptive interactions are important for treatment - high partitioning of contaminants onto organics can be expected which affects the use of sludge for land application or the selection of adsorbents. It appears as if organic adsorbents may be far more effective in removal of such contaminants if applied in forms with high specific surface areas (e.g. polymers).

To demonstrate this option, organic matter was added to a particle solution. Hematite was selected, as this colloid is spherical, monodispersed and well characterised. The adsorption of estrone increases considerably (see Figure 2) for all organics used (HA, FA and NOM) and over the pH 3 - 12 range. The largest increase of adsorption was provoked by HA at pH 3, which is the most aromatic, least soluble compound at conditions where the organics have the lowest charge. The increase due to the presence of natural organics can be explained with a modification of the colloid surface as a result of the interaction of estrone with the natural organics which adsorb on the hematite. Due to the high octanol-water partitioning coefficient the contaminants interact strongly with the natural organics.

Table 1. Interactions of estrone with naturally occurring particulates (pH 7-8, 100 ng/L estrone, 1mM NaHCO$_3$, 24h of adsorption)

Particle	Average Particle Diameter [μm]	Particle Concentration [g/L]	Estrone Adsorption [%]	Estrone Adsorption [ng/g*]
Hematite	0.075	$6.3 \cdot 10^{-3}$	0.74	150.64
Kaolin	0.1-4	0.1	8.87	140.46
Bentonite	0.1-10	0.1	13.53	175.84
Cellulose	20	0.1	8.35	113.96
Activated Sludge	100	5.4	22.38	6.15

* *dry weight*

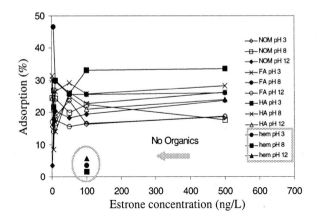

Figure 2. Impact of organic matter on adsorption of estrone by hematite

The implication of this is that trace pollutants in water and wastewater treatment systems are likely to be found associated with colloids as in natural systems most colloids have an organic coating. Further this effect can be taken advantage of to promote the removal of trace contaminants in water and wastewater treatment with the addition of adsorbents.

Removal of trace contaminants by particle addition

The addition of adsorbents is of advantage for the removal of trace contaminants, especially when membrane processes are used which do not retain trace contaminants. The objective of this study was a comparative investigation of common adsorbents used in the water and wastewater treatment industry, hence powdered activated carbon (PAC), ferric chloride coagulant ($FeCl_3$) and Magnetic Ion Exchange Resin (MIEX) were used.

Figure 3 shows the impact of pH and ferric chloride dosage on estrone removal in jar test experiments. The removal of such compounds is minimal during coagulation. This was expected as coagulation tends to favour the removal of large and hydrophobic compounds. Adsorption to the iron hydroxide precipitates is very low. This interaction and removal may change in the presence of natural organics and such experiments are yet to be conducted.

With powdered activated carbon the result is different. At relatively low concentrations of 5-10 mg/L PAC a substantial removal of estrone can be achieved as shown in Figure 4. Effects of competition are visible where adsorption in a 'clean' buffer solution is much higher than in surface water or secondary effluent. In surface water and effluent other organics compete for adsorption. This competition could be due to the blockage of pores and restriction of diffusion of trace contaminants to the adsorbent surface or due to the competition for adsorption sites. At higher dosages there is a large excess of adsorbent and hence the difference in matrix diminishes.

Magnetic ion exchange resin (MIEX®) has been developed for water treatment applications for enhanced natural organics removal (Slunjski *et al.*, 1999; Slunjski *et al.*, 2000). The resin preferentially adsorbs small and charged compounds from natural organics mixtures (Cook *et al.*, 2001). Figure 5 shows a light micrograph with particles on the order of 10-100 μm and an electron micrograph of a MIEX® particle with several small fractions broken off from particles. In higher resolution electron micrographs, the particles show a composition of the particles of very dense fibres, presumably magnetic iron compounds which give the resin its magnetic characteristics which lead to rapid aggregation if magnetised.

While the compounds studied are uncharged at neutral conditions, polar or hydrogen bonding interactions are responsible for a removal of up to 45 % of the

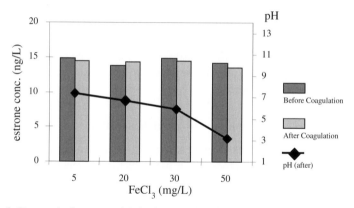

Figure 3. Removal of estrone with ferric chloride

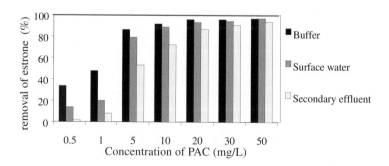

Figure 4. Removal of estrone with powdered activated carbon

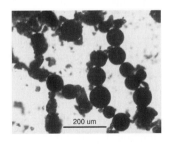

Figure 5. Light and electron micrograph of MIEX®

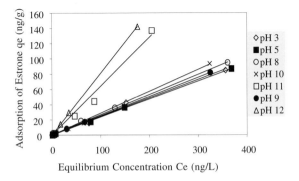

Figure 6. Removal of estrone with magnetic ion exchange resin (MIEX®)

estrone, which corresponds to an adsorption of 80-100 ng/g (Figure 6). This increases with pH - when the molecules are dissociated at a pH above 10.4, the removal increases drastically to about 70 %. This strong pH dependence can be explained with an additional ion exchange mechanism when the molecules are dissociated and carry a negative charge. Given the nature of those contaminants, the removal of small polar compounds is somewhat surprising. The interactions are attributed to hydrogen bonding or hydrophobic interactions.

There are a number of consequences from the above results. Both $FeCl_3$ and MIEX® are not very suited to remove the majority of the trace contaminants. PAC is better suited and appears to be the preferential choice when PAC is added in a sufficiently high dosage.

However, both MIEX® and PAC are both commonly used in water treatment. It is very likely that those particles will accumulate contaminants and to date very little is understood about the behaviour of such contaminants during regeneration and changes in feed water characteristics.

Removal of Trace Contaminants by Membranes

While microfiltration (MF) and ultrafiltration (UF) membranes are not expected to remove such small and polar compounds, most membranes tested showed a very similar removal of estrone at the start of operation. Removal was high at low and neutral pH, while it decreased substantially at a pH larger than 10.5. This could be attributed to adsorption effects, assumedly hydrogen bonding and hydrophobic sorption. This is shown for submerged MF (Memcor) and UF (Zenon) membranes in Figure 7. Those experiments were performed by adsorption tests of estrone on the membrane material without filtration. The adsorption of contaminants on hydrophobic membranes is higher than on hydrophilic materials.

Adsorption also dominates removal for some NF membranes. Some membranes remove estrone by size exclusion, others by adsorption. Figure 8 shows typical relationships between retention or permeate concentration and pH. As indicated previously the estrone molecule dissociates at pH 10.4 which leads to a drastic drop in retention. This behaviour suggests that the retention occurs due to polymer-contaminant interactions as opposed to size exclusion. When the molecules take a

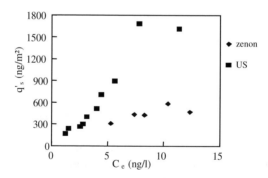

Figure 7. Adsorption of estrone on submerged MF and UF membranes

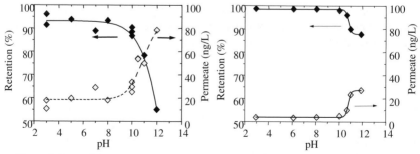

Figure 8. Retention and permeate concentration of NF (left, TFC-SR1) and RO (right, TFC-S) membranes

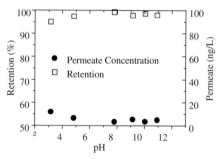

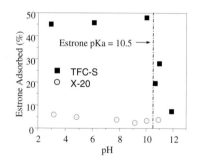

Figure 9. Retention and permeate concentration of X-20 membrane

Figure 10. Adsorption for the TFC-S membrane and X-20 membrane as a function of pH

negative charge at high pH, a repulsion between the negatively charged membrane and the organic anions causes reduced adsorption and facilitated transport through the membranes. For some very tight membranes (Figure 9) there is no effect of pH on retention, which indicates that the 'pores' are smaller than the contaminants and the membrane is an effective barrier independent of solution chemistry.

The adsorption as a function of pH for the tighter membranes is illustrated in Figure 10. Results reflect the shape of the retention curves in Figure 8 and Figure 9. While the materials are different (polyamide on polysulfon support (TFC-S) and polyamide-urea composite (X-20)) it appears that much of the adsorption of the TFC-S membrane is due to adsorption on the polysulfon support layer which offers a vast surface area. With the retention on the membrane surface of the X-20 membrane, the support is not available for adsorption and hence observed values are significantly smaller.

Those adsorption effects will only be operative during initial stages of filtration, but can cause the accumulation of a significant amount of trace contaminants and retention depends on the solution chemistry.

It is important to understand such retention and adsorption effects prior to membrane selection if the membrane is expected to act as a reliable barrier to contaminants. Such adsorption effects are also very important to understand the fate of pollutants in treatment systems and the possible desorption of contaminants during feed quality changes or cleaning operations.

Conclusions

While a broad range of results is presented in this paper, there is a common theme to the observations made regarding the interactions of trace contaminants, in this case a natural hormone estrone, and the materials used in treatment.

Powdered activated carbon clearly outperforms $FeCl_3$ and MIEX® in the removal of estrogens. Contaminants partition onto all particulates investigated in this study and interact with natural organics. Those findings are an essential contribution to the understanding of fate of contaminants in treatment systems as well as in the environment.

In membrane filtration, the retention mechanisms of some membranes are dominated by adsorptive effects while other membranes are capable of removing such small, polar compounds due to size exclusion. Adsorption decreases as the charge of the contaminants and the particle or membrane surface increases. During neutral charge conditions the adsorption of trace contaminants is maximised and this manifests itself with high intrinsic retention values. This retention, in the case of NF membranes, later shows a breakthrough behaviour (results not shown), unless the pore size of the membranes exhibits clear size exclusion phenomena. This was the case with tight RO membranes.

In water recycling, where a multiple barrier approach may be required to ensure low risk for water users, treatment is likely to be beyond economic feasibility and the product water achieved using such technology would approach ultrapure water. It is hence essential to plan water recycling with an integrated water cycle approach and determine the most sustainable water usage.

In the light of such results and a heated public debate it appears unreasonable to assume that direct reuse of water for personal consumption is a sensible solution. However, if the discharge of moderately treated wastewater persists one needs to realise that the global water cycle is closed and contaminants will reach our drinking water sources. It is essential that further developments incorporate reduced usage of synthetic chemicals and source separation, effective removal as well as destruction of the removed contaminants.

Acknowledgements

The authors wish to thank Dr Sheng Chang, Long Duc Nghiem, Maibritt Mastrup, Rikke Lund Jensen and Peter Ong for provision of experimental data and their contribution to the "Optimised Use of Membrane Hybrid Processes in Water Recycling" as well as Prof Tony Fane for useful discussions. The Australian Research Council and the Queensland Government are thanked for research funding through the SPIRT scheme.

References

Arcand-Hoy, L.D., Nimrod, A.C., Benson W.H., (1998) Endocrine modulating substances in the environment: estrogenic effects of pharmaceutical products, *International Journal of Toxicology* **17,**139-158
Colborn, T., Dumanoski, D., Myers, J. P. (1997) Our stolen future Abacus: London, UK
Cook, D., Chow, C., Drikas, M.(2001) in 19th Federal AWA Convention; Canberra, Australia
Council, N.R.(1999) Hormonally active agents in the environment, National Academy Press: Washington DC

Crisp, T.M., Clegg, E.D., Cooper, R.L., Wood, W.P., Anderson, D.G., Baetcke, K.P., Hoffmann, J.L., Morrow, M.S., Rodier, D.J., Schaeffer, J.E., Touart, L.W., Zeeman, M.G., Patel, Y.M. (1998) Endocrine disruption: an effects assessment and analysis, *Environmental Health Perspectives* **106**,11-56

EU (1999) Report of the Working Group on Endocrine Disruptors of the Scientific Committee on Toxicity, Ecotoxicity and the Environment (CSTEE) of Directorate General XXIV, Consumer Policy and Consumer Health Protection, p 96

Fawell, J.K., Sheahan, D., James, H.A., Hurst, M., Scott, S. (2000) Oestrogens and oestrogenic activity in raw and treated water in Severn Trent water, *Water Research* **35**, 1240-1244

Harland, L.(1999) Endocrine disrupting Chemicals. In 18th Federal AWWA Convention; AWWA, Adelaide, Australia

Johnson, A. C., Sumpter, J. P. (2002) Removal of Endocrine Disrupting Chemicals in Activated Sludge Treatment Works, *Environmental Science & Technology* **35**, 4697-4703

Johnson, A.C., Belfroid, A., Di Corcia A., (2000) Estimating steroid oestrogens inputs into activated sludge treatment works and observations on their removal from the effluent, *The Science of the Total Environment* **256**, 163-173

Jurgens, M.D., Williams, R.J., Johnson, A.C. (1999) Environment Agency, Bristol, UK.

Keith, L.H. (1998) Environmental Endocrine Disruptors, *Pure & Applied Chemistry* **70**, 2319-2326

Kirchner, M., Holsen, H., Norpoth, K. (1973) Fluorescence Spectroscopic Determination of Anti-Ovulatory Steroids in Water and Waste Water on the Thin Layer Chromatography Plate, *Zbl. Bakt. Hyg. I Abt. Orig. B* **157**, 44-52

Lucier, G. W. (1997) Dose-response relationships for endocrine disruptors: what we know and what we don't know, *Regulatory toxicology and pharmacology* **26**, 34-35

Manning, T. (2000) New South Wales Environment Protection Authority, Sydney

Mastrup, M., Jensen, R. L., Schäfer, A. I., Khan, S. (2001) in Schäfer, A.I., Sherman, P., Waite, T.D. (Eds) Recent Advances in Water Recycling Technologies, Brisbane, Australia, 103-112

Ohlenbusch, G., Kumke, M.U., Frimmel, F. H. (2000) Sorption of phenols to dissolved organic matter investigated by solid phase microextraction, *The Science of the Total Environment* **253**, 63-74

Ono, Y., Somiya, I., Kawaguchi, T., Mohri, S. (1996) Evaluation of Toxic Substances in Effluents from a Wastewater Treatment Plant, *Desalination* **106**, 255-261

Schäfer, A. I. (2001) Natural Organic Matter Removal using Membranes: Principles, Performance and Cost, Technomic: Lancaster

Seckler,D.,Barker, R., Amarasinghe, U.(1999) Water scarcity in the twenty-first century, *International Journal of Water Resources Development* **15** , 29-42

Slunjski, M., Bourke, M., Nguyen, H., Ballard, M., Morran, J., Bursill, D.(1999) in A.W.W. Association (Ed.) 18th Federal Convention, Adelaide, Australia

Slunjski, M., Cadee, K., Tattersall, J. (2000) in AquaTech; Amsterdam

Stahlschmidt-Allner, P., Allner, B., Römbke, J., Knacker, T. (1997) Endocrine disrupters in the aquatic environment, *Environmental Science and Pollution Research* **4**, 155-162

SWC Sydney Water Corporation (1999) Report, Sydney

Ternes, T.A., Kreckel, P., Mueller, J. (1999) Behaviour and occurrence of estrogens in municipal sewage treatment plants - II. Aerobic batch experiments with activated sludge, *The Science of the Total Environment* **225**, 91-99

Tyler, C. R., Routledge, E. J. (1998) Natural and anthropogenic environmental estrogens: the scientific basis for risk assessment. Estrogenic effects in fish in English rivers with evidence of their causation, *Pure Appl. Chem.* **70**, 1795-1804

U. EPA (1997)

U. EPA, US EPA (1997) Office of Research, Washington

H.H. Hahn, E. Hoffmann, H. Ødegaard (Eds.)
Chemical Water and Wastewater Treatment VII, pp. 331-338
© IWA Publishing, London
ISBN: 1 84339 009 4

Phosphorus Recovery from Sewage Sludge by Thermal Treatment and Use of Acids and Bases

K. Stark

Royal Institute of Technology (KTH), Div. of Water Resources Engineering, Stockholm, Sweden
email: stark@aom.kth.se

Introduction

In Sweden there are significant differences in opinions and policies between organisations and municipalities regarding the strategy for sludge management in the future. The national proposal of 75% phosphorus recovery from sewage sludge would make high demands on wastewater treatment plants around Sweden if realised. The proposal is at present under investigation by the Swedish Environmental Protection Agency (SEPA) as a commission from the government.

The demand of phosphorus recovery has led to sludge applications besides agricultural use. Sludge fractionation is considered to be a sustainable method for sludge handling where the sludge is seen as the raw material and products are recovered. A sludge handling system with product recovery may involve the following different steps (Hultman and Levlin, 1997). First, sludge is dissolved by different methods such as physical (heat or pressure), mechanical (use of ultra sonics), biological or chemical methods (use of acids and/or bases). The treated sludge may be used directly or the liquid phase is separated from the solid phase, where product recovery will continue.

In the Scandinavian countries phosphorus is mainly removed from wastewater by chemical precipitation. This causes the phosphorus to be strongly bound to metal ions and for phosphorus release, it may be necessary to use acids combined with high temperature and pressure.

Technologies that reduce both sludge mass and volume, whilst producing a reusable sludge product, are becoming increasingly favoured. This chapter describes some of these methods.

[331]

Phosphorus recovery system

Phosphorus recovery from digested sludge, where the phosphorus is mostly chemically bound, requires advanced treatment to release the phosphorus into a concentrated solution, which will be used to produce a phosphorus product. There are three commercial systems available in Sweden (KREPRO, BioCon and Aqua Reci), which can recover a high amount of the phosphorus in the sludge. They are described below.

KREPRO

The KREPRO sludge treatment process uses thermal hydrolysis and the addition of acid to lower the pH. Thermal hydrolysis means that the material is treated by high pressure and temperature, which destroys the larger organic molecules in the sludge. Four main products are recovered from the sludge: biofuel, phosphate, precipitant, and carbon source (Karlsson and Hansen, 2000). Both digested and raw sludge can be treated. The process is continuous and can be divided into seven main steps: thickening, acidification, thermal hydrolysis, biofuel separation, phosphorus precipitation, phosphate separation and recycling of precipitant and carbon source. The sludge is thickened to 5-7 % DS and acidified with sulphuric acid to a pH between 1 and 3. The coagulant, heavy metals and phosphorus are partly dissolved by this treatment. The KREPRO process has run as a pilot plant in Helsingborg and the city of Malmö in Sweden is planning to install a modified KREPRO process. The chemical demand for the process in Malmö is calculated to approximately 500 kg/ton DS.

BioCon

The BioCon process uses thermal drying and incineration followed by acid leaching of the produced ash, and product recovery through a system of ion exchangers. The process consists of three parts: a drying unit for the sludge, an incineration unit, and a recovery unit (Svensson, 2000). In the recovery unit the ash is ground in a mill and leached out with sulphuric acid, whereby the metals and the phosphate are dissolved in a solution. Thereafter the material is separated from the solution with an ion exchanger. The products from the process are phosphoric acid, precipitation agent $Fe_2(SO_4)_3$, and potassium hydrogen sulphate ($KHSO_4$). The residual sludge contains heavy metals.

The Swedish city Falun has installed the BioCon process and is planning to install a phosphorus recovery unit, which will be the first incineration plant in Sweden with phosphorus recovery. The calculated chemical demand for the plant in Falun is approximately 500 kg/ton DS where sulphuric acid, (H_2SO_4), hydrochloric acid (HCl), potassium chloride (KCl), and sodium hydroxide (NaOH) are used.

Aqua Reci

Aqua Reci is based on a process that uses supercritical water oxidation followed by phosphorus recovery. Supercritical water oxidation is a technology recently applied to oxidation of sludge from wastewater treatment plants. Full-scale experience with this technology is limited and the development of technology for phosphorus recovery is in the starting phase. Feralco (2001) showed that a high degree of leaching of phosphorus occurs when sodium hydroxide is applied at temperatures over 90 °C. Lime is added to obtain calcium phosphate as the product. Parallel experiments have been performed from the autumn 2001 at the Royal Institute of Technology (KTH), described later in the text.

Treatment of Sewage Sludge by Supercritical Water Oxidation

Supercritical water oxidation is an innovative and effective destruction method for organic wastewater and sludge. The process uses the unique properties of water in its supercritical state, where it behaves very differently from water under standard conditions. A major advantage of supercritical water oxidation systems is that they are capable of completely destroying organic compounds without producing harmful emissions (Patterson *et al.*, 2001). For instance, organic carbon is converted into carbon dioxide, organic and inorganic nitrogen into nitrogen gas (N_2), and halogenated organics and inorganics into their corresponding acids. Supercritical water oxidation is a radical oxidation reaction and is also exothermic. Once the feed organic is concentrated enough, energy maintains the reactor temperature.

The critical point of water is over 374 °C and 221 bar with the properties between those of a gas and a liquid. Compared to incineration in air, supercritical water oxidation takes place at much lower temperatures and may be seen as an alternative to drying and incineration. The process is described in Figure 1.

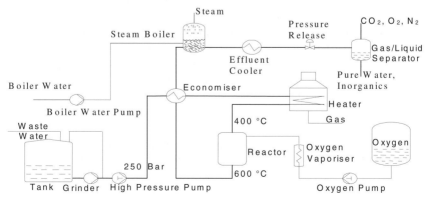

Figure 1. Principal flow sheet of the Aqua Critox® process (Gidner *et al.*, 2000)

A pilot plant with supercritical water oxidation has operated in Karlskoga, Sweden for 2 years and with a sludge capacity of 250 kg/h. Tests at this plant showed that organic material was totally oxidized, and inorganic material was nearly all in solid phase. The process has a fast reaction time: 30 - 90 s to destroy the organic material at a pressure of 275 bar and a temperature between 400 and 600 °C.

Supercritical water oxidation clearly has some direct environmental advantages compared to other treatment methods for sewage sludge. Svanström *et al.* (2001) showed that supercritical water oxidation is environmentally attractive also from a life cycle perspective.

Leaching experiments

Leaching experiments were carried out at the Department of Land and Water Resources Engineering at the Royal Institute of Technology, Stockholm, Sweden. Sludge from Bromma wastewater treatment plant, Stockholm, was run with supercritical water oxidation in the pilot plant in Karlskoga, Sweden. Hydrochloric acid (HCl) and sodium hydroxide (NaOH) were added at room temperature to the sludge product and mixed for 2 h. Up to 2M HCl (6.78 g HCl/g SS) and up to 5M NaOH (18.59 g NaOH/gSS) were used. The chemical demand is shown in Figure 2 for the acid and in Figure 3 for the base. The difference in amount of chemical used is due to the result of phosphorus release.

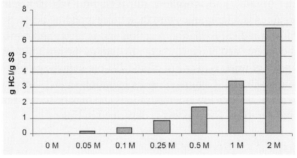

Figure 2. Acid demand for each set of mixing in the leaching experiment

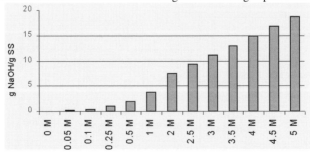

Figure 3. Base demand for each set of mixing in the leaching experiment

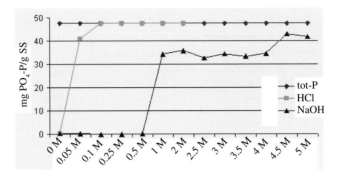

Figure 4. Phosphorus leaching from SCWO sludge using acid (HCl) and base (NaOH)

Phosphorus was leached out easier with hydrochloric acid than with sodium hydroxide at room temperature (Figure 4). At 0.1 M HCl, there was 100% phosphorus release while a ten-fold higher concentration of NaOH (1M) resulted in only 70% phosphorus release.

Phosphorus was leached out before iron, and iron was leached out more effectively with acid than base. The concentration of HCl that released 100% phosphate (0.1 M) leached only 1% of the iron, while 1 M NaOH leached 70 % of the phosphorus and only 0.04% of the iron (Figure 5 and 6). A very high

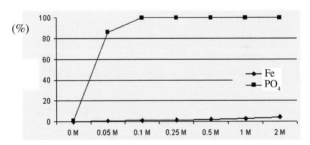

Figure 5. Leaching of phosphorus and iron with acid (HCl)

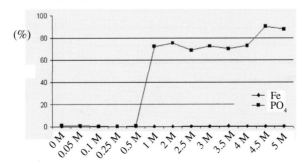

Figure 6. Leaching of phosphorus and iron with base (NaOH)

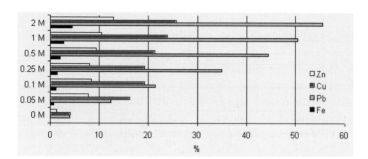

Figure 7. Heavy metals leached in acid conditions (Zn=Zinc, Cu=Copper, Pb=Lead, Fe=Iron)

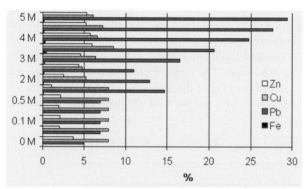

Figure 8. Heavy metals leached in basic conditions (Zn=Zinc, Cu=Copper, Pb=Lead, Fe=Iron)

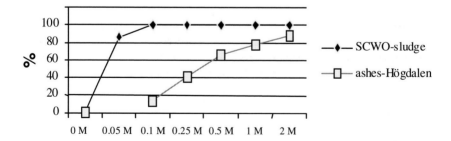

Figure 9. Phosphate leaching with acid from supercritical water oxidation sludge (SCWO-sludge) and from ashes from co-incineration of sewage sludge and household wastes

concentration of base (4.5 M NaOH) was required to release 90% of the phosphate and 0.2% of the iron (Figure 6). The iron may disturb the process in the step receiving a valuable clean phosphorus product.

Heavy metals, especially lead (Pb), are also leached to a higher percentage with acid than with base at room temperature (Figure 7 and 8).

The leaching results from supercritical water oxidation may be compared with co-incineration of sewage sludge and household wastes. Phosphorus was more easily leached with acid from the sludge residual of supercritical water oxidation than from the ash from traditional incineration (Figure 9).

Discussion

All three systems described in this chapter use thermal treatment and high chemicals consumption for the dissolution and recovery steps. An important difference between the systems is the end product. KREPRO produces ferric phosphate, while the BioCon process produces phosphoric acid and Aqua Reci calcium phosphate.

A disadvantage of the KREPRO and BioCon processes is that at the same time that phosphorus is dissolved from the sludge/ash, most of the inorganic materials are also leached out. Addition of precipitation agents leads to a high chemical demand to release the phosphate from the sludge and to recover the phosphorus product.

A disadvantage of the Aqua Reci process is lack of full-scale experiences. The process has only been tested in laboratory scale. Performed experiments will develop the process at a temperature over 90 °C. Leaching with sodium hydroxide and lime recovers calcium phosphate, but aluminum is leached out together with the phosphate.

While the KTH experiments showed that a low concentration of acid resulted in a high release of phosphorus, heavy metals were also leached under these conditions.

The supercritical water oxidation process may be seen as an interesting alternative to drying and incineration for sludge handling in Stockholm.

Acknowledgement

The author would like to thank MISTRA, the research programme of the Urban Water and Stockholm Water Co, for financial support and is thankful for valuable comments and help from Bengt Hultman, Erik Levlin, Monica Löwén and Agnes Mossakowska.

References

Feralco (2001) Pilot test of final sludge disposal from Stockholm Water´s water and wastewater treatment plants by use of the Aqua Reci-process, working document (in Swedish)

Gidner, A., Alemark, M., Stenmark, L., and Ekengren, Ö. (2000) Treatment of sewage sludge by supercritical water oxidation. IBC´s 6th Annual Conference on Sewage sludge, 16-17 Feb 2000, London, England

Hultman, B., Levlin, E., Löwén, M., Mossakowska, A., and Stark, K. (2001) Recovery of phosphorus and other products from sludge and ash, Stockholm Water Co, Report No.6, March 2001 (in Swedish)

Hultman, B., and Levlin, E. (1997) Sustainable sludge handling. Proceedings of a Polish-Swedish seminar, KTH, Stockholm, May 30, 1997. Advanced wastewater treatment, Report No. 2, TRITA-AMI REPORT 3045

Karlsson, I., and Hansen, B. (2000) Recycling of components separated from waste water sludge. WEFTEC, October 14-18, 2000, California, USA

Patterson, D.A., Stenmark, L.,and Hogan, F. (2001) Pilot-scale supercritical water oxidation of sewage sludge. 6th European Biosolids and Organic Residuals conference, 12-14 Nov 2001, UK

Svanström, M., Fröling, M., Modell, N., Peters, W.A., and Tester, J. (2001) Life cycle assessment of supercritical water oxidation of sewage sludge. 6th European Biosolids and Organic Residuals Conference, 12-14 Nov 2001, UK

Svensson, A. (2000). Phosphorus from sewage sludge – a study of KREPRO-process and BioCon process from a life cycle perspective. Masters Thesis, Chemical Environmental Science, Chalmers Institute of Technology, Göteborg, Sweden (in Swedish)

H.H. Hahn, E. Hoffmann, H. Ødegaard (Eds.)
Chemical Water and Wastewater Treatment VII, pp. 339-348
© IWA Publishing, London
ISBN: 1 84339 009 4

Chemical Treatment Effects on Phosphorus Availability in Sludges

S.K. Dentel* and J.T. Mah

*Dept. of Civil & Environmental Engineering, University of Delaware, USA
email: dentel@udel.edu

Introduction

Organic wastes such as animal manures and wastewater sludges are commonly applied to agricultural lands in many areas of the world. The practice is attractive from agronomic, economic, and logistical standpoints.

Neither sludge nor manure contains the high concentrations of nutrients such as nitrogen (N) and phosphorus (P) that are available in commercial fertilisers. Nonetheless, suitable mass applications can provide sufficient amounts of either N or P. In fact, the amount of P available through these wastes exceeds the amount applied as commercial fertilisers, suggesting a strong incentive for maximal utilisation of these materials (Table 1).

Unfortunately, the typical practice of determining manure or sludge applications by N requirements leads to a surfeit of P well beyond what the crop will assimilate. Table 2 illustrates this for the application of either poultry litter or digested sludge to corn, leading to overapplications of P by 420 % and 540 %, respectively.

Thus, if either sludge or manure is to be agronomically utilised as a "useful material," proper management of both N and P loadings are needed to assure that P will not be overapplied. Otherwise, the environmental harms to surface waters and shallow ground waters can be significant. P losses by erosion, surface runoff, and subsurface drainage from agricultural fields have been implicated in the degradation of water quality in many areas nationally and internationally. Most recently, excessive P levels have been associated with the accelerated growth of toxic *Pfiesteria* spp. in eutrophic waters of the eastern U.S. Given this, some states plan to require N and P management plans for agricultural systems that use organic wastes, including manures and sludges. This trend is apparent in other areas internationally.

[339]

Table 1. Annual U.S. generation of organic wastes and P from wastes (10^6 Mg/year)

	U.S. generation	P content
Manures (cattle, swine, poultry)	73	2.3
Municipal sludge	7.5	0.2
Commercial fertilisers		1.6

Table 2. P overloading when applications are based on N requirements for corn

Material	Application rate (kg ha^{-1})	Plant Available Nitrogen (PAN)			Phosphorus (P)		
		Percent (dry basis)	Required (kg ha^{-1})	Applied (kg ha^{-1})	Percent (dry basis)	Required (kg ha^{-1})	Applied (kg ha^{-1})
Poultry litter	8600 75% TS	2.1	135	135	1.6	25	105
Digested sludge	36000 25% TS	1.5	135	135	1.5	25	135

If controls are to be mandated on P releases, the possible risks of P application to soils should be systematically and scientifically assessed. This will allow development of best management practices (BMPs) that properly account for the numerous factors affecting P releases from land-applied sludges.

The major determinants of P mobility and the potential risk to surface waters are listed in Table 3, categorised according to sludge-related, soils-related, and locality-related factors. Although it is important to define and quantify all of these components, this paper primarily reports on recent work concerning the first category, with the objective of determining the effects of wastewater treatment processes, chemical additives used, and solids treatment processes on the resulting forms, solubility, and biological availability of phosphorus in municipal sludge. The findings will assist in developing a decision matrix for use in determining BMPs for sludge application to land.

Table 3. Factors affecting P releases from land-applied sludge

Sludge-related factors	Soils-related factors	Local factors
Wastewater type	P concentration in solid, soluble phases	Hydrology
Presence of water treatment residuals	P speciation and bioavailability	Topography
P-removal processes (BNR, chemical)	P-binding capacity (e.g. presence of Al, Fe)	Crop management practices
Chemical additives used (Fe, lime)	Erosion potential	Other nutrient applications
Stabilisation processes used	Porosity and subsurface drainage	Ditch & stream flow characteristics
		Proximity to surface waters
		Sensitivity of surface waters to elevated P levels

Previous Research

The literature on P mobility in various types of sludge is limited, but provides clear indications of the important factors. Table 4 summarises the most relevant published literature on P speciation and availability in sludges and in soil-amended sludges.

Essentially, previous research has suggested overall patterns and governing influences. The presence of Al and Fe in sludge has been found to play a key role in rendering P less available. Biological processes do not tie up phosphorus to a comparable extent. The presence of Ca is associated with greater P mobility, but whether this is simply a pH-related phenomenon has not been determined. Previous work has aimed at defining the stoichiometry of Al and Fe association with P during the removal processes, but has not been concerned with the subsequent speciation or availability of captured P. The form of Fe or Al added, and where it is added, could affect this stoichiometry.

In fact, even the analytical methods used for assessing P speciation and "availability" have been quite disparate, particularly when preparation and analysis are applied to liquid or dewatered sludge samples. We also found no reported method for long-term storage of sludge that took into account the preservation of P speciation. Some of our initial work was therefore in methods development. In addition, we considered it quite important to characterise the wastewater and

Table 4. Summary of results from previous characterisations of land-applied sludges

Authors & year	Conclusions	Comments
Chang et al. (1983)	P distribution unaffected by sludge type	No sludge used with high Ca, Al, or Fe
McCoy et al. (1986)	Most P associated with Al, Fe, Ca and unavailable; less than 5% organic P	Composted sludge with Al or Fe present
Kyle and McClintock (1995)	Fe and Al decreased P solubility more than BPR	
Jokinen (1990)	Al reduced P solubility; Ca increased soluble P	
Corey (1992)	Increased (Al+Fe):P ratio decreased P availability	Availability determined by soybean plant uptake
Wen et al. (1997)	Sludge did not increase P uptake by plants as much as manure.	Composting increased P extractability but not availability

Other studies consider forms and solubility of P in sludge (Cavalloro *et al.*, 1993; Cline *et al.*, 1986; Folle *et al.*, 1995; Frossard *et al.*, 1996; Furrer *et al*, 1984; James and Aschmann, 1992; Lindo *et al.*, 1993; Lund *et al.*, 1976; O'Connor *et al.*, 1986; Pierzynski *et al.*, 1990; Taylor *et al.*, 1978) but do not address the effects of treatment processes or soil characteristics

sludge treatment processes where samples were being obtained so that no spurious influences (e.g. the presence of water treatment residuals in a wastewater influent) would skew our results.

Procedures

The initial phase of this research consisted of the identification of candidate treatment facilities and completion of an initial field survey. Details are given in Dentel *et al.* (2000). Eight sites were then selected to provide an array of sludge types and chemical dosing practices. Large sludge samples were collected at these plants, including different sampling locations at some facilities. Specific procedures were developed for sludge storage as also described in Dentel *et al.* (2000).

Samples from selected sampling sites were characterised by total solids, pH, and P fractions. The soluble fraction was separated by 0.45 μm membrane filtration with glass fibre filter prefiltration and analysed for total P, soluble total P, total and soluble acid hydrolysable (condensed) P, total reactive P (ortho P), and soluble reactive P. Solid samples were analysed for total P, extractable total P, total acid hydrolysable P, extractable acid hydrolysable P, total reactive P (ortho P), and extractable reactive P. Table 5 defines these P fractions.

The analytical methods for these P fractions were adapted from Methods of Soil Analysis, Standard Methods, and EPA methods. Since the published methods were not developed for the analysis of sludges, considerable modification was required. As refined for this study, the fractions were determined as follows: total P was by nitric/perchloric acid digestion followed by colorimetric analysis; acid hydrolysable P by sulphuric mild acid hydrolysis followed by colorimetric analysis;

Table 5. Operationally defined phosphorus fractions

Dissolved Orthophosphate (DOrthoP)
  DOrthoP = Dissolved Reactive Phosphate (DRP)
Dissolved Condensed Polyphosphates (DCondP)
  DCondP = Dissolved Acid Hydrolysable P (DAHP) - DRP
Dissolved Organic Phosphates (DOrgP)
  DOrgP = Dissolved Total P (DTP) - DAHP = DTP - (DCondP+DOrthoP)
Total Orthophosphate (TOrthoP)
  TOrthoP = Total Reactive Phosphate (TRP)
Total Condensed Polyphosphates (TCondP)
  TCondP = Total Acid Hydrolysable P (TAHP) - TRP
Total Organic Phosphates (TOrgP)
  TOrgP = Total Phosphorus (TP) - (TAHP) = TP - (TCondP + TOrthoP)

---

reactive P by direct colorimetric analysis. Kinetic studies showed that it was often necessary to lengthen all extraction times for pulverised dried sludges due to their lesser wettability compared with soils; in addition, the higher P concentrations required modified reagent doses and sample quantities.

Laboratory experiments were also conducted to determine the effects of additive chemicals on P availability in sludges utilising a standard jar test with a sludge volume of 500 mL. The optimal doses of conditioning chemicals used were determined on the basis of preliminary jar tests followed by capillary suction time tests.

Results

The results provide much useful information on the transport of P through a wastewater treatment plant and its subsequent mobility. For example, comparisons of total and soluble P concentrations at two other facilities (Table 6) indicate large differences in P fractionation between biological and chemical removal of P. These distinctions are of clear relevance to the development of appropriate BMPs for land application of sludges.

Multi-plant comparisons: A set of 16 sludges were analysed for 6 P types (> 192 measurements), as well as extractant pH, and a variety of analyses and comparisons completed. Only the most significant findings are presented here.

Figure 1 provides the total phosphorus concentrations on a solids basis, grouped by site and sample. The P amounts are all between 1 % and 3 % of the solids mass. However, total P concentrations will change as wastewater concentrations vary, assuming a high percentage removal in treatment. The fate of phosphorus in land-applied sludges depends on the speciation of this P as well as the total loading. Figure 2 therefore presents P speciation, with concentrations divided by the total measured P concentration to give all fractions on a relative basis. Figure 2 also groups the results by phosphate species rather than by site or sample.

Table 6. Comparison of P fractions at the Parkway and Little Patuxent WWTPs

WWTP and Sample	TS (%)	pH	TP (mg/L)	SP (mg/L)	Comments
Parkway					
2E Return Activated Sludge	0.9	5.8	100-200	#5	Alum residuals in influent
Thickened 1E & 2E Solids	4.0	6.2	300	3	
Little Patuxent					
Thickened 1E Solids	5.0	6.0	400-600	40-60	Biological phosphorus removal
Thickened 2E Solids	4.0	6.3	1000-1100	600-700	

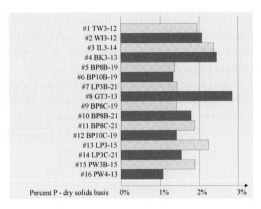

Figure 1. Percent total P levels on a solids basis for 16 sludge samples

Clearly, most of the P in all samples was in the total ortho-P form, whether associated with the liquid or solid phase. A smaller but significant amount was held as total organic P. This is true for all samples and sites. Several things are borne out:

- Over the range of eight samples that are very different in treatment type, additives and locations, the relative proportions of the phosphate species remain remarkably similar.

- There is very little dissolved phosphate relative to the insoluble forms. Of the insoluble P forms, most is ortho-P, with the remaining amount primarily organic P.

- Four samples indicate higher dissolved ortho-P concentrations than the others. One of these (Sample 3) is a sludge sample after 2-stage anaerobic digestion; the other three (Samples 7, 13, and 14) are from a plant using biological P removal.

- Several of the total condensed polyphosphate (TCondP) values are negative. As shown in Table 1, these values are obtained by subtracting total reactive P from total acid hydrolysable P (TAHP). It is likely that TAHP values were decreased through losses by precipitate formation on glass surfaces, causing the negative TCondP results. Modifications in the test method have corrected this difficulty and shown TCondP levels to be essentially negligible relative to the TP levels.

- A clear correlation exists between total ortho-P and total P, even from a variety of sludge types (Figure 3). Total organic P appears to be so weakly related to total P that it may be taken as an essentially constant fraction of the solids fraction. The intersection of the two trend lines may represent the minimum P content for microbial viability. Additional P is almost completely as ortho-P (1:1 slope on the graph).

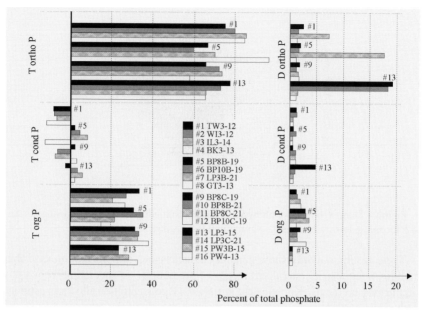

Figure 2. Phosphate speciation normalised by total P, organised by species

Results of a separate experiment are shown in Figure 4. In this experiment, ortho-P and pH were measured initially (including samples with an initial pH adjustment) and then after stoichiometric addition of ferric chloride based on the measured P concentration. The initial conditions are the indicated upper boxed points; the remaining points show a pattern that is indicative of solubility limits for ferric phosphate and ferric hydroxide.

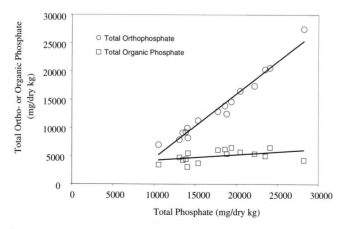

Figure 3. Total phosphate speciation as a function of total P concentration

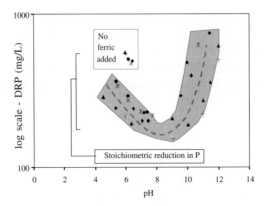

Figure 4. Dissolved reactive phosphorus (DRP), following various pH adjustments then stoichiometric ferric additions based on initial P concentrations

These results not only show general characteristics of P speciation (and thus mobility), but also a generalised means of predicting this speciation from a few simple measurements. Knowing the initial total P concentration allows an estimate of the organic and inorganic fractions through the use of the regression lines and equation shown on Figure 3 (and neglecting condensed P as justified by Figure 2). At intermediate pH, Figure 4 shows that the reaction of any Fe present is stoichiometric with the present P. When both Fe and Al are present, the molar ration ([P]/([Al] + [Fe]) readily provides the amount of P immobilisation, but the complete solubility problem must be solved at more extreme pH levels. The overall approach is shown in Figure 5.

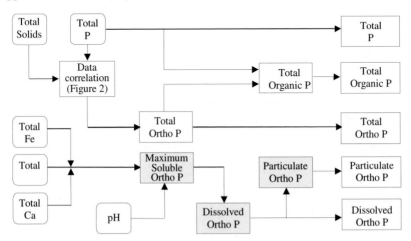

Figure 5. Schematic of procedure for determining major P species in sludges

These predictive steps are necessary if the land application of sludge is to be managed on a scientifically appropriate basis. The analytical results presented in this paper show a considerable variability in the fractionation of P in various sludges, and this should be accounted for in any rational means of managing sludge application to land.

Conclusions

Comparisons of numerous sludge samples indicated that, although the amount of total P in the sludges varied, its fractionation in the total sample remained relatively constant. However, total ortho-P increases in proportion to the total amount of P in the analysed samples. Sludges from treatment plants utilising biological nutrient removal displayed a much higher P release capability. Chemical treatment with ferric iron immobilises P in stoichiometric proportion if the pH is moderate, but if this is not the case, solubility considerations must be taken into account.

The importance of these factors suggests that sludge properties must be considered when determining the best management practices for utilising sludge as a fertiliser. A procedure for determining the P availability in sludges can be developed that is both logical and fundamentally supportable.

Acknowledgements

Support of this research by the Metropolitan Washington Council of Governments and the City of New York is gratefully acknowledged.

References

Cavallaro, N., Padilla, N., and Villarrubia, J. (1993) Sewage sludge effects on chemical properties of acid soils. *Soil Sci.* **156**, 63-70

Chang, A. C., Page, A. L., Sutherland, F. H., and Grgurevic, E. (1983) Fractionation of phosphorus in sludge-amended soils. *J. Environ. Qual.* **12**, 286-290

Cline, G. R., Lindemann, W. C., and Quintero, R. (1985) Dynamics of extractable phosphorus during nonsterile and sterile incubation of sludge-amended soil. *Soil Sci.* **140**, 98-104

Corey, R. B. (1992) Phosphorus regulations: Impact of sludge regulations. *Crop Soils* **20**, 5-10

Folle, F.J., Shuford,W., Taylor,R.W., Mehadi, A.A., and Tadesse, W. (1995) Effect of sludge treatment, heavy metals, phosphate rate, and pH on soil phosphorus. *Commun. Soil Sci. Plant Anal.* **26**, 1369-1381

Frossard, E., Sinaj, S. Zhang, L-M, and Morel, J. L. (1996) The fate of sludge phosphorus in soil-plant systems. *Soil Sci. Soc. Am. J.* **60**, 1248-1253

Furrer, O.J., Gupta, S.K., and Stauffer, W. (1984) Sewage sludge as a source of phosphorus and consequences of phosphorus accumulation in soils. pp. 279-294. In P. L'Hermite and H. Ott (eds.) *Processing and Use of Sewage Sludges*. D. Reidel Publ. Co., Dordrecht, Netherlands

James, B. R., and Aschmann, S. G. (1992) Soluble phosphorus in a forest soil Ap horizon amended with municipal wastewater sludge or compost. *Commun. Soil Sci. Plant Anal.* **23**, 861-875

Jokinen, R. (1990) Effect of phosphorus precipitation chemicals on characteristics and agricultural value of municipal sewage sludges. *Acta Agric. Scan.* **40**, 141-147

Kyle, M. A., and McClintock, S. A. (1995) The availability of phosphorus in municipal wastewater sludge as a function of the phosphorus removal process and the sludge treatment method. *Water Environ. Res.* **67**, 282-299

Lindo, P., Taylor, R.W., Shuford, J.W., and Adriano, D.C. (1993) Accumulation and movement of residual phosphorus in sludge-treated Decatur silty clay loam soil. *Commun. Soil Sci. Plant Anal.* **24**, 805-1816

Lund, L.J., Page, A.L., and Nelson, C.O. (1976) Nitrogen and phosphorus levels in soils beneath sewage disposal ponds. *J. Environ. Qual.* **5**, 26-30

McCoy, J.L., Sikora, L.J., and Weil R.R. (1986) Plant availability of phosphorus in sewage sludge compost. *J. Environ. Qual.* **15**, 404-409

O'Connor, G.A., Knudsten, K.L., and Connell, G.A. (1986) Phosphorus solubility in sludge-amended calcareous soil. *J. Environ. Qual.* **15**, 308-312

Pierzynski, G.M., Logan, T.J., and Traina, S.J. (1990) Phosphorus chemistry and mineralogy in excessively fertilised soils: Solubility equilibria. *Soil Sci. Soc. Am. J.* **54**, 1589-1595

Taylor, J.M., Sikora, L.J. Tester, C.F., and Parr, J.F. (1978) Decomposition of sewage sludge in soil: II. Phosphorus and sulfur transformations. *J. Environ. Qual.* **7**, 119-123

Wen, G., Bates, T.E. Voroney, R.P. Winter, J.P., and Schellenbert, M.P. (1997) Comparison of phosphorus availability with application of sewage sludge, sludge compost, and manure compost. *Commun. Soil Sci. Plant Anal.* **28**, 1481

H.H. Hahn, E. Hoffmann, H. Ødegaard (Eds.)
Chemical Water and Wastewater Treatment VII, pp. 349-362
© IWA Publishing, London
ISBN: 1 84339 009 4

Effects of pH and Ca/P Ratio on the Precipitation of Phosphate

Y. H. Song, H. H. Hahn and E. Hoffmann[*]

*Institute for Aquatic Environmental Engineering, University of Karlsruhe, Germany
email: Erhard.Hoffmann@isww.uka.de

Abstract

Phosphorus recovery from wastewater supports the principle of sustainability and may help meet the needs of the phosphate and water industries. Studies aiming at understanding the precipitation of calcium phosphate from wastewater are essential to improve the technologies of phosphorus recovery. In the present paper, a chemically defined system with a phosphate concentration of 20 mg P/L and an ionic strength, expressed as conductivity of 2.0 mS/cm, which was equivalent to that in wastewater, was designed to study the effects of pH and initial Ca/P ratio on the precipitation of calcium phosphate. A computer programme based on an iterative method was written to calculate the saturation index of the precipitation system. The precipitation experiments were undertaken in batch reactors with a pH range from 7.5 to 11.0, initial Ca/P ratios from 1.67 to 6.67, and a constant temperature of 21.5°C. It was found that both high pH values and high initial Ca/P ratios in the precipitation system favoured the precipitation efficiency and the precipitation rate of calcium phosphate. At initial Ca/P ratios of 1.67 and 3.33 the lowest pH values to achieve rapid precipitation were 9.0 and 8.5, respectively; at initial Ca/P ratios above 5.0, rapid precipitation could be achieved at pH 8.0. At pH values from 7.5 to 8.0, after a reaction time of 1.5 hrs a special acceleration phenomenon was observed, where high precipitation efficiencies could be achieved. From the variations of the concentrations of calcium and phosphate, it could be concluded that the final precipitates were hydroxyapatite in the experimental pH range, and this was confirmed by X-ray diffraction measurements of the precipitates. pH and initial Ca/P ratio may affect the precipitation of calcium phosphate due to the supersaturation of the precipitation system.

[349]

Introduction

Phosphorus is an essential element for all living organisms and its utilisation has greatly promoted the development of agriculture and industry. Modern society does not use phosphorus resources in a sustainable way, however. Phosphates are manufactured from phosphate-containing rocks, then are consumed in agriculture and industry, and finally go into soil, rivers and the sea. This, on the one hand, is exhausting the limited phosphate deposits that are unevenly distributed in the world, on the other hand, has caused eutrophication of water bodies. With the emphasis on sustainability, more stringent nutrient discharge limits, and more restrictions on sludge disposal, both the wastewater treatment industry and the phosphate industry have to consider the alternatives of conventional phosphorus removal technologies and opportunities for phosphorus recovery.

Considerable world-wide research projects have been undertaken on phosphorus recovery technologies, some of which are already at the pilot- or large-scale stage. From the point of view of availability, the crystallisation processes of calcium phosphate and magnesium ammonium phosphate hexahydrate (struvite) have been emphatically studied, and a number of technologies such as the DHV Crystalactor™ Pelletiser, the CSIR Fluidised Bed Crystallisation Column, and the Kurita Fixed Bed Crystallisation Column have been developed (Brett *et al.,* 1997). In these technologies, phosphate is extracted from wastewater by a technique such as Phostrip, and then is precipitated and/or crystallised in dedicated reactors such as fluidised bed or fixed bed reactor to produce useful end products. But there are still a lot of problems with the application and acceptance of these technologies. For example, the DHV Crystalactor™ Pelletiser is one of the technologies used to recover phosphate as calcium phosphate. In this technology, the presence of carbonate may affect the crystallisation, so concentrated sulphuric acid (96 %) is added to remove carbon dioxide. The addition of acid decreases the pH to 3, at which carbon dioxide is released, and then the pH must be increased to 9 to precipitate calcium phosphate. While this operation is an important step for obtaining good product, it inevitably increases the complexity and the cost of the process, especially in the area where the influent contains a large amount of carbonate. Besides carbonate, other components such as magnesium and organic matter may also affect the precipitation or crystallisation process of calcium phosphate. These examples illustrate the importance of understanding the effects of process conditions on the precipitation and crystallisation of phosphate in order to develop mature technologies for phosphorus recovery.

From the point of view of industry, to recover phosphorus as calcium phosphate is promising (Driver *et al.,* 1999). Calcium phosphate formation can be induced with a high calcium concentration and high pH by adding lime, which is inexpensive, readily available, and free of environmental problems. Calcium phosphate is just a component of phosphate rock, so it is a good raw material both for industry and agriculture if it is recovered in a suitable physical form.

Some of the chemical and physical-chemical factors that may influence the precipitation of calcium phosphate from wastewater include the composition (e.g. concentrations of phosphate, calcium, magnesium, organic matter and carbonate), the pH, and the temperature of the influent. The goal of the present work was to determine the effects of two operational factors, pH and initial Ca/P ratio on the precipitation of calcium phosphate. In the current study, a wider pH range (7.5 to 11.0) than that in traditional studies of calcium phosphate precipitation was examined, to simulate conditions in phosphate recovery from wastewater. Furthermore Ca/P ratios ranging from 1.67 to 6.67 were tested in a chemically defined system containing 20 mg P/L based on the characteristics of actual wastewaters.

Experiments

The precipitation of calcium phosphate was initiated by the rapid mixing of $CaCl_2$ and K_2HPO_4 or KH_2PO_4 solutions. Experiments were performed by adding 500 mL of 2.15, 4.30, 6.45, and 8.60 mM $CaCl_2$ to 500 mL of 1.29 mM K_2HPO_4 or KH_2PO_4 within 30 sec, in order to get Ca/P ratios of 1.67, 3.33, 5.00, and 6.67, respectively. The reverse order of addition was also used with identical results, so the former order of addition was used throughout the experiments. Prior to mixing, the pH of each reactant solution was adjusted to a value between 7.5 and 11.0 at intervals of 0.5 by adding NaOH or HCl, as needed. In order to keep the ionic strength of the precipitation system equivalent to that of wastewater, 1.01 g KNO_3 was added to the $CaCl_2$ solution prior to mixing, to arrive at a concentration of 0.01 M KNO_3 after mixing. The conductivity of the precipitation system reached a value of about 2.0 mS/cm. The initial mixing was done by strong stirring with a HI 200M magnetic stirrer (Hanna Instruments, Singapore) for 3 min; afterwards solutions were stirred at a constant reproducible rate. The change in pH with time, after mixing, was monitored with a PH 191 pH meter (WTW GmbH, Germany) and solutions were maintained at the aforementioned constant pH values by adding 0.45 M NaOH. Aliquots of 15 mL for calcium and phosphate analyses were removed at frequent intervals after mixing, and the samples were filtered with Standard Glassfibre Filters 13400 (Sartorius AG, Germany). All of the above reactant solutions were prepared from Analytical Grade reagents. The water used in the experiments was deionized water, which had a conductivity of 0.7 µS/cm. All the experiments were performed at $21.5 \pm 0.5°C$.

The calcium concentration of the filtrate was analysed using the complexometric method, and the phosphate concentration was analysed with the molybdenum heteropolyphosphate complex method at a wavelength of 700 nm with a Lambda 2 UV/VIS Spectrometer (Perkin Elmer, Germany).

After a reaction time of 3 hrs, the reaction mixture was filtered through glassfibre filters, and the precipitates were dried at a room temperature around 22°C. The powdered samples were placed in glass capillary holders with 0.5 mm diameter. X-ray diffraction (XRD) measurements of the precipitates were run on a Huber Guinier camera equipped with a Ge-monochromater and a Huber 670 image plate (Huber Diffraktionstechnik GmbH, Germany). Diffractograms were recorded with Cu-radiation (Cu $K_{\alpha 1}$ = 0.15406 nm) over an angular range of 8-100 °2θ with a step width of 0.005 °2θ per sec counting time.

Calculations of species concentration and supersaturation

Depending on the physical and chemical environment, different kinds of calcium phosphate phases (Table 1) may precipitate from saturated solutions.

Table 1. Calcium phosphate phases

Phases	Abbreviation	Composition	Ca/P
Dicalcium phosphate dihydrate	DCPD	$CaHPO_4.2H_2O$	1.00
Octacalcium phosphate	OCP	$Ca_4H(PO_4)_3.3H_2O$	1.33
Tricalcium phosphate	TCP	$Ca_3(PO_4)_2$	1.50
Hydroxyapatite	HAP	$Ca_5(PO_4)_3OH$	1.67

The supersaturation (S) of a system with respect to a given calcium phosphate phase indicates the thermodynamic driving force of the precipitation reaction. It is defined as

$$S = IAP/K_{sp} \tag{1}$$

where IAP is the ionic activity product and K_{sp} is the solubility product of the calcium phosphate precipitate. The IAPs of different calcium phosphates are defined as

$$IAP_{HAP} = (\ [Ca^{2+}]\ f_2\ )^5\ (\ [PO_4^{3-}]\ f_3\ )^3\ (\ K_w/[H^+])\ f_1 \tag{2}$$

$$IAP_{OCP} = (\ [Ca^{2+}]\ f_2\ )^4\ (\ [PO_4^{3-}]\ f_3\ )^3\ [H^+]\ f_1 \tag{3}$$

$$IAP_{TCP} = (\ [Ca^{2+}]\ f_2\ )^3\ (\ [PO_4^{3-}]\ f_3\ )^2 \tag{4}$$

$$IAP_{DCPD} = [Ca^{2+}]f_2[HPO_4^{2-}]f_2 \tag{5}$$

in which f_z denotes the activity coefficient of a z-valent ion and K_w refers to the ionic product of water. Moreover, a saturation index (SI) is defined as

$$SI = \log (S) \tag{6}$$

to facilitate further discussion.

The calculation of ionic activity product necessitates a detailed knowledge of the chemical speciation of ions in solution. Assuming that no precipitation has taken place just after mixing, the ions, ion-pairs and equilibria listed in Table 2 should be taken into account in the present precipitation system. As an approximation, the influence of the dissolved atmospheric CO_2 on the precipitation system is ignored. The thermodynamic dissociation and association constants used in the calculation are also given in Table 2. The calculations were performed by an iterative method (Feenstra, 1979; van Kemenade and De Bruyn, 1987) with a computer program written in FORTRAN, and the activity coefficients were calculated according to Davies equation

$$-\log f_z = AZ^2 \{I^{1/2}/ (1 + I^{1/2}) - 0.3I\} \qquad (7)$$

where I was the ionic strength, and A had a value of 0.505 at 21.5°C. The calculated values of supersaturation help to understand the precipitation of calcium phosphate.

Table 2. Species, equilibria and constants used in the calculation of supersaturation

Equilibrium	pK	Reference
$H_3PO_4 \leftrightarrow H_2PO_4^- + H^+$	2.1329	Bates 1951
$H_2PO_4^- \leftrightarrow HPO_4^{2-} + H^+$	7.2099	Bates and Acree 1943
$HPO_4^{2-} \leftrightarrow PO_4^{3-} + H^+$	12.42	Vanderzee and Quist 1961
$H_2O \leftrightarrow H^+ + OH^-$	14.115	Covington et al. 1977
$Ca^{2+} + PO_4^{3-} \leftrightarrow CaPO_4^-$	-6.462	Chughtai et al. 1986
$Ca^{2+} + HPO_4^{2-} \leftrightarrow CaHPO_4^0$	-2.3914	Gregory et al. 1970
$Ca^{2+} + H_2PO_4^- \leftrightarrow CaH_2PO_4^+$	-0.7350	Gregory et al. 1970
$Ca^{2+} + OH^- \leftrightarrow CaOH^+$	-1.356	Gimblett and Monk 1954
$Ca_5(PO_4)_3OH(s) \leftrightarrow 5Ca^{2+} +3PO_4^{3-} + OH^-$	58.501	McDowell et al. 1977
$Ca_4H(PO_4)_3.3H_2O(s) \leftrightarrow 4Ca^{2+} + 3PO_4^{3-} + H^+ + 3H_2O$	50.606	Christoffersen et al. 1989
$Ca_3(PO_4)_2(s) \leftrightarrow 3Ca^{2+} + 2PO_4^{3-}$	28.828	Gregory et al. 1974
$CaHPO_4.2H_2O(s) \leftrightarrow Ca^{2+} + HPO_4^{2-} + 2H_2O$	6.5898	Patel et al. 1974

Results

The effect of pH on the precipitation of calcium phosphate

For the convenience of discussion, a precipitation efficiency $\alpha(t)$, where $0 \leq \alpha(t) \leq 100\%$, is defined as

$$\alpha(t) = (C_0 - C_t) / C_0 \bullet 100\% \qquad (8)$$

where C_t is the concentration of phosphate at the reaction time t and C_0 is the initial concentration of phosphate in solution. This ratio of precipitated phosphate to the initial phosphate in solution also indicates the extent of the precipitation reaction.

Figure 1 shows the effect of pH on the precipitation of phosphate. In Figure 1a where the initial Ca/P ratio is 1.67, the precipitated phosphate is plotted as a function of the reaction time at different pH values ranging from 8.0 to 11.0. At a given pH value the precipitated phosphate increases as the reaction proceeds. As pH increases, the $\alpha(t)$ improves significantly. Within a reaction time of 3 hrs, the higher the reaction pH was, the higher the $\alpha(t)$ could be reached. Three different precipitation processes can be distinguished. At pH 8.0 the $\alpha(t)$ is so low that it is less than 8.0 % within a reaction time of 3 hrs. At pH 8.5 the precipitation reaction can be precisely described by an empirical kinetic equation as

$$C_0 - C_t = 7.659\ln(t) - 9.221 \qquad (9)$$

where C_0 (mM) is the initial concentration of phosphate in solution and C_t (mM) is the concentration of phosphate in solution at the reaction time t (min). A transformation is obtained by differentiating the above equation

$$-dC_t/dt = 7.659 \cdot 1/t \qquad (10)$$

where $-dC_t/dt$ is the precipitation rate of phosphate, and 7.659 is a constant related to the reaction conditions including supersaturation, pH, and temperature. Equation (10) indicates that the precipitation rate is inversely proportional to the reaction time. At pH $\geq$ 9.0, the precipitation of phosphate is very fast at the beginning, but afterwards the precipitation rate slows down although the $\alpha(t)$ still increases. Unlike the situation at pH 8.5, where the precipitation reaction advances gradually, the precipitation reactions at pH values from 9.0 to 11.0 are rapid reactions, which achieve maximum efficiencies in 0.17 hr, and the values of $\alpha(t)$ depend on the pH values. This can be seen more clearly in Figure 2, in which the values of $\alpha(t)$ at reaction times of 0.17 hr and 3 hrs are compared.

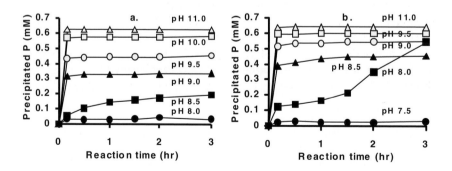

Figure 1. The effect of pH value on the precipitation of phosphate. (a) Initial Ca/P ratio 1.67; (b) Initial Ca/P ratio 3.33

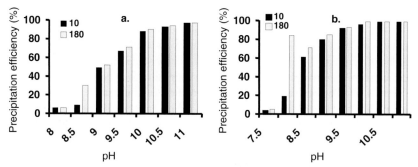

Figure 2. The comparison of precipitation efficiencies at reaction times of 0.17 hr and 3 hrs. (a) Initial Ca/P ratio 1.67; (b) Initial Ca/P ratio 3.33

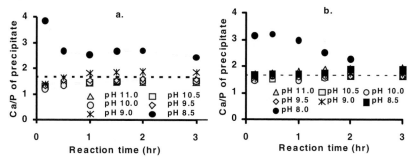

Figure 3. The Ca/P ratios of the precipitates. (a) Initial Ca/P ratio 1.67; (b) Initial Ca/P ratio 3.33

The precipitated phosphate is plotted as a function of the reaction time in Figure 1b, where the initial Ca/P ratio is 3.33 and the pH values range from 7.5 to 11.0. As pH increases, the $\alpha(t)$ also improves significantly. Here again three different situations can be distinguished. At pH 7.5 the $\alpha(t)$ is low, less than 5.0 % in a reaction time of 3 hrs. When pH $\geq$ 8.5, rapid precipitation reactions occur and the values of $\alpha(t)$ depend on the pH values. An interesting phenomenon occurred at pH 8.0: at first the precipitation advanced gradually, but after a reaction time of 1.5 hrs the precipitation reaction accelerated, so that the $\alpha(t)$ was nearly the same as that at pH 9.0 when the reaction time was 3 hrs. This phenomenon deserves special attention.

According to the precipitated calcium and phosphate, the Ca/P ratio of the precipitate can be calculated, and the results are shown in Figure 3. It shows that both at initial Ca/P ratio of 1.67 and pH $\geq$ 9.0, and at initial Ca/P ratio of 3.33 and pH $\geq$ 8.5, where rapid precipitations occurred, the precipitates had Ca/P ratios ranging from 1.2 to 2.0, and at the reaction time of 3 hrs got close to 1.67, the Ca/P ratio of HAP. At a given pH value, with the proceeding of the precipitation reaction, the Ca/P ratio of the precipitate has an increasing tendency. But below the pH value of rapid precipitation, *i.e.* where the precipitation reaction advances gradually and where acceleration phenomenon exists, the Ca/P ratio of the precipitate may be higher than 2.0, even above 3.0 at the early stage of the reaction. With the proceeding of the precipitation reaction, the Ca/P ratio goes down, and tends to get close to 1.67. The above evolution phenomenon of Ca/P ratio suggests that no matter what kind of precipitation process the reaction follows, the final precipitate tends to transform to thermodynamically most stable calcium phosphate, HAP.

The effect of the initial Ca/P ratio of solution on the precipitation of calcium phosphate

Figure 4 shows the effect of the initial Ca/P ratio of solution on the precipitation of calcium phosphate at pH values of 9.0, 8.0, and 7.5. In Figure 4a where the pH value is 9.0, the precipitation reactions are all fast. As the initial Ca/P ratio increases from 1.67 to 5.00, the precipitation efficiency $\alpha(t)$ improves. In Figure 4b where the pH value is 8.0, the precipitation reactions show the typical acceleration characteristics that were shown in Figure 1b, and the $\alpha(t)$ improves as the initial Ca/P ratio increases. In Figure 4c, when the pH is 7.5 and the initial Ca/P ratio of 6.67, an obvious acceleration phenomenon can be also observed. This is similar to the situation at pH 8.0.

The effect of the initial Ca/P ratio of solution on the Ca/P ratio of the precipitate is shown in Figure 5. Both at pH 9.0 and 8.0, the highest initial Ca/P ratios produced precipitates with the lowest Ca/P ratios. At the reaction time of 3 hrs, all the precipitates originating from different initial Ca/P ratios had Ca/P ratios close to that of HAP. The above results indicate that initial Ca/P ratio may influence the precipitation rate and efficiency, but the final precipitates tend to have stoichiometric compositions of HAP.

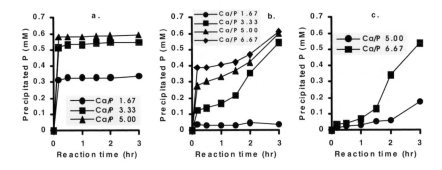

Figure 4. The effect of the initial Ca/P ratio of the solution on the precipitation of phosphate. (a) pH 9.0; (b) pH 8.0; (c) pH 7.5

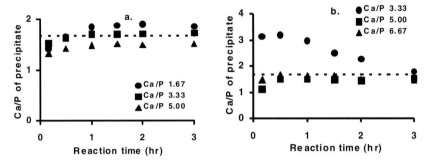

Figure 5. The effect of the initial Ca/P ratio of solution on the Ca/P ratio of the precipitate. (a) pH 9.0; (b) pH 8.0

X-ray diffraction patterns (XRD) of the precipitates

Representative XRD patterns for the precipitates originating from different pH values and initial Ca/P ratios are shown in Figure 6. In all three patterns the Bragg peaks correspond to HAP (ICDD 84-1998). The broad lines reflect small crystallite sizes in the precipitate. These data show that the variations in pH and initial Ca/P ratios do not affect the final formation of HAP.

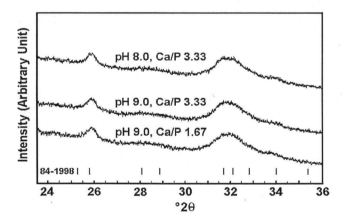

Figure 6. X-ray diffraction patterns of the precipitates

Discussion

The effect of pH on the supersaturation of calcium phosphate

The precipitation of calcium phosphate from solution is a base uptake process. Because the process is related to the equilibria listed in Table 2, the uptake of base promotes the dissociation of the hydrogen containing phosphate species, and then promotes the precipitation of calcium phosphate. In a given precipitation system without base addition, the pH will decrease as the precipitation reaction proceeds. In the present experiments, the pH value of the precipitation system was kept constant by base addition, so the pH value of the system affected the speciation of phosphate and the base uptake during the precipitation. The effect of pH value on the precipitation of calcium phosphate can be analysed by calculating the supersaturation of the precipitation system (Figure 7).

The calculations indicate that the present precipitation system is highly supersaturated with respect to HAP, OCP, and TCP, but not DCPD, which has a very low S value and the corresponding SI is not shown. In the present studies the

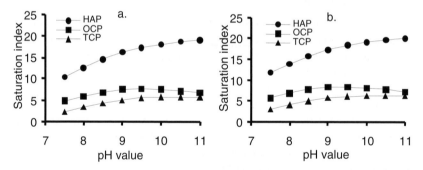

Figure 7. The effect of pH on supersaturation. (a) Initial Ca/P ratio 1.67; (b) Initial Ca/P ratio 3.33

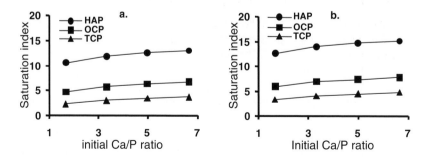

Figure 8. The effect of initial Ca/P ratio on the supersaturation. (a) pH 8.0; (b) pH 7.5

SI with respect to OCP is taken to discuss the precipitation reactions. Comparing Figure 7a with Figure 1a, at pH 8.0 the SI is only 5.95, so $\alpha(t)$ is low; at pH 9.0 the SI is 7.49 and rapid precipitation may occur; at pH 8.5 where SI is 6.86, precipitation advances gradually. Comparing Figure 7b with Figure 1b, at pH 7.5 where SI is 5.81, $\alpha(t)$ is low; at pH 8.5 where SI is 7.78, rapid precipitation may occur; at pH 8.0 where SI is 6.93, a special acceleration phenomenon appears.

The above comparisons indicate that the thermodynamic driving force, supersaturation, accounts for the occurrence and precipitation efficiency of calcium phosphate. After the rapid precipitation of the early stage, subsequent precipitation becomes very slow (Figure 2), because the supersaturation of the solution has decreased. The present experiment was carried out in a batch system without phosphate and calcium supplements during the reaction, so the rapid precipitation rate cannot continue. In a continuous system, however, rapid precipitation could be maintained. It can be concluded that the pH value of the precipitation system influences the supersaturation, then the precipitation rate and efficiency.

The effect of the initial Ca/P ratio of solution on the supersaturation of calcium phosphate

Figure 8 shows the effect of initial Ca/P ratio of the solution on the supersaturation of calcium phosphate. The SI values are directly proportional to the initial Ca/P ratio. Comparing Figure 8 with Figure 4b, when initial Ca/P ratios were 5.00 and 6.67, the SI values were high enough to promote rapid precipitations initially, but then the reactions proceeded slowly until a reaction time of 1.5 hrs, after which the acceleration phenomena appeared. At pH 7.5 and initial Ca/P ratios ≤ 3.33, both the SI values and $\alpha(t)$ are low; at initial Ca/P ratios of 5.00 and 6.67, where the SI values are 6.35 and 6.70, respectively, acceleration phenomena exist, too. Since an increase in the initial Ca/P ratio increases the supersaturation of the precipitation system, the thermodynamic driving force of the precipitation becomes stronger.

The acceleration phenomena at pH 8.0 and 7.5

As shown previously, when the pH of the precipitation system was 8.0 or 7.5, and if the SI value with respect to OCP was suitable, an acceleration phenomenon appeared after a reaction time of 1.5 hrs. In the present experiments, at pH 8.0 with initial Ca/P ratios from 3.33 to 6.67 (SI values from 6.93 to 7.77), and at pH 7.5 with initial Ca/P ratios from 5.00 to 6.67 (SI values from 6.35 to 6.67), acceleration phenomena were observed after a reaction time of 1.5 hrs. The final precipitation efficiency after a reaction time of 3 hrs may even approach or surpass the efficiencies of the reactions at higher pH values (Figure 1b). Although the above phenomena occur at certain SI values, they seem to be characteristics of this special pH range, from 7.5 to 8.0.

In a kinetic study on the precipitation of calcium phosphate, Van Kemenade and De Bruyn (1987) used relaxation time, t_R, to study the precipitation kinetics. In a single growing phase in a precipitation system, the relaxation time is the time from zero to the inflection point of the sigmoidal kinetic curve, where the precipitating rate reaches its maximum value. In a complex precipitation system where the precipitation involves the growth of more than one phase, the kinetic curve may have more than one inflection to indicate the formation of different phases, so relaxation time for each precipitating phase can be defined. In a given precipitation system, the relaxation time of a certain phase is highly reproducible. Analysing the kinetic curve of the present experiments at pH 8.0 or pH 7.5, where an acceleration phenomenon may occur after a reaction time of 1.5 hrs, precipitation was found to be a two-stage precipitation. The relaxation time for the second phase at pH 8.0 and initial Ca/P ratio 3.33 was 1.9 hrs. The second phase might also be a rapid growing phase, so the final precipitation efficiency was high. It is difficult to identify the phases in the process of precipitation only by the present Ca/P ratios of the precipitates, for the precipitates may adsorb some calcium and phosphate ions. Instrumental methods such as TEM and XRD are needed to distinguish different phases of the precipitates. If the system is designed properly, high precipitation efficiency can be obtained at pH values as low as 8.0 and 7.5.

The composition of the precipitate

As shown in Figures 3 and 5, the Ca/P ratio of the precipitate continuously changed as the precipitation proceeded. If the reaction is a rapid precipitation, the Ca/P ratio of the precipitate will increase slightly to arrive at a value close to 1.67; if the reaction advances gradually or accelerates after a second phase forms, the Ca/P ratio of the precipitate may be fairly high, but as the reaction proceeds, the Ca/P ratio decreases and tends to approach 1.67. The present precipitation system was complex, for several mineral phases with different Ca/P ratios might occur in the precipitation process, and the formed phases might transform to other phases. Thus the calculated Ca/P ratio was in fact a comprehensive reflection of the precipitates. In the precipitating process, the newly formed precipitate might have large surface area, especially in the gradually advanced reaction or the reaction with acceleration phenomenon, so excessive calcium ions could be adsorbed by the new precipitate, thus showing high Ca/P ratio in the early stage of the reaction. As the reaction proceeds, more phosphate might be combined into the solid phase, and then the Ca/P ratio of the precipitate got close to the stoichiometric value of HAP. XRD measurements of the precipitates confirmed that after a reaction time of 3 hrs, the precipitates were HAP although the precipitation mechanisms, rates and efficiencies might be different because of different reaction conditions, such as pH and initial Ca/P ratios.

Conclusions

In the present paper, a chemically defined system with a phosphate concentration of 20 mg P/L and an ionic strength equivalent to wastewater was designed to study the effects of the pH value and the initial Ca/P ratio of solution on the precipitation of calcium phosphate. High solution pH and high initial Ca/P ratio favour the precipitation efficiency and precipitation rate of calcium phosphate. The precipitation system with an initial Ca/P ratio of 1.67, the stoichiometric Ca/P ratio of HAP, had rapid precipitation at pH 9.0; if the initial Ca/P ratio was increased to 3.33, the pH needed to achieve rapid precipitation decreased to 8.5. At a given pH value, as initial Ca/P ratio increases, the precipitation efficiency and the precipitation rate also increase. When the initial Ca/P ratio reaches 5.00, the pH value needed to achieve rapid precipitation can be decreased to 8.0, and at an initial Ca/P ratio of 6.67, the precipitation efficiency at pH 7.5 may be as high as 83 % after a reaction time of 3 hrs. In the pH range from 7.5 to 8.0, the precipitation may accelerate after a reaction time of 1.5 hrs, and the final precipitation efficiency can reach a fairly high value. It indicates that a well-designed precipitation system at low pH values could be considered. This, on the one hand, can save the base and calcium needed for phosphate recovery, on the other hand, can avoid the problem of the strong basicity of the effluent after the phosphate recovery process.

The effects of pH and initial Ca/P ratio on the precipitation of calcium phosphate are due to their effects on the thermodynamic driving force of the precipitation reactions. No matter how the pH value and the initial Ca/P ratio of the precipitation system vary, after a specific reaction time the final precipitates tend to transform to HAP, and this has been confirmed by XRD measurements.

Acknowledgements

The authors would like to thank DFG-Graduiertenkolleg 366 "Grenzflächen-phänomene in aquatischen Systemen und wässrigen Phasen" for the financial support to Dr. Yong Hui Song. Thanks also go to Prof. Fritz H. Frimmel of the Engler-Bunte-Institute, Department of Water Chemistry of the University of Karlsruhe, Dr. Dietfried Donnert and Dr. Peter Weidler of the Institute for Technical Chemistry of the Karlsruhe Research Center Technology and Environment, and Dr. Ernst Antusch who was formerly a member of the authors' institute for their kind help with the research.

References

Bates, R.G. (1951) First dissociation constant of phosphoric acid from 0-60°C. Limitations of the electromotive force method for moderately strong acids. *J. Res. Nat. Bur. Stand.* **47**, 127-134

Bates, R.G., and Acree, S.F. (1943) pH values of certain phosphate-chloride mixtures and the second dissociation constant of phosphoric acid from 0-60°C. *J. Res. Nat. Bur. Stand.* **30**, 129-155

Brett, S., Guy, J., Morse, G.K., and Lester, J.N. (1997) *Phosphorus Removal and Recovery Technologies.* Selper Publications, London, pp. 31-44

Christoffersen, J., Christoffersen, M.R., Kibalczyc, W., and Andersen, F.A. (1989) A contribution to the understanding of the formation of calcium phosphates. *J. Crystal Growth* **94**, 767-777

Chughtai, A., Marshall, R., and Nancollas, G.H. (1968) Complexes in calcium phosphate solutions. *J. Phys. Chem.* **72**(1), 208-211

Covington, A.K., Ferra, M.A., and Robinson, R.A. (1977) Ionic product and enthalpy of ionization of water from electromotive force measurements. *J. Chem. Soc. Faraday Trans. 1* **73**, 1721-1730

Driver, J., Lijmbach, D., and Steen, I. (1999) Why recover phosphorus for recycling, and how? *Environ. Technol.* **20**(7), 651-662

Feenstra, T.P. (1979) A note on the calculation of concentrations in the case of many simultaneous equilibria. *J. Chem. Edu.* **56**(2), 104-105

Gimblett, F.G.R., and Monk, C.B. (1954) E.M.F. studies of electrolytic dissociation. Part 7-some alkali and alkaline earth metal hydroxides in water. *Trans. Faraday Soc.* **50** 965-972

Gregory, T.M., Moreno, E.C., and Brown, W.E. (1970) Solubility of $CaHPO_4.2H_2O$ in the system $Ca(OH)_2$-H_3PO_4-H_2O at 5, 15, 25, and 37.5°C. *J. Res. Nat. Bur. Stand.* **74A**(4) 461-475

Gregory, T.M., Moreno, E.C., Patel, J.M. and Brown, W.E. (1974) Solubility of β-$Ca_3(PO_4)_2$ in the system $Ca(OH)_2$-H_3PO_4-H_2O at 5, 15, 25, and 37°C. *J. Res. Nat. Bur. Stand.* **78A**(6), 667-674

McDowell, H., Gregory, T.M., and Brown, W.E. (1977) Solubility of $Ca_5(PO_4)_3OH$ in the system $Ca(OH)_2$-H_3PO_4-H_2O at 5, 15, 25, and 37°C. *J. Res. Nat. Bur. Stand.* **81A**(2, 3), 273-281

Patel, J.M., Gregory, T.M., and Brown, W.E. (1974) Solubility of $CaHPO_4.2H_2O$ in the quaternary system $Ca(OH)_2$-H_3PO_4-$NaCl$-H_2O at 25°C. *J. Res. Nat. Bur. Stand.* **78A**(6), 675-681

van Kemenade, M.J.J.M., and De Bruyn, P.L. (1987) A kinetic study of precipitation from supersaturated calcium phosphate solutions. *J. Colloid Interface Sci.,* **118**(2), 564-585.

Vanderzee, C.E., and Quist, A.S. (1961) The third dissociation constant of orthophosphoric acid. *J. Phys. Chem.* **65**(1), 118-123

H.H. Hahn, E. Hoffmann, H. Ødegaard (Eds.)
Chemical Water and Wastewater Treatment VII, pp. 363-368
© IWA Publishing, London
ISBN: 1 84339 009 4

Recycled Iron Phosphates in the Fertiliser Industry

*R. Puska and P. Ylinen**

*Kemira, Espoo, Finland
email: paula.ylinen@kemira.com

Introduction

Phosphorous ore is the main P source of the fertiliser industry today and the main commercial phosphate deposits are in the United States, Russia, Morocco and China. After mining and beneficiating the phosphate minerals, mainly hydroxo and/or fluorapatites ($Ca_5(OH,F)$ $(PO_4)_3$) are processed to phosphoric acid and fertilisers.

Substantial amounts of phosphates are also available in the side streams from various sources, such as sewage treatment, the food industry and cattle manure. The purpose of our investigation was to explore whether iron phosphates from sewage treatment could be exploited in the fertiliser manufacturing industry.

Phosphorus in Fertilisers

Mineral fertilisers are applied on cultivated soils to ensure that an adequate supply of P and other nutrients are available to crops. Today there is a wide range of phosphatic fertilisers available. They may contain only phosphorus, but more often also nitrogen and potassium as well as some micronutrients.

Soon after its application, fertiliser phosphorus undergoes various reactions in the soil, some of which transform the phosphorus into forms less usable by plants. Therefore, the repeated applications of phosphatic fertilisers and animal manures lead to the accumulation of phosphorus in the soil reserves. The high content of phosphorus in soil increases the risk of P runoff into surrounding waterways. Therefore, in many countries authorities control the phosphorus application rates in order to sustain the soil phosphorus balance.

[363]

Criteria for phosphatic raw materials

The total phosphorus content is one of the main criteria considered for the raw materials contained in fertilisers. Another important criterion is the ratio of impurities to the total phosphorus content. Certain elements, such as Ca, Fe, Al, Mg, F and Si define how well phosphatic minerals suit the production processes for phosphoric acid and fertilisers. Other more undesirable elements, such as heavy metals and radioactive elements, affect the purity level of the phosphatic end products and thus their suitability in the food production chain. Especially the cadmium concentration of phosphate rocks is widely discussed in connection with phosphatic fertilisers, because the main sources of Cd in cultivated soils are fertilisers and animal manures together with the atmospheric deposition (Louekari *et al.*, 2000). All these quality issues need to be carefully evaluated when recycled phosphates are considered for use as a raw material in the fertiliser industry.

Nutritional efficacy

In terms of a fertiliser product, the most important issue is that the nutrients in the fertilisers are available for use by plants and crops. There are many factors that define the efficacy of fertiliser phosphorus in cultivated soil. The most important ones are a crop's P-utilization capacity, soil pH, and reactive Fe-, Al- and Ca-content in the soil. Soil texture, soil organic matter content, and soil phosphorus status also play important roles in the possible bioavailability of fertiliser phosphorus (Laegreid *et al.*, 1999).

The solubility properties of the nutrient compounds in fertilisers are used to predict the availability of these nutrients for plants. In terms of phosphate compounds, there are various methods for assessing the phosphorus reactivity in soil solutions. The reactivity is conventionally estimated by measuring the solubility of the phosphate in extracting solutions, such as water, neutral ammonium citrate, 2 % citric acid, 2 % formic acid, and acidic and alkaline citrate solutions (Chien, 1993).

In agronomy, iron phosphates as chemical compounds are generally considered to be almost totally unavailable to plants and crops especially in slightly acidic and neutral soils (Sutton, 1985). Actually iron and aluminium phosphates are considered to be the ultimate end products of phosphorus reactions in certain soils, and these are unavailable to plants. However, there are studies where certain iron phosphates have been found to be more bioavailable than others. One factor is the level of crystallinity (Huffman, 1962; Lindsay, 1979; Richards and Johnston, 2001, Willis *et al.*, 1999). Amorphous hydrated iron phosphates have been reported to be significantly more available to plants than crystalline iron phosphate minerals, such as strengite. These facts and findings encouraged us to study how well plants can take up phosphorus from iron phosphate recovered from sewage treatment.

Iron phosphates from sewage sludge treatment as fertiliser raw materials

Environmental issues, such as surface water eutrophication, demand that phosphorus should be removed from waste waters. Removal is carried out either by chemical precipitation and crystallisation or by biological methods. Inorganic phosphates, such as calcium phosphates, iron phosphates and struvites are currently the main products from phosphorus recovery processes (Morse *et al.* 1998). Especially struvites have been proposed as suitable products for fertiliser applications (Booker *et al.*, 1999).

Kemira Kemwater´s iron coagulants are used to separate phosphorus from the sewage streams. Iron binds phosphates efficiently with high recovery when reacting in phosphorus-containing liqueurs. Kemwater´s sludge recycling process KREPRO recovers iron phosphate as an acidic sludge with 30-40 % dry matter content. This amorphous product is free from heavy metals and organic contaminants (Gonzales and Karlsson, 2001). Phosphorus content is approx. 10 % as P and iron content 24 % as Fe. Our preliminary solubility tests indicated rather good bioavailability. Iron phosphate from Kemwater´s KREPRO process has no water-soluble phosphorus, but its P is 100 % soluble in neutral ammonium citrate and 30 % soluble in 2 % citric acid.

The aim of our growth experiments was to study if $FePO_4$ from Kemwater´s KREPRO process could be utilized as a fertiliser raw material. Two different uses of the fertiliser were tested, one for forestry and the other for agriculture.

Applications for forestry

To encourage tree growth on acidic, mineral-poor peatlands, PK fertiliser is needed. Iron deficiency is also reported in these areas, so the application of $FePO_4$ allowed us to introduce both phosphorus and iron into the fertiliser formula. We produced test granules by compacting muriate of potash with three citrate-soluble P sources: 100 % apatite, 50 % apatite + 50 % iron phosphate and 100 % iron phosphate. For comparison we also produced PK samples from the 100 % water-soluble single super phosphate (SSP). The efficacy of the trial fertilisers was tested on birch seedlings. The plants were grown in controlled conditions in pots with an acidic, mineral-poor peat as the growth substrate. Fertilisers were spread on the substrate surface. Reference pots received K but no P. Nitrogen demand was fulfilled by regular N additions to all the treatments. There were seven replicates. Plant growth was followed for 204 days and water samples were taken regularly during the test to monitor the P leaching risk.

Fertilisers that were produced either from apatite rock or water-soluble SSP clearly gave the best growth results. The combination of apatite-P and iron phosphate-P released phosphorus slowly for use by the plants and hence the growth was slow at the beginning but continued longer and was well balanced later on. However, plants supplied only with $FePO_4$-P had great difficulties in starting their growth.

As expected, the leaching tendency of phosphorus was the highest with the PK fertiliser produced from water-soluble SSP. Some leaching of P also occurred from the apatite-based fertiliser treatments in these acidic soil conditions. However, if only half of the phosphorus came from apatite and half from KREPRO's $FePO_4$, then phosphorus leaching was as effectively prevented as with the fertiliser produced from 100 % $FePO_4$.

Applications for agriculture

The nutritional performance of the $FePO_4$-P was tested both on a Scandinavian soil (pH 5.5, P content 3.6 mg/kg) and on a Mediterranean soil (calcareous, pH 8.7, P content 13 mg/kg). Ryegrass was selected in this greenhouse experiment for the test plant because it allows successive harvesting and thus the possibility of monitoring the availability dynamics of phosphorus. Phosphorus was applied either as 100 % water-soluble triple super phosphate (TSP) or as pure $FePO_4$ originating from the KREPRO process. Reference pots received no P. All the treatments had sufficient and equal amounts of N and K available throughout the experiment. There were five replicates.

On both soils TSP allowed the best growth and $FePO_4$ only partly fulfilled the phosphorus need of the ryegrass. $FePO_4$-P growth was better on the acid soil than on the calcareous soil. The native P reserves of the acid soil were able to supply the ryegrass only 3 mg P total for six successive harvests. For $FePO_4$- and TSP-fertilisers, 7.5 mg P and 11 mg P, respectively, were taken up by the grass. During the same growing period, calcareous soil was able to provide 3.5 mg P for the grass without any P; $FePO_4$ provided 7 mg and TSP 14.5 mg of P.

Conclusions

Based on the results of our study, we conclude that recycled iron phosphate may have potential in fertiliser production. The target soil and crop must be well defined, however, and the purity level of the recycled material must be extremely well guaranteed. In many countries, legislation regulates the fertiliser life cycle from raw materials to manufactured products and further to the application rates and practises on the farm. To meet the demands of sustainable agriculture, raw material purity and environmental safety questions play a key role when new fertiliser products are formulated and developed or when new raw materials are selected for exploitation.

Additionally, there are logistical and production technology issues that limit the type of recycled raw materials, and all this means that a lot of research and development work is still needed before the practical applications can be realized.

References

Louekari, K., Mäkelä-Kurtto, R., Pasanen, J., Virtanen, V., Sippola, J. and Malm, J. (2000) Cadmium in fertilisers, risks to human health and the environment, Ministry of Agriculture and Forestry in Finland, Publication 4/2000

Laegreid, M., Bockman, O.C. and Kaarstad, O. (1999) *Agriculture, Fertilisers & the Environment,* CABI Publishing in association with Norsk Hydro ASA, p. 150

Chien, S.H. (1993) Solubility assessment for fertiliser containing phosphate rock, *Fertilizer Research* **35**, 93-99

Sutton, P. (1985) Feasibility of using iron and aluminium phosphates as phosphatic fertilizer sources on calcareous soils, Dissertation thesis, University of Nebraska-Lincoln, Nebraska, USA

Huffman, E.O. (1962) Reactions of phosphate in soils, recent research by TVA, *The Fertilizer Society, Proceeding* No. 71

Lindsay, W.L. (1979) Chemical equilibria in soils, Reprint of the first edition, The Blackburn Press, Caldwell, New Jersey, USA

Willis, R.B. and Allen, P.R. (1999) Measurement of amorphous ferric phosphate to assess iron bioavailability in diets and diet ingredients, *Analyst* **124**, 425-430

Richards, I.R. and Johnston, A.E. (2001) The effectiveness of different precipitated phosphates as sources of phosphorus for plants, Report on work undertaken for CEEP, EFMA, Anglian Water UK, Thomas Water UK and Berlin Wasser Betriebe

Morse, G.K., Brett, S.W., Guy, J.A. and Lester, J.N. (1998) Review: Phosphorus removal and recovery technologies, *The Science of the Total Environment* **212**, 69-81

Booker, N.A., Priestley, A.J. and Fraser, I.H. (1998) Struvite formation in wastewater treatment plants: Opportunities for nutrient recovery, *Environmental Technology* **20**, 777-782

Gonzales, J. and Karlsson, I. (2001) Sustainable, safe and economically viable, *World Water & Environmental Engineering,* **24** (6) 12

Authors Index